**Books are to be returned on or before
the last date below**

19 DEC 1977

-6 APR 1978

20 MAY 1978

30 AUG 1978

23 OCT 1978

1 4 APR 1998

LIBREX—

Measurement of man at work

Measurement of man at work

An appraisal of physiological and psychological criteria in man-machine systems

Edited by
W T Singleton *University of Aston in Birmingham*
J G Fox *University of Birmingham*
D Whitfield *University of Aston in Birmingham*

Based on papers presented to a
symposium held in Amsterdam,
September 1969, and sponsored by the
International Ergonomics Association

Taylor and Francis Ltd. London   1973

First published 1971 by Taylor & Francis Ltd,
10/14 Macklin Street, London WC2B 5NF.

Reprinted 1973

© 1973 Taylor & Francis Ltd

ISBN 0 85066 041 6 (cloth)
ISBN 0 85066 071 8 (paper)

Printed and bound in Great Britain by
Taylor & Francis Ltd, 10/14 Macklin Street
London WC2B 5NF.

List of contributors

Bonjer F. H.
Sweilandstraat 16
Warmond
The Netherlands
(Papers 1 :4, 2 :4)

Borg G.
The Institute of Applied Psychology
University of Stockholm
Råsundavågen 101
171 37 Solna
Sweden
(Paper 2 :7)

Burrows A. A.
Douglas Aircraft Company
Long Beach
California
USA
(Paper 3 :3)

Chapanis A.
Department of Psychology
The Johns Hopkins University
Baltimore
Maryland 21218
USA
(Introductory paper)

Chiles W. D.
Psychology Laboratory
Civil Aeromedical Institute
P.O. Box 25082
Oklahoma City
Oklahoma 73125
USA
(Paper 2 :11)

Christensen J. M.
Human Engineering Division
Aerospace Medical Research Laboratory
Wright Patterson Air Force Base
Ohio
USA
(Paper 3 :2)

Defayolle M.
Division de Psychologie
Centre de Recherches du Service
de Santé des Armées
108 *Boulevard Pinel*
69 *Lyon* 3[e]
France
(Paper 2 :3)

Dinand J. P.
Division de Psychologie
Centre de Recherches du Service
de Santé des Armées
108 *Boulevard Pinel*
69 *Lyon* 3[e]
France
(Paper 2 :3)

Edwards E.
Department of Ergonomics and Cybernetics
University of Technology
Loughborough
Leicestershire
England
(Paper 2 :8)

Fairbank B. A.
Institute for Perception RVO-TNO
Kampweg 5
Soesterberg
The Netherlands
(Paper 3 :5)

Fox J. G.
Department of Engineering Production
The University of Birmingham
Birmingham 15
England
(Paper 3 :6, Editor)

Gentil M. T.
Centre de Recherches du Service
de Santé des Armées
108 Boulevard Pinel
69 Lyon 3^e
France
(Paper 2 :3)

Groll-Knapp E.
Hygiene Institute University of Vienna
Department of Environmental Hygiene
A–1095 Vienna/Austria
Kinderspitalgasse 15
(Paper 2 :2)

Hartman B. O.
Neuropsychiatry Branch
Clinical Sciences Division
USAF School of Aerospace Medicine
Brooks AFB
Texas
USA
(Paper 3 : 4)

de Jong J. R.
de Genestetlaan 25
Bilthoven
The Netherlands
(Paper 3 : 1)

Kalsbeek J. W. H.
Laboratory of Ergonomic Psychology-TNO
Zuiderzeeweg 10
Amsterdam-oost
The Netherlands
(Paper 2 : 5)

Leplat J.
Laboratoire de Psychologie du Travail
41 rue Gay-Lussac
75 Paris 5^e
France
(Paper 1 : 6)

Michon J. A.
Institute for Perception RVO-TNO
Kampweg 5
Soesterberg
The Netherlands
(Paper 3 : 5)

Murrell K. F. H.
University of Wales Institute of Science
and Technology
Occupational Psychology Department
8 Cathedral Road
Cardiff
Wales
(Paper 1 : 5)

Pailhous J.
Laboratoire de Psychologie du Travail
41 rue Gay-Lussac
75 Paris 5^e
France
(Paper 1 : 6)

Parrot J.
Centre d' Etudes Bioclimatiques du C.N.R.S.
21 rue Becquerel
67 Strasbourg
France
(Paper 1 : 2)

Rabideau G. F.
The Bunker-Ramo Corporation
15–1/2 East Main Street
Fairborn
Ohio 45324
USA
(Paper 2 : 10)

Rey P.
Institut de Physiologie
20 rue de l'Ecole-de-Médecine
1211 Genève 4
Switzerland
(Paper 2 : 6)

Rolfe J. M.
Institute of Aviation Medicine
Royal Air Force
Farnborough
Hants.
England
(Paper 2 : 9)

Shackel B.
Department of Ergonomics and Cybernetics
University of Technology
Loughborough
Leicestershire
England
(Paper 3 : 7)

Sinaiko H. W.
Institute for Defense Analyses
400 Army Navy Drive
Arlington
Virginia 22202
USA
(Concluding paper)

Singleton W. T.
Department of Applied Psychology
University of Aston in Birmingham
Birmingham 4
England
(Paper 1 : 1, Editor)

Streimer I.
San Fernando Valley State College
18111 Nordhoff
Northridge
California 91324
USA
(Paper 1 : 3)

Whitfield D.
Department of Applied Psychology
University of Aston in Birmingham
Birmingham 4
England
(Editor)

Wilkins W. L.
Navy Medical Neuropsychiatric Research Unit
San Diego
California 92152
USA
(Concluding paper)

Wisner A.
Laboratoire de Physiologie du Travail et
d'Ergonomie du Conservatoire National
des Arts et Metiers
41 rue Gay-Lussac
75 Paris 5ᵉ
France
(Paper 2 : 1)

Contents

Preface

This book began with a comment by A. Chapanis at a meeting of the Council of the International Ergonomics Association. He thought that it would be interesting to take a serious look at the differences between American 'human factors engineering' and European 'ergonomics'; in particular their relative emphasis on, and utilization of, physiological and psychological methods and techniques. The proposal was accepted that a Symposium should be held in 1969 on the topic 'Psychological versus physiological criteria in man–machine systems'.

This happened in April 1968, during the annual meeting of the Ergonomics Research Society held at the University of Sussex. At the same meeting a committee to organize the conference was formed. The Chairman was J.R. de Jong (Netherlands) and the members were F.H. Bonjer (Netherlands), A. Chapanis (USA), W.T. Singleton (UK) and A.Y. Wisner (France). The invitation to hold the meeting in the Netherlands was accepted by the I.E.A.   H.P. Ruffell Smith (I.E.A. President) and E. Grandjean (I.E.A. Secretary) were *ex-officio* members of the committee and played a useful part in the early stages. The committee met a number of times during 1968/69 always in Holland and usually in Amsterdam. The committee was fortunate in having the services of Dr. H.A.W. Klinkhamer and Miss J.W. Gijsbers from the Netherlands Institute of Preventative Medicine as a Secretariat.

These meetings generated an interesting debate on the problems and structure of ergonomics and the possible strategies in approaching the question posed by Chapanis. The original intention was to try to answer the following questions.

1. For what kind of ergonomics problems is it more appropriate to use physiological criteria, psychological criteria, a combination of the two?
2. When both physiological and psychological criteria are used in an experiment, how well do they agree?
3. When principles of equipment design are based on findings achieved in experiments using physiological criteria, do they agree with principles arrived at using psychological criteria?

From a close examination of these questions the first point requiring clarification is the distinction between physiology and psychology. Obviously there is a considerable difference between the basic training of the two kinds of specialist but even this is diminishing in that most university psychology courses contain a large component of physiology and vice versa. At the research and application level the distinction is even more difficult to make, perhaps particularly so in relation to the problems of man at work. The question is further discussed in several papers within this book. For the present it is adequate to think only in terms of measurement. Physiological techniques are those which concentrate on electrical, chemical or mechanical data from the body or its waste products. Psychological techniques are those which concentrate on either opinions or performance.

Given that the main purpose was to examine and compare physiological and psychological techniques it would clearly be undesirable to separate the Symposium into sections on these lines.

After extensive discussion it emerged that, following a general introduction, there should be three main sections called respectively, Man, Techniques and Applications. The 'Application' section was easiest to understand and define, its purpose being to consider methodologies and strategies which are in use in relation to real problems and to illustrate these by recent case-studies. The distinction between 'Man' and 'Techniques' is not an easy one, but it is important and is worthy of detailed explanation. The expertise of every technologist is twofold: there is some topic about which he claims to have extensive knowledge and there is a repertoire of specialist techniques relevant to this topic in which he is a skilled practitioner. In the case of ergonomics the topic is normal, healthy man functioning in relation to some rational purpose in a particular environment; the techniques are those concerned with the measurement of man or of man–machine performance. In a modern technological society at any time man usually has some relationship with a machine, e.g. an air conditioning unit, a vehicle, a machine-tool. Clearly man has many descriptors and there must be a wide range of relevant techniques. The descriptors can each be approached by a variety of techniques but also a technique may provide evidence relevant to a variety of descriptors. For example, 'arousal' is a descriptor, two relevant techniques are postural measurement and heart rate measurement. Postural measurement is a technique relevant to arousal and also to physical size. Heart rate measurement is a technique relevant to arousal and also to energy expenditure. In this sense 'Man' and 'Techniques' are two separate dimensions. The ergonomist is professionally interested in descriptors of man which are actually or potentially useful in relation to man–machine interaction. This was the topic of discussion in the 'Man' section, in particular current trends in measurement and assessment. The ergonomist is equally interested in data acquisition procedures about man at work—this was the topic of discussion in the 'Techniques' section, in particular, recently developed techniques, their advantages and limitations. This sectional division seemed to provide a useful structure to the Symposium and it has been retained in this book.

Having established the philosophy of the Symposium, the Organizing Committee took up the more administrative but still taxing problems of determining roughly what should be covered within each section and also who should be invited to provide the coverage. It was necessary to restrict the topics to those which seemed either to be currently developing or to have made such a powerful contribution in the past that they could not be ignored. Thus the Symposium and the book are not intended to be comprehensive and exhaustive; the objective was rather to be timely, interesting and stimulating. It was also necessary to restrict attendance at the Symposium; it was agreed that about forty would be the optimum in that any number less than this could not hope to be a reasonable cross section of current workers and any number greater than this would be likely to reduce the quality and intensity of interaction between contributors. The problem of acquiring such a small sample across so many countries, disciplines and problems was an impossible one and in fact the Symposium has some obvious gaps and weaknesses. It has been possible to attempt some slight improvements in the book and for this reason there are a few papers here which were not presented at the meeting. Nevertheless, some gaps remain and we are aware that justice has not been done to certain countries and to certain problems. For example, the book is weak on biochemistry and anatomy, partly because the original orientation was towards psychology and physiology and partly because certain key workers in these fields could not attend.

Professor Chapanis conducted a survey by questionnaire towards the end of the Symposium and it emerged that almost everyone thought that it had been worthwhile and that further symposia of this kind should be sponsored by the I.E.A. There were a number of excellent supporting facets. The Symposium was held at the Royal Tropical Institute, Amsterdam, which proved very suitable for the purpose in terms of both working and domestic accommodation. The Dutch organizers, Dr. de Jong, Dr. Bonjer, Dr. Klinkhamer and Miss Gijsbers (secretary to Dr. Bonjer) provided most efficient preparatory services. In the actual support of the Conference, Miss Gijsbers was joined by my own secretary, Miss Love.

By this time my two editorial colleagues had become involved, their first major task being to record and clarify the discussion which took place following each paper. We have decided that this is best communicated not by a literal reproduction of the comments of each discussant but by incorporation in the epilogue which follows each section of the book. We hope that no discussant will feel personally aggrieved when he recognizes a point he made at Amsterdam recorded in this book without direct acknowledgment to the individual source. We consider that, since all participants were discussants and all contributed to the undoubted stimulation of the atmosphere then any credit which reflects from these discussions reflects equally on all participants.

It remains for me to acknowledge our debt to the I.E.A. Council, who supported this Conference, in particular Professor Chapanis who provided the original inspiration and Dr. de Jong who guided the deliberations of the Organizing Committee; to the members of the committee who provided such stimulating sessions in Amsterdam before the Symposium took place; to the participants who were in Amsterdam from 16th–18th September 1969; and to all the authors who have contributed written papers since then.

> W. T. Singleton,
> University of Aston in Birmingham,
> England.
> March 1970.

Professor Chapanis holds degrees in psychology. He has
worked on human factors research and applications in
industry and in the services, and now is Professor of
Psychology at Johns Hopkins University, Baltimore,
U.S.A. He has had numerous consulting assignments
with industrial companies and official agencies.

The search for relevance in applied research

Alphonse Chapanis

1. Human factors engineering and ergonomics

Human factors engineering, or its equivalent, human engineering, is a term used almost exclusively on the North American continent. Everywhere else, with the possible exception of the U.S.S.R., *ergonomics* is the word that most closely approximates its American analogue. Although these two terms, *human factors engineering* and *ergonomics*, look and sound worlds apart, the fields they represent are more nearly alike than they are different. Both are concerned with designing for human use. More specifically, both disciplines apply information about human characteristics, capacities, and limitations to the design of human tasks, machines, machine systems, and environments so that people can work safely, comfortably, and effectively.

For all that, there is one way in which ergonomics differs from its sister discipline. This is a difference that has been observed, discussed, and described informally and in writing (Fowler 1969, and Singleton *et al.* 1967) on both sides of the Atlantic : ergonomics seems to be more physiologically-oriented than does its American counterpart.

2. The general purpose of this symposium

It's a good thing that the study of man–machine systems has developed along somewhat different lines in Europe and America. Both ergonomics and human factors engineering are relatively new disciplines, and, as new disciplines, they are not always sure of themselves, of their goals and their methods. The different emphases that we have used in our work provide us with a unique opportunity to compare and evaluate the merits of our respective approaches.

This symposium was conceived as a way for a number of us to get together and, through mutual discussions, to develop more powerful and more useful strategies for solving problems of man at work. I think of this symposium as being dedicated entirely to methodological problems, to an exploration of tactics and techniques. By pooling our collective experiences and findings, and through our evaluations of them, I am confident that we will strengthen and further the technology to which we are bound by our common interests.

3. The purpose of this paper

It would be presumptuous even to guess, at the start of this symposium, what we shall conclude, or what we shall and shall not be able to agree on. For one thing, I am not entirely sure of many answers and I myself have come to this symposium in a spirit of search and inquiry. What I can do, however, is to structure the problem as I see it. In so doing I hope that we can at least all start our deliberations from some common ground of understanding even though we may have arrived here from different directions. With this in mind, then, I shall try to do two things.

First, I shall formulate several elementary propositions about the conduct of research in general, and of man–machine research in particular. It is possible that you will not find my statements profound because I may be merely stating things that you know, believe, and take for granted. Yet there is, I think, some merit in getting these thoughts down on paper. Some of the points I shall make may be so obvious that we tend to forget them when we become enmeshed in the intricate details of our own research.

1

The second thing I shall try to do is to state clearly and simply the exact nature of the problem that has brought us all here together.

4. Some propositions about the research process

Research is a many-sided activity and what we might say about it depends, to some extent, on which face we happen to be looking at. Let me remind you, however, that the title of our symposium directs our attention to *criteria*.

4.1. *Our basic problem is one of criteria and dependent variables*

An experiment may be described simply as a deliberately arranged situation in which the experimenter:

- varies some factors (the *independent* variables), while
- minimizing the influence of some other factors that are not of interest to him at the moment (the *controlled* variables),
- in order that he may measure changes in behaviour (the *dependent* variables) that are the result of, or are produced by, the independent variables.

To start with, let us all agree that in this symposium we are primarily concerned with the selection and choice of *dependent* variables in our research. Although the title of our symposium then goes on to specify *physiological* and *psychological criteria*, I would like to frame the purpose of our symposium in a somewhat larger context. It is not simply a matter of deciding about physiological versus psychological variables, although that is certainly an important decision that the experimenter must make. In my opinion, the real problem can be put in this way, 'How do we decide what kind of output, index, or criterion to measure when we do an experiment?' I hope that each of us will keep this question in mind throughout the remainder of our meetings.

4.2. *Any experiment can be designed with many possible dependent variables*

My next proposition is that any competent research man can think of a great many different dependent variables for any experiment. Let us take a relatively simple and circumscribed problem: the problem of evaluating the legibility of type. In their comprehensive review of the literature on this topic, Cornog and Rose (1967) found that over the years investigators have used an astonishing variety of dependent variables in studying this problem. Some of these dependent variables are:

- the amount of meaningful information retained after reading connected prose or text,
- the time taken to read a sample of connected text,
- errors made in recognizing individual symbols,
- the number of symbols recognized per unit time,
- the number of words or symbols that can be read on either side of a fixation point,
- recognition thresholds measured in terms of the distance at which symbols first become identifiable,
- recognition thresholds in terms of the amount of light at which symbols first become identifiable,
- speed in transcribing symbols into a machine,
- errors in transcribing symbols into a machine,
- eye-blink rate during reading,
- heart rate during reading, and
- nervous muscular tension during reading.

This list by no means exhausts the dependent variables that have been used in this kind of research, but it will perhaps convey some idea of the diversity of measures that can be used with even a fairly limited kind of problem. Remember that the independent variable was in every case some variation in typography.

If we turn to a more general kind of problem, for example, the effects of noise on human behaviour, we find that the number of dependent measures increases greatly. Table 1 gives a partial list of some that have been used in studies of this kind. This list can, of course, be multiplied several times by subdividing each of the variables in it. For example, one can take at least two separate measures, speed and accuracy (or errors), in tasks such as cancelling *c*'s, name checking, and number checking. Time estimation can easily be divided into four different measures: the estimation of short time intervals, the estimation of long time intervals, the estimation of filled time, and the estimation of unfilled time. As you know, each of these tends to give results that differ from each other (Woodrow 1951). EEG records yield several different measures; and so on.

Table 1. A partial list of dependent variables that have been used in studying the effects of noise on human behaviour. (From Plutchik 1959.)

Annoyance value of the noise	Minnesota Clerical Test
Auditory fatigue	Minnesota Form Board Test
Blood oxygen saturation	Monitoring lights
Blood pressure	Monitoring steam pressure gauges
Cancelling *c*'s	Muscle tension
Clock-watching	Name checking
Critical flicker fusion	Number checking
EEG	Palmer sweating
EMG	Peristaltic contractions
Extrapolating the movement of a visual target	Pulse rate
Finger volume	Respiration amplitude
Flow of gastric juices	Serial reaction time
Hearing loss	Somatic complaints
Inserting a stylus into a moving tape containing irregularly-sized and irregularly-spaced holes	Subjective feelings such as irritation and distraction
	Time estimation
	Trembling
	Word fluency

Let us look at this in still another way. Imagine that you are designing an experiment and that you want to use some sort of a measure of psychomotor performance as a dependent variable. How many different measures do you have to pick from? No one really knows. When Fleishman and Hempel (1954) made their factor analytic study of dexterity they used 15 different tests. In a later study of physical proficiency and manipulative skill (Hempel and Fleishman 1955) they used 46 different tests; and in a still later article on complex psychomotor performance (Fleishman and Hempel 1956) they used 23 different tests. Collectively these add up to 84 separate tests, each of which claims to be a measure of some sort of psychomotor performance. You could use any one of them as a dependent variable.

These examples are enough to establish the point. Any clever research man can easily find a large number of different dependent variables for any particular experiment he has in mind.

4.3. *Dependent measures sometimes give the same results*

Sometimes apparently diverse dependent measures give essentially the same results. When this happens the experimenter can consider himself fortunate. An illustration of such a happy coincidence can be found in a study by Jenkins and Connor (1949). Among other things, these investigators were concerned with what is now referred to as the C/D, or control-display, ratio. This is the gearing ratio between the movements of a control and the movements of some element on a display that is linked to the control.

The experimental apparatus used by Jenkins and Connor was simple. The subject grasped with his right hand a control knob located at waist level in front of him. Rotating the knob moved a pointer along a linear scale directly in front of the subject's eyes and at a normal reading distance from them. Any one of ten gearing ratios could be introduced between the knob and the pointer. At one extreme, the pointer moved only 0·22 in. for a complete revolution of the knob; at the other extreme it moved 34 in. The subject's task was also simple: at a signal he moved the pointer until it matched, within very close tolerances, a particular position on the scale.

Jenkins and Connor used two dependent measures: (a) the total time to make a setting (a psychological, or performance, measure); and (b) accumulated action potentials taken from the subject's active forearm (a physiological measure). Since the experimenters did not collect data for all subjects under all conditions, I show in Figure 1 average data for only three subjects for whom comparable data are available at the gear ratios shown. These are the subjects identified

Figure 1. Average performance of three subjects in setting a pointer on a linear scale as a function of the gearing ratio between the movements of the pointer and the knob. (After Jenkins and Connor 1949.)

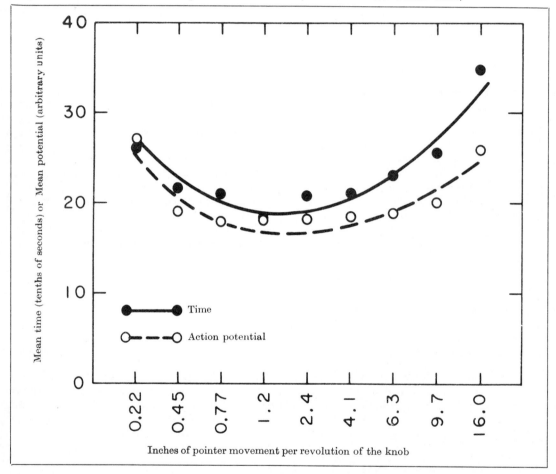

as DMS, HWQ, and JKD in their Table 1. Everything considered the agreement between the two dependent measures is remarkable. Both curves show that there is a middle range of gear ratios, from about 0·77 to 4·1 in. of pointer movement per revolution of the knob, at which performance is best. Performance is poorer beyond this range in either direction. All in all, these findings represent a comforting and reassuring correspondence.

Another example of the close correspondence that can be found between apparently diverse dependent measures is shown in Figure 2. These are data on heart rate level, arithmetic computation, skin resistance level, and probability monitoring taken at periodic intervals throughout the day over a period of 15 days. The data are averages for two teams of man (11 in all) who were isolated in a special chamber simulating a kind of space vehicle. Although all four measures show regular fluctuations corresponding to earthly day and night cycles, the men were actually

Figure 2. Average performance of 11 subjects during a 15-day confinement. (After Adams and Chiles 1961.)

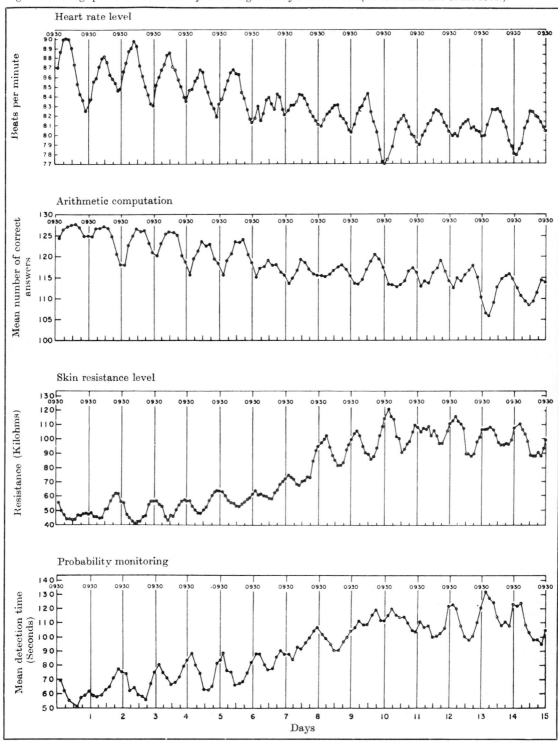

not on a normal earth schedule. Throughout this entire period the men alternated four hours of work with two hours of rest, a schedule that might be used in certain types of missions where equipment has to be monitored around the clock.

Although the four measures in Figure 2 are quite different, the curves are strikingly similar, not only in long-term trends, but even in the small daily cycles that are imposed on the long-term trends. Note that the daily fluctuations for the lower two curves are inverted as compared with the two upper ones. This is a trivial matter for it is only a reflection of the way the measures are expressed. The important thing is that the curves are almost exact mirror images of each other. I have had two of my assistants compute the intercorrelations among these four measures in two ways, first without making any correction for long-term trends, and second, after eliminating, or partialling out, the effect of the long-term trends. The results of these computations are shown in Table 2. The correlation coefficients for the raw data are surprisingly high. Eliminating the effects of the overall trends results in correlations between the daily cycles only. This procedure reduces the magnitudes of all the correlation coefficients although some of them still remain high.

Table 2. Intercorrelations among the four measures shown in Figure 2. The upper value in each case is the correlation coefficient for the data as they stand, uncorrected for long-term trends. The lower value is the correlation coefficient remaining when the long-term trend has been eliminated, or partialled out, for each set of data.

	Arithmetic computation	Skin resistance	Probability monitoring
Heart rate	$+0.77$	-0.85	-0.84
	$+0.22$	-0.49	-0.35
Arithmetic computation		-0.83	-0.88
		-0.44	-0.61
Skin resistance			$+0.95$
			$+0.58$

Examples such as these are not easy to find. When they do occur they usually bring joy to a research man's heart because they simplify greatly his task of summarizing what his data mean. In any case, it is important for us to note that dependent variables, sometimes even quite different dependent variables, may yield essentially identical results in an experiment.

4.4. *Dependent variables may be correlated but may have different sensitivities*

Sometimes dependent variables may be correlated, that is, they may all show effects in response to changes in independent variables, but the dependent variables may have different sensitivities. By differential sensitivity I mean that one dependent measure may show significant effects sooner than another, for example, with fewer data, or with smaller variations in the independent variable.

A simple example of what I mean is in the correlation between time and errors. There is now a substantial body of evidence to show that in human factors work, time and errors are positively correlated. Let me clarify. Assume that we have tested N pieces of equipment (or arrangements, or layouts, or work methods) and that for each of the N items we have an average time and an average error rate. The correlation between these average times and errors for the N items will almost certainly be positive, that is, those designs or arrangements of apparatus that can be used fastest are also the ones that can be used with fewest errors. Hence, the correlations I am referring to here are not correlations of time and errors between or within individuals. They are correlations of time and errors for pieces of equipment.

Even though time and errors are so related, it is also abundantly clear that time is a more sensitive measure than errors. (See, e.g., Chapanis and Lindenbaum 1949, Chapanis and Lockhead 1965, Chapanis and Mankin 1967, Chapanis and Scarpa 1967, Scales and Chapanis 1954, Shackel 1959, Whitfield 1964, and Wright *et al.* 1969.) When both measures are collected in the same experiment, on the same subjects, and on the same variations of apparatus, it is easier to find significant results with the time measures than with the error measures.

Another quite different example can be found in the effects of anoxia on performance. Anoxia is a condition of oxygen deprivation that occurs most commonly on ascent to high altitudes. The dark-adapted threshold of the human eye is one of the first bodily functions to show a significant effect due to anoxia and, indeed, increases in the dark-adapted threshold have been demonstrated at altitudes as low as 2,260 m. By contrast, significant decrements in arm–hand steadiness, pursuit tracking, and mental computations are not demonstrable below altitudes of about 3,660 m and similar decrements in simple memory tasks and the ability to name colours do not appear until altitudes of 4,570 m have been reached. Still other psycho-motor tasks, for example, reaction time, show no significant decrements below altitudes of about 5,490 m (Ernsting 1965).

It is important to note that these large variations in sensitivity to the effects of anoxia are not confined exclusively to the physiological or the psychological measures. Nor can one assert that one class of measures is more sensitive than the other. For example, although vision seems to be particularly sensitive to oxygen lack, hearing is particularly resistant to it (Ernsting *op. cit.*). Although changes in pulse rate are detectable with only mild anoxia, changes in arterial blood pressure do not occur until the individual is suffering from severe oxygen deprivation (Green 1965 and McFarland 1938).

4.5. *Sensitivity is not necessarily a virtue in a dependent variable*

My next proposition follows from the preceding one. It is that the more sensitive of two dependent measures is not necessarily the better one. The issue here concerns the distinction between what is called the statistical significance of a result and its practical significance (Chapanis 1959, Chapanis 1967, and Leplat 1969). In striving for statistical significance, or sensitivity (used here in the special way I defined it in Section 4.4), in a dependent variable, the experimenter runs the risk of finding results that are of trivial importance from a practical standpoint.

In basic research the experimenter is typically unconcerned with the practical significance of his findings. In ergonomics work, however, we cannot be so cavalier. If the results of an experiment show that the better of two methods, devices, or procedures, can account for only one per cent more of the total variability in performance, we will have squandered valuable time and resources in running that experiment. We cannot ignore the practical consequences of our findings if we are to make substantial advances in our science and technology.

4.6. *Dependent variables may be unrelated*

My next proposition is that many dependent variables are unrelated to each other. Examples are so numerous that you merely need to scan through abstracts and summaries of research studies in almost any issue of any journal to find them. The literature abounds with experiments in which results obtained with one dependent variable seem to be unrelated to those obtained with some other dependent variable. Below are four quotations from four different sources to show the kind of thing I mean.

- This experiment was designed to compare the information-handling performance of Ss in making verbal and motor responses to two sets of Arabic numerals.... When verbal responses were made, the conventional numerals were consistently superior.... No such clear superiority was evidenced for either set of numerals when the motor responses were made. (Alluisi and Martin 1958.)
- Measurements of simple motor performance during whole-body vibration have yielded varied results, depending more upon the nature of the tests than upon the quality of the environmental stress. (Guignard 1965.)
- A very large decrement in detection performance was observed, but it was unrelated to changes in skin conductance levels.... There was no significant relationship between overall conductance level and performance. (McGrath *et al* 1959.)
- An experiment ... tested the effects of a 90 db, 8000 cps tone on the four measures of critical flicker fusion, cancelling *c*'s, word fluency, and trembling. Only the last measure showed any effect.... (Plutchik 1959.)

Statements such as these are all the more interesting because when an experimenter begins an experiment he makes a deliberate selection from among all the possible dependent variables available to him. What he tries to do is to select those dependent variables that, in his opinion, are likely to give him significant results. His goal, after all, is to publish a paper, and non-significant results do not usually make publishable material. It is instructive, therefore, that one can find so many non-significant relationships among dependent variables even when those variables have been deliberately selected to give positive results.

I suspect, although I cannot prove, that if we could do a gigantic cross-correlational study of all conceivable human dependent variables we would find that most of the correlations would be zero. Some support for my viewpoint comes from the three factor analytic studies to which I have already referred. Fleishman and Hempel discovered five orthogonal (that is, independent) factors among their 15 tests of dexterity; 15 orthogonal factors among their 46 tests of physical proficiency and manipulative skill; and 9 orthogonal factors among their 23 complex psychomotor tests. This is a total of 29 unrelated and independent factors all falling under the general heading of psychomotor performance.

The non-relatedness of many dependent variables in human research has two important implications that are worth dwelling on for a moment. First, although investigators may use the same words in describing their respective experiments, this does not mean that they are really talking about the same things. So when you see a number of experiments that purport to have measured the effects of something on psychomotor performance, or on complex decision making, or on vigilance, or on cardio-vascular efficiency, it is important to look closely at the exact dependent measures that have been used in those experiments. The data may or may not refer to the same human functions. The second implication is that we have to choose carefully and wisely when we plan applied research. If our experiments are to be meaningful, our dependent measures must have some relation to the tasks in the real-life problem that we are investigating.

4.7. *Sometimes negative findings are valuable*

In a certain sense it is unfortunate that the customs of science dictate our research strategies. Dependent variables should be picked for their relevance to some practical problem and not only because they are likely to yield significant results. In ergonomics and human factors work, it is sometimes important to know that some things *do not* make a difference. If you can tell an engineer with confidence that it does not make any difference which of two devices, or methods, he uses, you give him much greater freedom in his design work.

A good example comes from our early work on weightlessness. Before man had actually flown into outer space, there was a great deal of concern and worry about how well he could perform under zero-g conditions. Early experiments on the psychomotor performance of human operators under zero-g conditions quickly dispelled all these early fears (Simons 1964). So long as an operator was tethered, or held in place, there was no demonstrable effect of zero-g on psychomotor performance. Although the finding was negative, its practical impact was great. It meant that engineers and designers could ignore this problem in their design work. As you know, our numerous subsequent flights into space have amply demonstrated that this was indeed a needless worry. To sum up my point, then, in our studies of the tactics of applied research we must not overlook the value of negative findings to properly phrased practical questions.

5. The particular purpose of our symposium

Having stated some propositions that I think are important background considerations for our symposium, let me turn now to the question, 'What exactly is the problem that brings us here together?'

5.1. *Our interests are in applied research, not in basic research*

My next proposition is that in this symposium we are concerned primarily with applied research and not with basic research. Let me make it absolutely clear right now that I believe wholeheartedly in basic research. I think basic research is good, we need more of it, and it should be supported everywhere. Nonetheless, I think that we need to keep clearly in mind the purpose of this symposium. As ergonomists or human factors engineers we are not primarily concerned with the study of man simply for the knowledge itself. Nor are we primarily concerned with the study of man only to prove or to disprove theories about his behaviour and functioning. Ours is a practical goal : we are interested in studying man primarily so that we can design better man–machine systems, and better evaluate how these systems perform (Kraft 1958, Wood 1958).

I think of the sum total of human knowledge, the *Body of Knowledge,* as a gigantic amorphous mass that is constantly changing in size and shape as new knowledge extends the boundaries first in one place, then in another. There are, of course, gaps, holes, and other discontinuities on the surface and inside the mass. The mass itself is made up of an intertwined and overlapping nexus of facts, generalizations, principles, and theories. These originate from the various sciences : astrophysics, biology, chemistry, geography, geology, physics, physiology, psychology, sociology, and so on.

The basic scientist is free to wander anywhere inside this mass and to investigate anything that excites his curiosity. Indeed the basic scientist is limited only by his own resources and energies. The ergonomist and human factors engineer, on the other hand, have to provide answers to practical questions. These questions are not self-generated : they come from the problem itself. Only a very small percentage (much smaller than a fraction of one per cent) of all the information inside this *Body of Knowledge* can be used in the solution of any practical problem. I have already written about how little so-called basic research is useful even when the research was ostensibly undertaken to solve some real-life problem (Chapanis 1967) : a viewpoint that is supported by Murrell's recent paper (1968). One of the most difficult tasks for the ergonomist or human factors engineer is to find and identify that very small percentage of information that will really contribute to the solution of whatever problem he may have at hand. When the information does not already exist, it will not come from basic research carried out in complete freedom. It will come from research that has been carefully planned and directed with the specific applied question in mind. In fact, I am prepared to argue that in psychology at least the best basic research has had its origins in some applied problem ; in some attempt to grapple with and find a solution for a real-world question. But that is the subject for another paper.

Even if you do not agree on this point, I hope that we can at least agree that our primary goal is the solution of practical problems. This symposium is not concerned with the accumulation of basic information *per se.*

5.2. *On levels of explanation*

There is another facet to the distinction between basic and applied research. This has to do with the level of understanding that is required for applied work as compared with the level of understanding that the basic scientist tries to achieve. Some of you will point out that there is no genuine discontinuity between basic and applied research. Applications depend on a storehouse of good basic information. I agree. But I would also argue that there are several levels of explanation in any science. For most practical purposes one does not need to dig deeply into these layers of explanation to find important, correct, and useful information.

The electronic engineer who is designing a circuit for an amplifier does not need to know the exact molecular composition and structure of the transistors that he designs into his circuits. He needs only to know the input and output characteristics of these transistors. Similarly, the systems programmer who is designing a computer into an automatic accounting system for a bank does not need to know all about the wires in the computer or the way the magnetic core memory units work. That is much too microscopic a level of information for his needs. Not only is such detailed information unnecessary but it may do more harm than good because it clutters up his mind with details that he cannot possibly use in his work. By analogy, I think that there is a correct level of understanding for the applied behavioural scientist as well. Let me illustrate with a couple of examples.

As you know, about eight per cent of otherwise normal men have some sort of colour vision defect. There are several levels at which one can comprehend this phenomenon. One may comprehend it at a relatively simple descriptive level, for example, that most people with defective colour vision confuse red, yellow, and green colours, and that they rarely confuse yellow and blue colours. These simple statements can, of course, be quantified with much greater precision by providing the exact physical specifications of those colours that are commonly confused and those that are seldom confused.

At another, more basic level of understanding, one may classify colour vision defects into anomalous trichromatism, dichromatism, and monochromatism. One can also describe completely the colour-mixture matching and other psychophysical curves that are characteristic of each of these kinds of colour vision defects. Further, one can work out precise hereditary patterns that yield these colour vision defects, and finally, one can speculate with some reasonable certainty about the photochemical, receptor, and ganglion cell abnormalities that account for defective colour vision.

I claim that for applied purposes the first descriptive level of understanding is enough. The man who is designing a man–machine system needs only to know that if (a) he is going to use colours for identification or coding purposes, and if (b) the operators will be unselected in so far as their visual characteristics are concerned, he must be careful about his choice of colours. I argue that it does the designer of equipment no good at all to know about rods, cones, electro-retinograms, photochemicals, or sex-linked recessive inheritance patterns. Although these things may be all important to the basic scientist, they are irrelevant to the practising ergonomist, human factors engineer, and design engineer.

I feel so strongly about this point that I would like to illustrate with one more example. As you know, most handbooks of ergonomics and human factors engineering contain information about so-called compatible and non-compatible ways in which displays and controls should move. For example, people read dials and scales most easily if the numbers increase in a clockwise direction around the face of the dial. Similarly, most people expect that a clockwise rotation of a knob should produce a clockwise movement of a pointer on a scale. And so on. Figure 3 shows examples of some simple compatible and non-compatible relationships that apply to the movements of a knob and pointer on a scale.

An interesting basic question, and one that I am frequently asked, is this one: 'Are such compatible movement relationships learned through years of experience with the devices around us, or are they the result of some kind of innate, biological mechanism that pervades our perceptual systems?' However interesting this question may be, I contend that the answer to it is of little use to the practical man who has to design something right now. We can show that errors and accidents are more likely to occur when compatible movement relationships are violated in the design of machine systems. It is true that people can, with sufficient training, learn to use non-compatible movement relationships. But when such people are stressed they tend to forget the non-compatible patterns they have learned and regress to the more compatible ones.

Given these facts, I claim that it makes no difference whether these compatibility relationships are learned or innate. The man who is designing automobiles to be used next year, or the engineer who is designing a space capsule for the next flight to the moon, would be foolhardy to design controls and displays that moved in non-compatible ways. The answer to the basic question has nothing to do with the practical design decision.

All that I have been trying to say about levels of explanation can be summarized in this way: I believe that it is possible to have a proper and valid systems science based on principles of *system* functioning. For the design and evaluation of systems these principles do not have to be phrased in molecular terms. Similarly, I believe that for the solution of man–machine systems

Figure 3. Some compatible knob–pointer movement relationships on the left and some non-compatible movement relationships on the right. (After Chapanis 1965.)

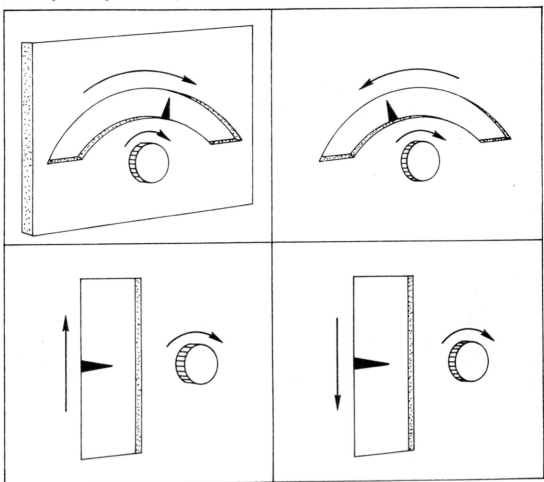

problems it is possible to generate valid principles of human behaviour that do not go into the neurophysiology of body functioning. To formulate useful statements about how the eye receives information from radar scopes it is not necessary to know all about rods, cones, and ganglion cells. To understand human decision making in practical situations, it is not necessary to know about EEG's and cortical responses. To design space capsules for operation under conditions of zero-g it is not really necessary to know about semicircular canals and otolith organs.

5.3. *On systems criteria*

So far I have been talking entirely from the experimenter's point of view. It is time to turn the problem around and to look at it from the standpoint of the engineer and the systems designer.

Let us approach the problem this way. When you are trying to decide what kind of a new automobile to buy, what factors do you consider? Do you judge an automobile by what it will do to your blood pressure, respiration rate, or brain wave pattern? I doubt it. I am equally sure that you are not at all concerned about what the automobile might do to your critical flicker fusion threshold, your reaction time, or to that hypothetical quantity that has been called your 'spare mental capacity' (Brown and Poulton 1961). On the contrary, I strongly suspect that you will make your decision on the basis of one or more of the following: appearance, availability of service and parts, cost, ease of handling, economy of operation, frequency of repair record, luggage storage capacity, riding qualities, safety features, size, speed.

11

To generalize, the value or worth of a system is normally judged by several criteria. These criteria are not necessarily all compatible. In the case of automobiles, for example, speed and economy of operation are, in general, incompatible. Faster, high performance cars are usually more costly to operate than those that have less power. Criteria also vary greatly from system to system and many criteria are specific to particular systems. Among the criteria that might be used to evaluate an automated assembly line in a factory are (a) the number of units produced per unit time, and (b) the amount of raw materials wasted during the manufacturing process. Criteria that are specific to telephonic communication systems are (a) the naturalness of the transmitted voice signals, and (b) freedom from crosstalk. Military systems are frequently evaluated by a criterion of kill probability. Systems criteria are not necessarily obvious and easy to agree on. Nonetheless, anyone who designs, builds, or buys a system does so according to certain systems criteria, criteria that are either explicit or implicit.

Table 3. Some common dependent measures (or criteria) used in ergonomic and human factors research (in the column on the left) and some general systems criteria (in the column on the right).

Experimental criteria	Systems criteria
Accuracy (or, conversely, errors)	Anticipated life of the system
Cardiovascular responses	Appearance
Critical flicker fusion	Comfort
EEG	Convenience
Energy expenditure	Ease of operation or use
Muscle tension	Familiarity
Psychophysical thresholds	Initial cost
Ratings	Maintainability
(e.g., of annoyance, comfort, etc.)	(e.g., mean time to repair)
Reaction time	Manpower requirements
Respiratory responses	Operating cost
Spare mental capacity	Reliability
Speed	(e.g., mean time to failure)
Trials to learn	Safety
	Training requirements

In Table 3 I have listed a 'baker's dozen' of important criteria (in the right-hand column) that apply to many, if not most, man–machine systems. These criteria are self-explanatory. Other things equal, the better of two systems is the one that (a) has a longer anticipated life, (b) has the more pleasing appearance, (c) is more comfortable for the people who use it, (d) is more convenient, (e) is easier to operate or use, (f) is more familiar, that is, more similar to systems in existence, (g) is cheaper to buy, (h) is quicker to repair, (i) requires fewer operators, (j) is cheaper to operate, (k) breaks down less often, (l) is safer, and (m) can be used with less highly trained personnel. Criteria like these, I contend, are the real indicators of system performance. These are the criteria with which we must come face-to-face if we are to do our jobs as proper ergonomists and human factors engineers.

I do not think that I need to prove to you that ergonomics or human factors considerations enter into every one of these systems criteria. What may not be so obvious, however, is that the human considerations almost always get involved in compromises of one sort of another : in what engineers frequently refer to as 'trade-offs'. A simple example of such a compromise is the decision about what sizes of doors and accesses to provide in an aircraft. On the one hand, the designer would like to have doors and windows as large as possible so that, in the event of an emergency, passengers can be evacuated as quickly as possible. Human factors engineers have, in fact, studied this very problem. But the design problem is not solved that easily. Designing aircraft with large doors and windows decreases the structural strength of the aircraft, and thereby increases the probability that it will be seriously damaged in storms, turbulent air, and on impact with the ground. The engineer is, therefore, faced with the necessity of arriving at a delicate balance between two conflicting requirements. He must make doors big enough so that people can get out in a reasonable period of time, but not so big that he unduly weakens the aircraft. Safety and survivability are complex criteria.

The ergonomist or human factors engineer working in his laboratory is often unaware of these conflicting requirements, but they are there nonetheless. The necessity for making such compromises often explains why abstract laboratory findings cannot be accepted without modification in the practical world of systems design. Systems criteria are not neat and simple linear transformations of the outcomes of basic research.

5.4. *How do we relate experimental criteria to systems criteria?*

We come at last to the heart of my paper and to what I think is the real purpose of this symposium. In the left-hand column of Table 3, a few representative dependent variables or experimental criteria are listed as they appear in ergonomics and human factors research. This, of course, is a highly abbreviated list. Still, it is a set of dependent variables that you can find used over and over again in our journals.

As you look at the two columns of this table, I think you will have to agree that there seems to be little correspondence between them. It's difficult to see what some of the experimental criteria in the left-hand column have to do with any of the systems criteria in the right-hand column. Moreover, a number of systems criteria do not seem to be measured by any of the experimental criteria.

The question I would like to leave with all of you is the following: 'How can the variables in the left-hand column be matched to those in the right-hand column?' An alternative way of posing the problem is as follows: 'Of all the dependent variables that a research man could use in an experiment, how can he pick those that will have the greatest amount of transfer to the criteria that will be used in the design and evaluation of a system?'

There is one last point. In our attempts to relate experimental criteria to systems criteria, I do not think that we should be persuaded by logic, intuition, or appeals to our common sense. If there are connections between our experimental variables and the things that we want to measure about systems, these connections should themselves be demonstrable and measurable. In what follows in this symposium I hope that we shall not only hear about procedures and methods. I hope that we shall also hear about data that support those procedures and methods as defensible approaches to the solution of the increasing pile of man–machine problems that remain to be tackled.

This paper was prepared under Contract Nonr 4010(03) between the Office of Naval Research and The Johns Hopkins University. It is Report Number 22 under that contract. Reproduction in whole or in part is permitted for any purpose of the United States Government.

References

ADAMS O.S., and CHILES W.D., 1961, Human performance as a function of the work-rest ratio during prolonged confinement. *U.S. Air Force, Wright-Patterson Air Force Base, Ohio, Aerospace Medical Laboratory ASD Technical Report* 61–720.
ALLUISI E.A., and MARTIN H.B., 1958, An information analysis of verbal and motor responses to symbolic and conventional Arabic numerals. *Journal of Applied Psychology*, **42**, 79–84.
BROWN I.D., and POULTON E.C., 1961, Measuring the spare 'mental capacity' of car drivers by a subsidiary task. *Ergonomics*, **4**, 35–40.
CHAPANIS A., 1959, *Research Techniques in Human Engineering* (Baltimore, Maryland: The Johns Hopkins Press).
CHAPANIS A., 1965, *Man–Machine Engineering* (Belmont, California: Wadsworth Publishing Co., Inc.).
CHAPANIS A., 1967, The relevance of laboratory studies to practical situations. *Ergonomics*, **10**, 557–577.
CHAPANIS A., and LINDENBAUM L.E., 1949, A reaction time study of four control-display linkages. *Human Factors*, **1**, 1–7.
CHAPANIS A., and LOCKHEAD G.R., 1965, A test of the effectiveness of sensor lines showing linkages between displays and controls. *Human Factors*, **7**, 219–229.
CHAPANIS A., and MANKIN D.A., 1967, Tests of ten control-display linkages. *Human Factors*, **9**, 119–126.
CHAPANIS A., and SCARPA L.C., 1967, Readability of dials at different distances with constant visual angle. *Human Factors*, **9**, 419–425.
CORNOG D.Y., and ROSE F.C., 1967, *Legibility of Alphanumeric Characters and Other Symbols: II. A Reference Handbook* (National Bureau of Standards Miscellaneous 262–2, Washington, D.C.: U.S. Government Printing Office).

ERNSTING J., 1965, The effects of anoxia on the central nervous system. In *A Textbook of Aviation Physiology* (Edited by J.A. GILLIES) (London: Pergamon Press).

FLEISHMAN E.A., and HEMPEL W.E. Jr., 1954, A factor analysis of dexterity tests. *Personnel Psychology*, **7**, 15–32.

FLEISHMAN E.A., and HEMPEL W.E. Jr., 1956, Factorial analysis of complex psychomotor performance and related skills. *Journal of Applied Psychology*, **40**, 96–104.

FOWLER R.D., 1969, An overview of human factors in Europe. *Human Factors*, **11**, 91–94.

GREEN I.D., 1965, The circulation in anoxia. In *A Textbook of Aviation Physiology* (Edited by J.A. GILLIES) (London: Pergamon Press).

GUIGNARD J.C., 1965, Vibration. In *A Textbook of Aviation Physiology* (Edited by J.A. GILLIES) (London: Pergamon Press).

HEMPEL W.E. Jr., and FLEISHMAN E.A., 1955, A factor analysis of physical proficiency and manipulative skill. *Journal of Applied Psychology*, **39**, 12–16.

JENKINS W.L., and CONNOR M.B., 1949, Some design factors in making settings on a linear scale. *Journal of Applied Psychology*, **33**, 395–409.

KRAFT J.A., 1958, Industrial approaches to human engineering in America. *Ergonomics*, **1**, 301–306.

LEPLAT J., 1968–1969, La méthode expérimentale en ergonomie. *Bulletin de Psychologie*, **22**, 775–781.

McFARLAND R.A., 1938, The effects of oxygen deprivation (high altitude) on the human organism. *Washington, D.C., Civil Aeronautics Authority Technical Development Report No.* 11.

McGRATH J.J., HARABEDIAN A., and BUCKNER D.N., 1959, Review and critique of the literature on vigilance performance. *Los Angeles, California, Human Factors Research, Inc., Technical Report* 206–1.

MURRELL K.F.H., 1968, On the validity of ergonomic data. *Occupational Psychology*, **42**, 71–76.

PLUTCHIK R., 1959, The effects of high intensity intermittent sound on performance, feeling, and physiology. *Psychological Bulletin*, **56**, 133–151.

SCALES E.M., and CHAPANIS A., 1954, The effect on performance of tilting the toll-operator's keyset. *Journal of Applied Psychology*, **38**, 452–456.

SHACKEL B., 1959, A note on panel layout for numbers of identical items. *Ergonomics*, **2**, 247–253.

SIMONS J.C., 1964, An introduction to surface-free behaviour. *Ergonomics*, **7**, 23–36.

SINGLETON W.T., EASTERBY R.S., and WHITFIELD D. (Editors), 1967, Preface. In *The Human Operator in Complex Systems* (London: Taylor and Francis).

WHITFIELD D., 1964, Validating the application of ergonomics to equipment design: a case study. *Ergonomics*, **7**, 165–174.

WOOD C.C., 1958, Human factors engineering: an aircraft company chief engineer's viewpoint. *Ergonomics*, **1**, 294–300.

WOODROW H., 1951, Time perception. In *Handbook of Experimental Psychology* (Edited by S.S. STEVENS) (New York: John Wiley and Sons, Inc.).

WRIGHT P., HULL A.J., and CONRAD R., 1969, Performance tests with non-circular coins. *Ergonomics*, **12**, 1–10.

Section 1 - Man

Prologue

The measurement of man is not an area of knowledge where sudden insights and rapid advances are common. To approach even a moderate comprehension of the problem requires a level of cerebration to which even research workers rarely rise. The general scientific situation has been summarized by Russell (1948), and a correspondingly clear statement of the stage of knowledge reached by the professional specialist in human sciences is provided by Stevens (1951). However, within ergonomics there is a job to be got on with. In common with the engineer and the clinician we cannot refuse to cope with problems on the grounds that our solutions will be less than perfect because of the limited state of knowledge.

Our general problem can be stated thus. Any task makes demands on the man required to carry it out. What can we find out about these demands which will lead to better selection and training, better machine design, better work space design, better job design, better environmental design? In each case what do we mean by 'better' and how do we measure our level of success?

Since he has been educated as a scientist the first reaction of an ergonomist faced with a new problem is that he must measure something. But what and how and why? A measure is a datum obtained using a standard procedure under a standard set of conditions which hopefully results in adequate reliability. That is, when the procedure is repeated under the same conditions the result obtained is sufficiently similar to that obtained previously to justify treating the set of such repeated measures as a statistical entity. Section 2 will deal with recent advances in techniques which are providing meaningful measures.

The meaningfulness of a measure depends on its relationship to one or more descriptors. A descriptor is a dimension of man which might be interesting for one of two reasons. Firstly, it might be directly relevant to some practical aspects of ergonomics, e.g. anthropometric measures are relevant to the physical dimensions of work-spaces. Secondly, it might be sensitive to changes in the state of a given man working at a given task, e.g. flicker fusion frequency changes in conditions of exacting visual work.

Thus, the reason for taking a measure is that it relates to a descriptor, sometimes indirectly, sometimes directly. For example, a measure such as loss of weight with suitably controlled conditions can be translated directly into a sweat-rate which again with appropriate provisos can be used as a measure of heat stress which may or may not have an associated heat strain which finally is the relevant descriptor. This, which by ergonomic standards is a direct measure involves three logical or empirically based steps in the chain of argument. It is true that there are some more direct measures such as those obtained by measuring tapes or rulers in relation to the class of descriptors to do with physical size. Unfortunately most of the measures can only be connected to descriptors by an argument which is tenuous both logically and empirically; for example, changes in heart rate regularity as a measure of effort. These arguments often get so involved that it is not clear whether the measure is being used to support the existence of the descriptor or whether the descriptor is used as a justification for the measure; for example, in the use of performance on paper and pencil tests as a measure of intelligence.

The weakness of the present position is one reason why there seems to be so much ergonomics effort devoted to laboratory studies which apparently have little connection with man at work.

These studies are usually aimed at strengthening the evidence to support relationships between measures and descriptors. Ideally these relationships are demonstrated by direct causal connections, e.g. laboratory environments and body temperatures but frequently we have to be satisfied with reliable correlations, e.g. E.E.G. changes as a measure of depth of sleep.

The importance of clarifying the underlying assumptions and hypotheses in the relationship between measures and descriptors is most clearly seen in the context of environmental measurements. The problem arises because the importance of the physical parameters of the environment rests ultimately on their effect on man. Thus the reaction of man must be the basic referent and yet this is so variable as to be very difficult to quantify even in probabilistic terms. There is a conflict between the natural tendency to rely on physical measures which are well defined and unambiguous and their inappropriateness as single variables in relation to human reactions, e.g. temperature or bits of information.

From this point of view, the different techniques, conceptual models and favourite problems of the physiologists and psychologists may, by contrasts, provide us with some insight into these extremely difficult problems. The papers in this section were selected in this context and with particular reference to topics which active groups of workers in various countries seem to find interesting at the present time. A number of these problems are discussed in several papers, for instance arousal, stress and strain, physical and mental load. The importance of time of activity is also discussed in detail in two papers.

References

RUSSELL B., 1948, *Human Knowledge: Its Scope and Limits* (London: Allen & Unwin).
STEVENS S.S., 1951, Mathematics, measurement and psychophysics. In *Handbook of Experimental Psychology* (Edited by S.S. STEVENS) (New York: Wiley).

Professor Singleton holds psychology degrees. After skills research at Cambridge, he worked on ergonomics and training problems in the shoe industry. Subsequently, he taught ergonomics and now is head of the Applied Psychology Department at the University of Aston, Birmingham, England.

The measurement of man at work with particular reference to arousal

W. T. Singleton

1. Introduction

The operational aims of studies of man working are to improve the efficiency of the system in which the man functions and to ensure that his health is not in jeopardy.

The inherent difficulty is that all measures are partial and indirect. Partial in that health is itself a multidimensional concept and the role of the man as a contributor to system efficiency is also multidimensional. Indirect in that there is never an unambiguous relationship between the physiological and psychological data obtained by the investigator and the norms of acceptable health risk and required productivity which are essentially socioeconomic in origin. The difficulty would be reduced if there were more consistency between different measures, different individuals and different occasions but these are all serious sources of variability which add to the problems of interpretation and extrapolation. To use psychological jargon : criteria, validity and reliability are always suspect and subject to argument for all possible measures.

2. Psychological measures

Given so many uncertainties there is some justification for a behaviourist approach. Even though we have difficulty in interpreting our measures in biological terms we can at least choose measures which are direct indicators of performance such as speed and error rate and we can try to relate these to particular kinds of environment and task design. In other words, we can measure inputs and outputs and look at their interrelationship without concerning ourselves with the intermediate processes.

Unfortunately there are at least three difficulties about this approach. Firstly, speed and error data from an experimental comparison do not provide sufficient evidence to make decisions about potentially expensive changes in real equipment and tasks. Secondly, such data often contain no evidence on biological cost, an operator may be performing better in one of two tasks compared in an experiment because he is trying harder rather than because one task is easier than the other. Thirdly, unless the data are related to some conceptual model of human behaviour they will remain specific to the experimental situation and will make no contribution to more general knowledge.

On the practical limitations of speed and error data Chapanis (1961) has discussed the conflicting requirements of rigour and realism, the discrepancy between ease of measurement and systems relevance and the needs for probabilistic data. Wulfeck and Zeitlin (1962) propound a similar argument in terms of the importance of cost/value trade-off data and discuss the consequent need for more psychophysical data in the form of probabilistic error functions and the need to design experiments to provide data over broad ranges of variables with optimum interpolation accuracy. If these requirements can be met data in the form of empirical input/ output relationships are obviously relevant and justifiable but still they are only useful within the limits of the particular class of situations.

The problem of separating performance and effort will be raised again later. In this context it increases suspicions about the ambiguity of simple performance measures. This often occurs in the form of similar performance in two working situations which are obviously different because the operator sees it as his task to achieve a particular output rate and he adjusts his effort accordingly. This can be detected by various refinements of performance data (Singleton and Simister 1957) but it remains true that straightforward speed-error data will not easily reveal difference of effort. An objection at the more fundamental level has recently been raised by Deese (1969). He points out that using speed as a measure of complexity makes unwarranted assumptions about the essentially sequential rather than parallel nature of human information processing.

The importance of relating data to theory has been stressed continuously throughout the history of science, e.g. Poincaré (1913) 'There is a hierarchy of facts : some have no reach : they teach us nothing but themselves'. In our case we have the example of our predecessors in work study who concentrated entirely on techniques and profitability with the result that their discipline remained almost unchanged for decades.

It is obviously not easy to achieve the optimum compromise between practical utility and the development of abstract theory. The psychologist interested in man at work seems to have concentrated very much on empirical data of direct relevance to some real class of problems, e.g. in the study of population stereotypes (Loveless 1962). The employment of rigorous method-ology of experimental design has been used as an excuse to avoid thinking conceptually about the total problem. The result is lots of data which may be reliable in restricted laboratory situations but which is useless for the prediction of actual error rates in real situations and is equally valueless in its contribution to human information processing models.

3. Physiological measures

From World War I there has been continuous, if erratic, progress in the measurement and prediction of the effects of the thermal environment (Bedford 1964). During World War II the functional anatomist made his contribution to the design of workspaces (Le Gros Clark 1954). In the early post-war years there was a proper concentration on the design of simple workspaces such as chairs and tables (Akerblom 1954), typewriters (Lundervold 1958) and vehicles (McFarland 1954). During the early post-war period the redevelopment of European heavy industries and the contemporary food scarcities resulted in the rapid development of work physiology (Muller 1953, Christensen 1954 and Lehmann 1954).

The highly developed administrative and clinical skills of these early workers were such as to obscure for some time the real difficulties of taking measurements in these fields. Unfortunately it seems to be in the nature of covert behaviour sensing devices that their use causes some discomfort as well as some interference with ordinary activity. Since the subject has to be persuaded to submit to their application his reaction will be partly dependent on the skill of the persuader. Skill is needed to apply all physiological instrumentation, even the relatively simple apparatus associated with the measurement of various parameters of heat stress are such as to demand considerable experience in transportation, placement and reading. The problems of persuading workers to wear the clumsy apparatus required for respiration measurements and the tedious repetitive nature of pulse rate and body temperature measurements are usually mentioned only casually in the literature. Nevertheless, it was presumably for this reason that samples of operators were often small and individual differences are given little attention.

In all of these areas : energy expenditure, posture, light, heat, cold, noise and vibration the physiologist is attempting to do two things : to provide a method of measurement and to provide a set of standards. It is of no consequence to be able to determine that a particular task requires an energy expenditure of 4 Cal/min unless one has also a standard by which to assess whether 4 Cal/min is light or heavy, tolerable or intolerable.

The provision of standards is even more complex than the acquisition of data. Terms such as low/high, weak/strong, light/heavy, can only be defined arbitrarily and often the definitions must be changed when the circumstances change. For example, the definition of heavy work in Northern Europe cannot be used in India (Christensen 1964). This is partly because of differ-ences in average diet and body build but also because of the combination of effects of work and effects of climate.

This last reason is one illustration of another general difficulty about physiological measures. Stress on the operator tends to have similar effects whether it is due to work, to climate or to fear. Thus, in spite of many attempts (Weybrew 1967) it is not possible to distinguish these effects physiologically with any precision, nor is it possible to either separate the effects or predict combinations of effects. Most of the recent work on heat stress (e.g. W.H.O. 1968) has concentrated on extending the use of heat stress indices to cover differences of clothing and of work-load.

The combined effects of psychological and physiological stress are bound to appear even if the only psychological stress is due to the actual taking of measurements.

4. Physiological concepts

At first sight the physiologist would appear to be in a highly advantageous position in that he has the support of several centuries of study of human anatomy and physiology and an even greater comprehension of the physics of the environment. In practice this wealth of scientific background is not easily applicable to real problems of man at work.

Most academic physiology is too restricted to the micro level to be of any value in predicting the behaviour of the intact healthy worker who, in any case, rarely functions near the limit of his capacities. The relationship between simple physical measures of the environment such as temperature or sound energy level and the corresponding effects on workers invariably turns out to be a multidimensional problem with dominant influences from immeasurable variables such as attitude and motivation.

The physiologist can, of course, explore limits in the form of damage risk levels. In doing so he usually begins with a very simple conceptual model and then confounds the issues by pointing out that the simple model is not really adequate. For example, in the measurement of physical work, Figure 1, the simple model of food being burned to generate energy is complicated by the variable efficiency and capacity of the motor system depending on which muscles and joints are in use and on skill and training, the internal storage of energy in many different forms, the capacity to generate an oxygen debt and so on. Figure 2 shows the similar complexities which arise in relation to heat stress. The simple concept of a black body in equilibrium with the surroundings is complicated by physical activity, clothing, posture, acclimatisation, core/surface differences and so on.

These cases have been outlined to illustrate two points which appear to be generally true in relation to all problems where the physiologist studies some apparently straightforward physical aspect of man at work. Firstly it emerges that there are important, often dominant, variables which cannot be measured. Secondly the vast structure of micro-physiology provides no assistance.

Figure 1. Models for energy generation: (*a*) simple model; (*b*) more complex model.

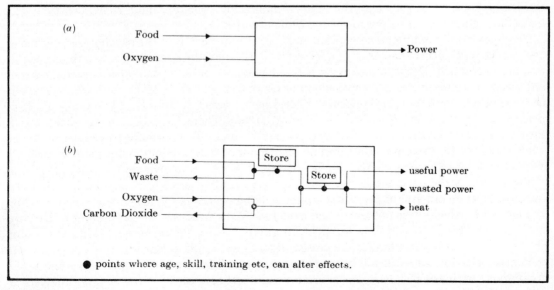

Figure 2. Models for heat balance: (*a*) simple model; (*b*) more complex model.

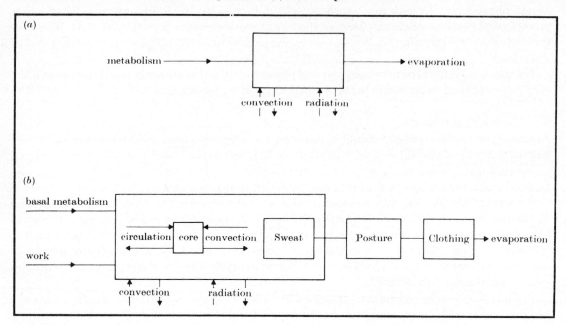

5. Psychological concepts

It used to be possible to distinguish between the approaches of physiologists and psychologists on the grounds that the former base their conceptual models on physical entities such as neurons and muscle fibres, whereas the latter think in terms of functional activities such as perception and memory. However, the models of physical work and heat exchange described in Figures 1 and 2 are broadly functional rather than physical in character but they remain physiological. Another possible distinction is to allow the physiologist to operate in the domain of energy exchange while the psychologist could restrict his activities to information exchange. This distinction also is by means no exact but it forms a convenient method of separation.

The pursuit of limits within information handling performance has proved even more elusive than for situations involving inputs and outputs of energy. The difficulty of defining information inputs to the human operator has forced the psychologist to rely excessively on simple artificial laboratory tasks where the limits obtained are probably more characteristic of the tasks than of human behaviour in general (Singleton 1969).

The complexities of human perception and learning are such that the psychologist has never been able to deal, other than at an empirical level, with variables associated with the presentation of information. Decision making which is dependent on both past experience and current objectives as well as status information is another area of ignorance. Much more is known about the output of information in the form of skilled performance. The state of knowledge in these fields has recently been described by Welford (1968). There have been many attempts to formulate comprehensive functional diagrams of human information processing (Broadbent 1958, Gagné 1962, Crossman 1964 and de Montmollin 1967), but in the design of work little seems to be gained at present by going beyond the two very simple models shown in Figure 3.

Figure 3(a) is the basic model for Task Analysis, that is procedures to identify what the operator must do in terms of accepting inputs, making decisions and generating outputs, while in Figure 3(b) (the inverse of 3(a)), is the model for Skills Analysis, the procedures to identify how the operator does his job in terms of information cues, the perceptual skills, the motor skills and the motor outputs. For the present it looks as though progress in conceptual models of human behaviour at work will be made by greater activities in the Task Analysis and Skills Analysis of ordinary people doing ordinary work rather than by more minute analysis of artificial laboratory tasks. These studies would appear to have little physiological content or background except that analysis of real tasks may require techniques for the detection of arousal levels.

Figure 3. Models for information processing: (*a*) Task Analysis model; (*b*) Skills Analysis model.

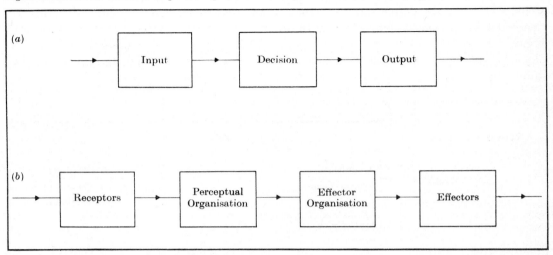

6. Arousal

In the post-war period there has been a revival of interest in level of alertness as a variable affecting performance. This was initiated on the physiological side by studies of the reticular system (Magoun 1954) and on the psychological side by studies of vigilance (Mackworth 1959). This work has been closely associated with the concept of stress as a descriptor of the load on the performer. It could be said that there has simply been a change in fashion founded on the optimistic belief that avoidance of terms such as fatigue and effort might result in avoiding also some of the disillusion which followed extensive work in these fields. This belief would have been more reasonable if the newer concepts had been defined more exactly in either conceptual or operational terms. Unfortunately, this was not so and the same disillusion is now in evidence in relation to stress and arousal studies.

The origin of these difficulties lies in inconsistencies and ambiguities of meaning. A measure of stress is sometimes a description of the conditions causing the subject's reaction, e.g. a heat stress index, sometimes a description of some aspect of the reaction itself, e.g. change in alpha frequency in the EEG, and sometimes a performance change thought to result from the conditions and the reaction to the conditions, e.g. increase in errors on an inspection task. Arousal can be introduced as an intervening variable which increases generalizing power, e.g. noise increases arousal, heat decreases arousal, or as a general descriptor of physiological changes, e.g. circadian rhythms, or as an additional factor which accounts for changes in behaviour when there are no changes in the environment or the task, e.g. fall-off in continued performance.

Some clarification is achieved by adopting the engineering convention of using stress in the sense of the stressor and strain for the reaction of the stressee. There are then two research problems : the study of relationships between stresses and strains and the study of the relationship between physiological and psychological measures of strain. It then follows that there are a number of possible but different justifications for the introduction of the concept of arousal. If relationships between stresses and strains are consistent and additive then arousal can be used to describe the intervening state of the organism and it can be measured directly by either the total stress or the total strain. If relationships between stresses and strains are inconsistent and integrate in peculiar ways then it can be hypothesized that discrepancies are due to a partially independent variable but, of course, neither the stress nor the strain can be used as a measure of this variable and the term becomes a cloak for ignorance. Similarly, if there are consistent relationships between psychological and physiological measures then arousal can be measured by each or both but if relationships are not consistent then the term must be restricted to one or the other or used for some masking factor which again cannot be measured by either.

The spectra of stresses and strains used in recent studies are shown in Figure 4; a more detailed schema of this kind has been produced by Weybrew (1967). Figure 4 is the result of an analysis of 140 papers dated from 1964 onwards and classified by the Defense Documentation Center, Office of Scientific and Technical Information as relevant to the 'arousal phenomenon'. Figure 5 shows an analysis of the same papers which supports the general impression that physiological as opposed to psychological measures of strain are more popular in Europe than in America.

Figure 4. Stresses and strains.

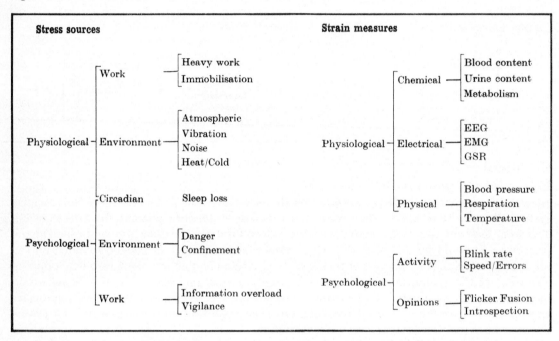

The increasing use of multiple criteria regardless of their origin in either psychology or physiology is to be welcomed. One way in which to explore their interactions is in the study of circadian rhythms and sleep. Kleitman (1939) distinguished between wakefulness of necessity (sub-cortical) and wakefulness of choice (cortical). Sleep studies are now being conducted with neurological (Johnson *et al*. 1968), biochemical (Rubin *et al*. 1969) and performance studies (Lubin 1967) from the same institute—the Neuropsychiatric Research Unit, San Diego.

Figure 5. Recent proportions of American/European studies.

Stress sources

American	Physiological	Psychological	Both
European	Physiological	Psychological	Both

Strain measures

American	Physiological	Psychological	Both
European	Physiological	Psychological	Both

Similar cross-discipline studies are being conducted in Europe (Frankenhaeuser *et al.* 1967). Another recent study by Thompson (1967) looked for phase differences between physiological and psychological circadian rhythms. There were considerable individual differences but some suggestion that somatic changes (e.g. metabolic rate) precede cortical changes (e.g. EEG) which precede performance changes (e.g. reaction time).

In a comprehensive review of many interdisciplinary studies Lacey (1967) suggests that the three complexes of the arousal process: electrocortical, autonomic and behavioural normally occur together but need not necessarily do so. Inconsistencies are most evident in studies of reaction to drugs where some combinations of drugs can lead to low arousal electrophysiological indicators and high arousal type behaviour and vice-versa.

In this context the use of concepts such as inverted U-shaped curves of arousal against performance (Malmo 1959) needs to be carefully defined. Unless such speculations are based on adequate measurement it is too easy to explain away any observed interaction of arousal change and performance but it is correspondingly difficult to predict such interactions.

Another arousal indicator which has not received very much recent attention is posture (Branton 1969). It seems that the classical teaching dictum 'sit up straight and pay attention' may be justified. Evaluations of seats used on long journeys and in schools together with laboratory studies of stabilization oscillations suggest that there may well be useful indicators of arousal available from this area.

The inconsistencies in definitions and in results within the general field of arousal should not be allowed to obscure the fact which is available introspectively and observationally that this is an important parameter of behaviour. Since the human body is full of feed-back systems it is probably meaningless to speculate about the basic origin of the phenomenon or to hope that there will ever be a single measure which provides a reliable and valid indicator. Arousal is a continuous dynamic variable with cyclical variations in time on which are superimposed irregular changes due to variations in inputs and outputs of energy and information. These inputs and outputs can be matched by combinations of neural and somatic activities which differ between tasks, times and individuals.

7. Discussion

These conclusions about arousal studies can be generalized to the total problem of measurement of man at work. Although it has always been recognized, e.g. in the rules of the Ergonomics Research Society, that both psychologists and physiologists have relevant concepts and skills, there has been a tendency to think of some problems, e.g. energy expenditure, as physiological and others, e.g. vigilance, as psychological.

Figure 6. Measures of human performance.

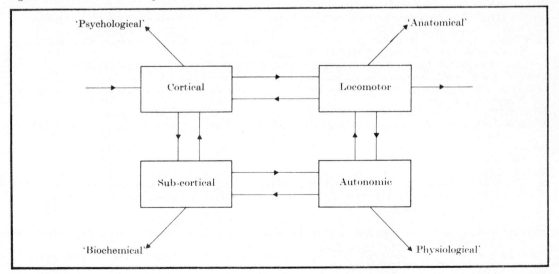

It now begins to look as though the complexities and interactions within the human body are such that neither discipline is adequate alone for the study of any problem of man at work. The physiologist (Figures 1 and 2) can be accused of too narrow an approach in treating the human operator as just another higher animal with insufficient regard for cortical dominance which can override endocrinal and autonomic parameters. Similarly the psychologist (Figure 3) can be accused of treating the human operator as too pure an information processing device without sufficient regard for sub-cortical and somatic factors which clearly influence performance.

It has been mentioned already that it would be possible to ignore either psychological or physiological variables if there were a consistent relationship between them under all conditions but this is obviously not so. There are at least four partially independent systems the cortical, the sub-cortical, the autonomic and the locomotor (Figure 6) which interact and which presumably can function in many different patterns of states which all result in the same input/output activity. To put it another way, a given job of work can be done by the human operator by a variety of different internal controls and adjustments which may never settle into a characteristic static pattern. Thus, to obtain a comprehensive measure of the man at any one time it is necessary to measure simultaneously psychological, biochemical, physiological and anatomical parameters.

The author is indebted to the Defense Documentation Center, Cameron Station, Alexandria, Virginia for the report bibliography on arousal phenomena mentioned in the text; and to Commander N. H. Berry, Office of Naval Research, who arranged for this search.

References

AKERBLOM B., 1954, Chairs and sitting. In *Symposium on Human Factors in Equipment Design* (Edited by W.F. FLOYD and A.T. WELFORD) (London: Lewis).

BEDFORD T., 1964, *Basic Principles of Ventilation and Heating* (London: Lewis).

BRANTON P., 1969, Posture and arousal (*Personal communication*).

BROADBENT D. E., 1958, *Perception and Communication* (London: Pergamon).

CHAPANIS A., 1961, On some relations between human engineering, operations research and systems engineering. In *Systems Research and Design* (Edited by D.P. ECKMAN) (New York: Wiley).

CHRISTENSEN E.H., 1953, Psychological valuation of work. In *Symposium on Fatigue* (Edited by W.F. FLOYD and A.T. WELFORD) (London: Lewis).

CHRISTENSEN E.H., 1964, Man at work: Studies on the application of physiology to working conditions in a sub-tropical country. *D.C.C. Health and Safety Series*, No. 4 (Geneva: I.L.O.).

CROSSMAN E.R.F.W., 1964, Information processes in human skill. *British Medical Bulletin*, **20**, 32–37.

DEESE J., 1969, Behaviour and fact. *American Psychologist*, **24**, 5.

FRANKENHAEUSER M., FROBERG J., HAGDAHL R., RISSLER A., BJORKVALL C., WOLFF B., 1967, Physiological, behavioural and subjective indices of habituation to psychological stress. *Physiology and Behavior*, **2**, 229–237.

GAGNÉ R.M., 1962, *Psychological Principles in System Development* (New York: Holt, Rinehart and Winston).

JOHNSON L., LUBIN A., NAITOH P., NUTE C., and AUSTIN M., 1969, Spectral analysis of the EEG of dominant and non-dominant alpha subjects during waking and sleeping. *Electroenceph. Clinical Neurophysiology*, **26**, 361.

KLEITMAN N., 1939, *Sleep and Wakefulness* (Chicago: Chicago University Press).

LACEY J. I., 1967, Somatic response patterning and stress : some revisions of activation theory. In *Psychological Stress* (Edited by M. H. APPLEY and R. TRUMBULL) (New York : Appleton Century Crofts).

LE GROS CLARK W.E., 1954, The anatomy of work. In *Symposium on Human Factors in Equipment Design* (Edited by W.F. FLOYD and A.T. WELFORD) (London: Lewis).

LEHMANN G., 1954, *Praktische Arbeits Physiologie* (Stuttgart: Georg-Thiemer Verlag) (2nd ed. 1962).

LOVELESS N.E., 1962, Direction of motion stereotypes: a review. *Ergonomics*, **5**, 357.

LUBIN A., 1967, *Performance under Sleep Loss and Fatigue in Sleep and Altered States of Consciousness* (Baltimore: Williams and Wilkins).

LUNDERVOLD A., 1958, Electromyographic investigations during typewriting. *Ergonomics*, **1**, 226.

MACKWORTH N.H., 1950, Research on the measurement of human performance. *M.R.C. Special Report* 268 (London: H.M.S.O.).

MAGOUN H.W., 1954, The ascending reticular system and wakefulness. In *Brain Mechanisms and Consciousness* (Edited by J.F. DELAFRESNAYE) (Springfield ICC: C.C. Thomas).

MALMO R.B., 1959, Activation: a neuropsychological dimension. *Psychological Review*, **66**, 367.

McFARLAND R.A., DAMON A., STANDT H.W., MOSELEY A.C., DUNLAP J.W., and HALL W.A., 1954, *Human Body-Size and Capabilities in the Design and Operation of Vehicular Equipment* (Boston: Harvard School of Public Health Monograph).

DE MONTMOLLIN M., 1967, *Les Systemes Hommes–Machines* (Paris: Presses Universitaires de France).

MULLER E.A., 1953, The physiological basis of rest pauses in heavy work. *Quarterly Journal of Experimental Physiology*, **38**, 205.

POINCARÉ H., 1913, The foundations of science. Quoted in STEVENS S.S., 1951, *Handbook of Experimental Psychology* (New York: Wiley).

RUBIN R.T., KOLLAR E.J., SLATER G.G., and CLARK B.R., 1969, Excretion of 17-hydroxycorticosteroids and vanillylmandelic acid during 205 hours of sleep deprivation in man. *Psychosomatic Medicine*, **31**, 1, 68.

SINGLETON W.T., 1969, Psychological limiting factors in human performance. In *Encyclopaedia of Linguistics Information and Control* (Edited by A. R. MEETHAM and R. A. HUDSON) (London: Pergamon).

SINGLETON W.T., and SIMISTER R., 1957, The design and layout of machinery for industrial operatives. *Occupational Psychology*, **31**, 26–34.

THOMPSON C.R.S., 1967, An investigation of the daily activity in the normal electroencephalogram and its relation to physiological and psychological circadian rhythms. *Unpublished dissertation, Department of Applied Psychology, University of Aston in Birmingham.*

WELFORD A.T., 1968, *Fundamentals of Skill* (London: Methuen).

WEYBREW B.B., 1967, Psychophysiological response to military stress. In *Psychological Stress* (Edited by M.H. APPLEY and R. TRUMBULL) (New York: Appleton Century Crofts).

WORLD HEALTH ORGANISATION, 1968, Health factors involved in working under heat stress. *Technical Report Series* (Geneva: W.H.O.).

WULFECK J.W., and ZEITLIN L.R., 1962, Human capabilities and limitations. In *Psychological Principles in System Development* (Edited by R.M. GAGNÉ) (New York: Holt, Rinehart and Winston).

Dr. Parrot is a psychologist, specializing in psychophysiology and applied psychology. He is a member of
the research staff of the Bioclimatic Research Centre at
the University of Strasbourg, with special interests
in the effects of environmental factors and drugs.

The measurement of stress and strain

J. Parrot

1. Human activity

In order to describe the situation of an operator performing a task, three sets of variables require consideration : input, internal and output (Figure 1).

Figure 1.   Variables relevant to operator performance.

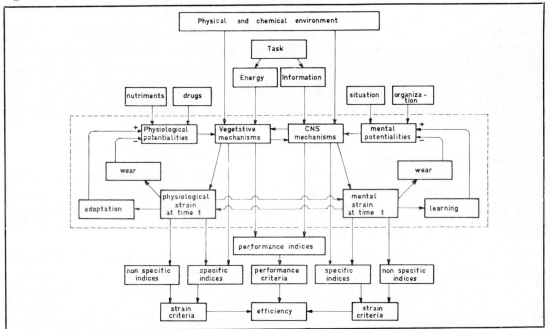

1.1. *Input variables*

This set is represented in the upper part of Figure 1 and includes the following.

■ Environmental conditions provided by physical agents : atmospheric pressure, gravity, light, air temperature, humidity, radiant heat flow, air movement, noise, vibration, atmospheric ionization, dust etc ; and by chemical agents : gaseous, liquid, or solid.

■ Input variables of the task to be performed. In any task it is always possible to distinguish between an energy component and an information component, their relative importance differing from one task to another. For an operator working at a soaking-pit, for example, the first one is particularly important and implies a high metabolic and thermal load. For an operator monitoring radar under normal environmental conditions, the second component is dominant.

■ External sources of energy or of substances influencing the energy turnover : diet, water and salt intake and also other types of ingesta such as drugs or alcohol.

■ 'Situation' factors depending upon socio-economic status : salaries, incentives or knowledge of results.

■ 'Organization' factors such as shift-work, schedule of rest-pauses, pace of assembly-line, and also housing conditions for diurnal or nocturnal sleep.

1.2. *Internal variables*

The second set of variables involves mechanisms which are specifically brought into play either by the task or by the environment or both. It is convenient here to distinguish between the following.

■ CNS mechanisms, not only cortical and subcortical, but neurological mechanisms in general which are responsible for the transmission and processing of information. Stress stimuli impinge upon these mechanisms by way of senses.

■ Vegetative mechanisms responsible for homeostasis and energy maintenance of the biological system. Stress agents interact here directly with the physical and chemical components of the body.

■ Physiological and mental potentialities on which these mechanisms depend ; e.g. constitutional factors such as sex, age, body build, etc, and functional factors such as acclimatization, training and learning methods.

Within each of these categories, interindividual variability comes into play. Between physiological and mental potentialities interaction processes are represented.

Because response processes induced within the system occur and change in time, strain in Figure 1 includes a temporal connotation. This temporal reference involves many aspects such as

■ the duration of strain, which may be prolonged for operators engaged in shift work ;

■ the rhythm for performing the task, as when the operator adopts his own tempo when working at a free-paced task or the organization of the operator's response within temporal limits when the task is paced ;

■ the possibility of interaction between stress and the rhythmic activity of the operator considered as a biological system.

In the figure, a feed-back loop has been sketched between strain and potentialities which may be influenced by external energy sources or recovery allowances ; e.g. for a man performing hard work at high environmental temperature, performance will be achieved with a minimal displacement of his physiological equilibrium only if he is provided with water and salt to replace that lost through sweating.

The general term 'wear' on this short-term loop may mean fatigue, or progressive difficulties in neuro-muscular transmission, or deterioration of an organ etc.

Another important effect has been added to express training or adaptation possibilities, in the positive sense of the term, when repeated exposure to stress results in a more economical and broader homeostasis. For instance in Canadian fishermen working for many hours every day with their bare hands immersed in cold water, blood pressure response and skin vasoconstriction are reduced to a mimum (Leblanc 1966). This effect is denoted by a positive sign in Figure 1.

Despite appearances, there is no crude symmetry between vegetative and metabolic mechanisms on the one hand and neural or communication mechanisms on the other, but exchange goes on continually between them. If mental strain or load results in fatigue, drugs may relieve the strain in some manner by way of physiological mechanisms. This is true also for food, both the look of it and its taste ; e.g. the way in which the prospect of a steak can raise one's 'spirits'.

1.3. *Output variables*

There are the different forms of measurable events produced by the system: performance events such as work done, quantity and quality of product, information transmitted per unit time etc, and physiological or psychophysiological phenomena which can be recorded, heart rate, temperature, sweat, EEG, EMG, galvanic skin responses, eye movements, blink-rate etc. All these events can be considered as indices of the activity of the system. A distinction can be made here between specific and non-specific indices. By specific indices are meant those which are specific to the mechanisms brought into play by the input forces, according to the nature of the task and the environmental factor; e.g. in a visual detection task such as radar monitoring, eye movements are specific indices, but heart rate and rectal temperature constitute non-specific ones. In a noisy environment, temporary threshold shifts of auditory sensibility for different frequencies are specific indices, but spontaneous whole body mobility and critical flicker fusion are non-specific.

This does not mean that non-specific indices are of lesser interest in relation to the strain induced by the task or the environment. Of all these indices, some are more pertinent than others, in a given 'situation'. If we could assume that we have really found good criteria of strain it would be possible to derive an evaluation of the efficiency of the man–machine system not from performance only, but from the ratio of performance or useful work to the organic cost paid for it.

2. Stress and strain

The word 'stress', in English engineering terminology, refers more to the external force acting upon a system, and 'strain' refers more to the effort this system has to provide and to the cost it has to pay in order to resist input stress forces. For instance one speaks of the compressive, or shearing, or torsional stress of a force acting upon a solid, and accordingly, of the strain induced within it. In French, the translation of 'stress' would be 'contrainte' and the translation of 'strain' would be 'astreinte'. These words contrainte and astreinte have recently been introduced by Metz (1967) into French environmental physiology and occupational health terminology relative to the effects of thermal factors. 'Contrainte thermique' gives the quantity of heat eliminated by the biological system in order to maintain its thermal equilibrium, and 'astreinte thermique' denotes the physiological or pathological changes resulting from thermal stress. The manifestation of strain appears particularly in some output variables: increase of internal temperature, heart rate and sweating; and in some cases in more dramatic results such as heat collapse due to breakdown of the thermoregulation or of water balance.

To sum up: the main point is not that the term 'stress' has to be dedramatized from its post-Selye (1959) meaning, but that stress as such is linked with strain. A stimulus constitutes a stress for a particular system if a strain ensues within that system. In other words, under some threshold, a given factor may be a stress only for those individuals in whom it results in strain.

2.1. *Strain criteria*

Defining strain criteria is the most difficult problem for each type of stress context. Sensitivity of an output index does not provide a guarantee for its validity as a criterion. For the same quantity of muscular work to perform their task, a woodman and a clerk may show the same oxygen consumption, but their respective heart-rates will greatly differ. Although a linear function of physical work load, oxygen consumption does not provide a good criterion of strain here, but heart rate does, if it appears, as seems likely, that this index is not only valid for different degrees of work load, but is also a good predictor of long-term physiological wear.

But in many cases a physiological criterion is on the final common path of several processes and varies as a result of these. This is precisely the case for heart-rate, which, as was shown by Vogt (1964), is additively influenced by metabolic and thermal stress. An appropriate laboratory analysis of the physiological transfer function involved in the overall control of heart rate was necessary to provide a practical methodology for its use as a multifactor criterion in actual work situations.

All these problems are physiological ones, and some still wait to be solved. For instance, since the respective roles of adrenal corticoids and catecholamines have been studied separately in different countries, it is difficult to find studies devoted to interrelations between adrenal cortex and adrenal medulla in response to different types of stresses.

Turning to a more practical point of view, the question of interacting stresses in relation to the study of man–machine systems may be formulated as follows : if we assume that the task results in strain within the operator, is this operational strain modified by another intervening stress? Environmental factors may induce strain in the operator, and as a result impair man–machine system efficiency. If any operational strain takes place, can ambient stresses provide a means for the study, design and validation of man–machine systems? These questions are difficult to answer if we do not content ourselves with assumptions or value judgments of our own choice in order to reduce an impracticable complexity to a practical form.

Two attitudes are possible : either performance only is considered, or physiological indices of strain are taken into account together with performance.

When performance only is considered, great possibilities for progress are already available. It is possible to answer questions like the following. If noise is added, or if temperature is raised, or if sleep recovery is not available, is the man–machine system efficiency markedly impaired or not? If one temperature has been found tolerable for work, does it remain tolerable if another stress such as noise is added? Can the effects of different stresses be summated or multiplied, or is there any possibility that they cancel each other out? In answer to some of these questions, the Cambridge research, for example, provided a great deal of information (Broadbent 1963, Wilkinson 1969). As far as they relate to experiments of the same type, our results in Strasbourg agree with those of Cambridge (Wittersheim *et al.* 1958, Grivel *et al.* 1959, Baumstimler 1963, 1968, Wittersheim 1968 ; as examples see Figures 2 and 3).

Figure 2. Mean response time (upper part of graph) and mean number of errors (bottom) on a multiple-choice paced task performed for 60 minutes in 3 trials P1, P2 and P3, in comfortable ambience (dry bulb, 22°C, relative humidity 50%) (open dots) and in severe heat (dry bulb, 40°C, relative humidity 50%) (black dots). Each 60 minutes' trial is divided into 4 quarters of an hour numbered 1 to 4 using 36 subjects. During successive trials P1, P2 and P3, mean RT increases regularly when Ss are working in comfortable ambience. Heat seems to prevent such an evolution, so that for P2 and P3 trials, mean RT are slightly shorter than in comfortable ambience.
Error rate is significantly increased when Ss are working in heat, despite learning effects between P1 and P3 trials (Grivel 1969, unpublished results).

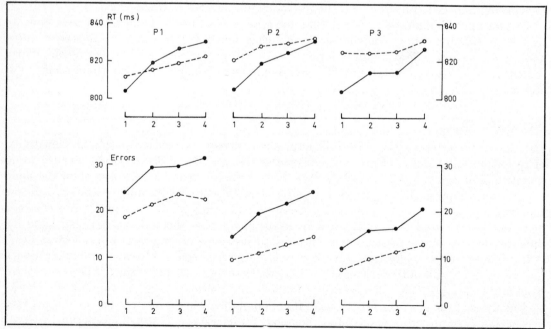

If physiological indices are taken into account as criteria of strain, their possible statistical correlation with performance indices should perhaps be handled with care. In fact, the use of physiological criteria of strain in relation to the man–machine system depends also on the objectives of the system or more crudely, on the goal which is aimed at by the owner of the machine. It may be, for instance, that the efficiency of a man–machine system would not be markedly impaired by noise, but that in the long term this condition would result in the operator's deafness. In a majority of cases, the operator has to adapt himself to the machine. If the machine in its environmental context cannot be modified, it must be monitored by another more suitable operator. In other words, the use of physiological criteria of strain depends upon the respective importance attributed to the operator's health or comfort and to the operations performed by him. Indeed in Figure 1 two sets of data are lacking: the investment made by the owner of the machine and the cost he is ready to pay in order for the man–machine system to remain operational; subjective impressions of the operator deriving from his somatic impressions, from knowledge of his performance, and also from 'situation' factors providing motivation.

Figure 3. Effect of motor-noise steps (80–105 dB SPL) (squares) and of shipyard noise (97 dB SPL at 50% of time) (triangles) on performance at a multiple-choice paced task, and corresponding performance in quiet (open dots): 50 minutes' trials. Q0 morning preperiod trial and Q5 next morning postperiod trial are performed in quiet, Q1 to Q4 afternoon trials are performed in experimental conditions. 9 subjects.
Left: mean RT (corrected values for covariance) are significantly reduced during 4 trials in noise (Q2 and Q3 data not available completely). Significant residual effect next morning.
Right: mean frequency of percentage errors of responses (corrected values for covariance). No significant effect of noise on mean error rate for 4 trials. Performance improved during first 50 minutes of exposure, in 2 noisy conditions, then deteriorated. Difference between respective slopes of deterioration in quiet and in shipyard noise is highly significant (Parrot 1969).

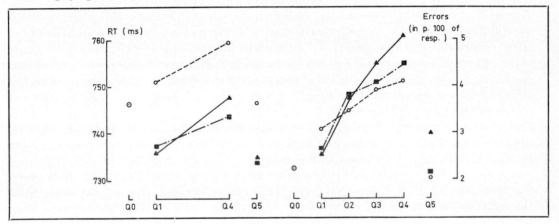

3. Recommendations

The experimental as well as empirical data presently available suggest the following.

- We must at first reach as complete a knowledge as possible of the operational strain itself when a task is performed in so-called normal conditions. This is particularly important from the point of view of inter-individual differences, in order to check whether these inter-individual differences can still be controlled when supplementary stresses are brought into play. In other words: if inter-individual differences are observed in so-called normal conditions, are they significantly modified when another stress factor is added to the situation?
- Since normal conditions can only be defined on arbitrary bases, it seems more realistic to explore the same system at different levels of possible stresses, including the so-called normal conditions, with at least one level below and one level above the expected level of these normal conditions. Factorial planning of experiments provides adequate procedures to systematize such investigations.

The same advantages can be expected in investigating a given environmental stress which will benefit from an experiment in which one or more other factors are systematically varied. For example: the cardiovascular effect of sudden noise may disappear under muscular work stress, or environmental heat stress. This means that heart-response is only of a contingent nature. If on the other hand heart-response remains unaffected by muscular work or heat stress, this makes this response a permanent feature of general physiological strain under acoustic stress.

■ Owing to the fact that some environmental stresses drastically influence the performance of some man–machine systems, research on limits of tolerance, relative to exposure duration, is required so that physiological or psychological responses do not reach unacceptable levels. Much work has to be done in this field, for example on the effects of impulse noises and noises involving transients.

■ Recent experiments in Strasbourg (Schwertz *et al.* 1968, Parrot and Wittersheim 1968, Parrot and Miller 1968) suggest that when applying repetitive transient stress stimuli instead of continuous ones, that is, stimuli inducing transient responses of the biological system, we have better opportunities for clarifying some internal response mechanisms, especially by changing input values and parameters of the transient stimulus. When such repetitive transient stress stimuli are produced according to a systematic time schedule, their transient effects are suitable for averaging procedures of the evoked potential type.

■ Concerning physiological and behavioural indices in relation to performance, it seems that, from the point of view of the design and validation of the man–machine system, micro-methods of study are fruitful. By micro-methods is meant thorough analysis of the different components of the operator's monitoring, particularly from the temporal point of view. For example Schmidtke's (1968) study of eye movements in radar monitoring by which he established a relation between amplitude of eye movements and detection probability. This finding might lead to an optimal search strategy and to better training methods. By recording eye movements in subjects monitoring a response key-set Parrot and Cubaynes (1968) observed that this is significantly more difficult to perform when forward movements of the right arm are requested. Further Hennemann (1967) who, by means of a cinematographic technique, analysed the different components of hand movements in monitoring the same key-set, between onset of the signal and achievement of response, showed the part and the meaning of preparation processes in this response.

■ Careful study of training processes, by use of such micro-methods, in subjects learning how to monitor a new machine, is also necessary, not only to discover how new voluntary motor patterns are acquired, in relation with dispositions or stereotypes, but in view of further comparisons with long duration motor performance, with or without environmental stress. Until recently, insufficient stress has been given to these learning processes, especially from the point of view of the genesis of errors and wrong motor patterns. What sort of relationship is there between fatigue or accident behaviour and wrong responses during learning?

These six points are not exhaustive; further investigation is clearly required to answer the difficult problem of interacting stresses in relation to the efficiency of man–machine systems.

References

BAUMSTIMLER Y., 1963, Analyse expérimentale des processus de vigilance. *Centre d'Etudes de Physiologie Appliquée au Travail, Strasbourg.*

BAUMSTIMLER Y., 1968, Commutation et généralisation de l'inhibition dans une activité volontaire. *Centre d'Etudes Bioclimatiques du C.N.R.S., Strasbourg.*

BROADBENT D.E., 1963, Differences and interactions between stresses. *Quarterly Journal of Experimental Psychology,* **15,** 205–211.

GRIVEL F., BAUMSTIMLER Y., WITTERSHEIM G., et METZ B., 1959, Facilitation des performances à une épreuve de choix multiple par le travail musculaire chez l'Homme. *C.R. Soc. Biol.,* **153,** 1598–1600.

GRIVEL F., 1969, Activités nerveuses supérieures chez l'Homme normal astreint à des ambiances thermiques sévères. *Centre d'Etudes Bioclimatiques, Strasbourg* (à paraître).

HENNEMANN M.C., 1967, Manifestations motrices de la préparation dans une activité sérielle de choix. *Centre d'Etudes de Physiologie Appliquée au Travail, Strasbourg.*

LEBLANC J., 1966, Adaptive mechanisms in human variation. *Annals of the New York Academy of Sciences,* **134,** 721–732.

METZ B.G., 1967, Paper to Scientific Group on Health factors involved in working under conditions of heat stress. *World Health Organization.* In *W.H.O. Report T.R.*412.

PARROT J., et WITTERSHEIM G., 1968, Effets transitoires d'échelons de bruit sur la motilité palpébrale et divers critères de performance au cours de l'exécution d'une tâche psychomotrice de longue durée. In *Problèmes Actuels de la Recherche en Ergonomie* (Edited by MORIN et PAILLARD) (Paris: Dunod), 166–173.

PARROT J., et CUBAYNES J.P., 1968, A study of kinesthetic familiarization with a response key-set by recording eye-movements. In *Tri-National Symposium on Information Feed-In and Information Processing by Man* (Edited by SCHOLZ) (Mondorf-bei-Bonn: Krupinski Verlag), 2 vols.

PARROT J., and MILLER J.C., 1968, Effect of alternating 80–105 dB noise-steps on spontaneous blink-rate and error-rate in a multiple-choice paced task. *Ergonomics*, **11**, 84.

PARROT J., 1969, Effets physiologiques et psychophysiologiques de bruits comportant des transitoires sur l'homme normal. *Centre d'Etudes Bioclimatiques du C.N.R.S., Strasbourg.*

SCHMIDTKE H., 1968, Beitrag zur Theorie der Signalentdeckung. *Psychologische Beiträge*, **10**, 464–468.

SCHWERTZ TH., GEBER M., et MARBACH G., 1968, Effets globaux de deux types de bruits et effets transitoires d'échelons de bruit sur la fréquence cardiaque enregistrée au cours d'une tâche psychomotrice de longue durée. In *Problèmes Actuels de la Recherche en Ergonomie* (Edited by MORIN et PAILLARD) (Paris: Dunod), 160–165.

SELYE H., 1959, Perspectives in stress research. *Perspectives of Biological Medicine*, **2**, 403–416.

VOGT J.J., 1964, Repérage de la contrainte thermique au moyen de critères physiologiques. *Centre d'Etudes de Physiologie Appliquée au Travail, Strasbourg.*

WILKINSON R., 1969, Some factors influencing the effect of environmental stressors upon performance. *Psychological Bulletin*, **72**, 260–272.

WITTERSHEIM G., GRIVEL F., et METZ, B., 1958, Application d'une épreuve de choix multiple avec enregistrement continu des réponses, des erreurs et des temps de réaction, à l'étude des effets sensori-moteurs de l'inversion du rythme nycthéméral chez l'Homme normal. *C.R. Soc. Biol.*, **152**, 1194–1198.

WITTERSHEIM G., 1968, Etude expérimentale des effets de l'alcool, de la caféine et du méprobamate sur l'activité psychomotrice de l'Homme normal. *Centre d'Etudes Bioclimatiques du C.N.R.S., Strasbourg.*

Dr. Streimer has psychology and physiology degrees. He has worked on human performance studies at the Boeing and North American Rockwell companies, and now combines the latter with a post in the Department of Psychology at San Fernando Valley State College, California.

Considerations of energy investment as determinants of behaviour

I. Streimer

1. Introduction

Discussions of measurement and assessment technique trends in ergonomics and human factors must take cognizance of dissimilarities in emphasis existing in the European and American approaches to these fields. A striking difference is seen in the apparent European concentration on the measurement of physiological phenomena as contrasted with the American concentration on psychological measurements. Chapanis (1969) attributed this emphasis difference to the impetus given the developing sciences by the academic training of the early practitioners on the two continents. European ergonomics moved in the direction of 'man-rating' tasks in terms of physical work demand loadings imposed on workers in operational situations. Conversely, the relative abundance of machine power afforded the American worker reduced physical labour demands and shifted research effort into those decision-making processes attendant to machine and process control. Essentially, it appears that practitioners on the two continents have responded to certain environmentally derived factors in shaping the direction and emphasis of their measurement and assessment practices. A major shaping influence, environmentally derived, appears to have been the technological response capability of the sponsoring society in terms of its ability to provide non-human energy input resources. Any considerations of measurement mode criteria and trends should be similarly guided in terms of interactions existing between operational requirements, human activity levels and the energy characteristics of the system under consideration.

Human assignment in man–machine systems may vary within a broad activity range from the almost purely physical to the almost purely mental work input modes. Functional assignment of men or machines may be determined by considerations of interactions between human response capabilities and characteristics, operational requirements and system energies available for allocation to human or machine operation. Typically, machine usage is preferred whenever human response capabilities are exceeded or overloaded, when the task is sufficiently repetitive to enable human replacement by a machine yielding more reliable cost-effective performance, or when the environment is too hazardous for humans. Conversely, humans are employed where these considerations do not pertain or when system resources do not permit the provision of non-human energy inputs.

2. Measurement of physical work

The examination and correlation of human expenditures with work output via physiological measurements has been the subject of much previous study by a number of investigators including Bink (1962), Müller (1962), Lehmann (1963), Streimer and Springer (1963) and Snook and Irvine (1969). The major findings indicate that in self-paced aerobic work which is not thermally, environmentally or psychologically stressed, direct and linear relationships exist between energy input levels and work output levels, Lehmann (1958), Poulsen and Asmussen (1962), Andrews (1967) and Streimer *et al.* (1969). It has been further demonstrated that in such physical work tasks, work efficiency changes produced by biomechanical considerations do not tend to change absolute energy input levels but rather to produce compensatory work rate alterations; in essence, a self-paced regulatory process prevails which limits energy expenditure rates, Snook and Irvine (1969), Streimer *et al.* (1967, 1969). The work output/energy input relationship has been exhaustively examined employing both direct and indirect measurement techniques. For the purposes of this paper, only indirect techniques less disruptive of normal operations will be discussed.

A number of investigators have employed heart rate as a measure of energy expenditure rate and work output rate in non-thermally stressed self-paced aerobic work: Poulsen and Asmussen (1962), Astrand *et al.* (1964), Andrews (1967). The reliability of these relationships has been such that, despite alterations produced in the oxygen-pulse ratio by various muscle group involvements, a number of relationships have been posited which may be employed in the evaluation of such factors as work load levels, relative task difficulties, operator physical condition, equipment comparisons: Henry and Farmer (1938), Tuttle and Dickenson (1938), Poulsen and Asmussen (1962), Malhotra *et al.* (1963). Heart rate has been successfully employed as an indicator of maximum oxygen processing capabilities (Maritz *et al.* (1961), Astrand *et al.* (1964)) and it has been found reliable in assessing human work producing capabilities. Brouha and Maxfield (1962) have suggested a technique which allows heart rate to be used as an indicator of work cost in thermally stressed environments and Tichauer (1963) has successfully employed pulse rate/energy cost relationships in the evaluation of the costs of high altitude manual work. Streimer *et al.* (1968, 1969) have found a high correlation between heart rate and energy expenditures during the performance of manual work at shallow depths. It may be generally stated that heart rate may be easily and validly employed as a measurement technique in the study of work costs during the performance of self-paced physical work in non-stressed environments.

Various facets of the respiratory processes have been employed as research tools with varying levels of success, in the measurement of energy expenditure during the execution of physical tasks. Hildebrandt and Cuntze (1965) examined expiratory peak flow–work load interactions and found a reliable relationship existed between the bronchomotor reaction and work load. Gautier *et al.* (1965) found stable relationships between attention levels and certain aspects of respiratory patterns during the performance of primarily sedentary tasks. Malhotra *et al.* (1962), Naimark *et al.* (1964) and Datta and Ramanathan (1969) examined the relationships between such factors as minute ventilation, pulmonary ventilation and ventilatory exchange ratios and energy expenditure rates and found positive correlations enabling prediction of work loads. Sharkey *et al.* (1966) found ventilatory rates could be used as predictors of performance energy costs in physical work and noted that ventilatory rates were a more accurate measurement than heart rate when the work required near maximal isometric muscular efforts. Although other techniques such as indirect calorimetry, time and motion studies and direct observation of body limb segment velocities have also been employed to assess energy input/work output relationships, to date, most investigations of human performance characteristics have been confined to heart rate and/or respiratory process measurement.

3. Measurement of mental work

Human response characteristics and capabilities during the performance of primarily mental work has also been the subject of much previous study. The bulk of these efforts have been directed to the examination of humans as information receivers and processors and as input elements to various control mechanisms. Current thinking posits the human in this process as a dynamic system continually engaged in decision-making activities through input structuring and output selection (Szafran 1966). The ability of humans to execute this process with respect to both the speed and number of information units processed per unit time has been pondered by many investigators. The effect of various stressors upon mental performance has been examined including the ability to make choice reactions, Hick (1952), stimuli presentation rate of perceptual integration ability, Edwards (1963), uncertainty on response time, Hilgendorf (1966), and others too numerous to mention. In all instances psychologists have examined the characteristics and capabilities of the mental processes indirectly through measurement of mental process behavioural outputs. Output measurement itself is subject to confounding because of various sensory-motor interactions and further confounding is possible through the effects of display control–sensory-motor interactions. The early books of Fitts (1947) and Chapanis *et al.* (1949), revealed the profound importance of machine and task design interaction with various aspects of sensory-motor functional outputs.

Various investigators have attempted to directly measure the physiological correlates of mental performance. Opton (1964) examined E.E.G. patterns as a function of subject age in an attention task and found relationships which were reliable. Other investigators, including Benson *et al.* (1965), Kiriakov (1964), Davis and Edwards (1965), Gautier *et al.* (1965), Kalsbeek and Ettema (1964), Gibson and Hall (1966), and McNulty and Noseworthy (1966), have employed a variety of measurements including galvanic skin response, heart rate, respiratory patterns, palmar skin resistance, myography, blood pressure, critical flicker fusion and other techniques as indices of mental activity with varying levels of success. The apparent lack of success in the application of physiological measurements to psychological processes stems from the impact of stress upon these measurements. It is somewhat para-doxical that the very element, stress, that lessens the reliability of physiological measuring techniques when applied to psychological phenomena is itself of sufficient importance in man–machine system operation to warrant the application of these techniques to the examination of stress.

4. Effects of stress

The application of physiological measurement techniques to the study of primarily mental tasks becomes increasingly important whenever task requirement–environment interactions are liable to produce stress in operators. Stress development may be a consequence of many factors including work load imposition rates, perception of work load, and duration of exposure to the stress producing situation. Most importantly, exposure to stress can reduce, over time, operator response characteristics and capabilities to levels inadequate to satisfactory task execution. Performance failure may be produced via reductions in operator ability to handle peak impulse loads, via reductions in the output characteristic levels of operators performing self-paced tasks or via elevation of operator perceptual thresholds to levels inappropriate to operational success.

Data exist which reveal that individuals executing self-paced mental or physical work will select work rates which are comfortable in the sense of being non-overly demanding. Wiener (1948) suggested that man be considered a self-regulating mechanism with respect to his information processing capabilities. Studies including those of Bahrick *et al.* (1954), Benson *et al.* (1965, 1966), Poulton (1962), and Adams and Creamer (1962) have examined various aspects of information processing capabilities and have noted the alterations in primary channel processing capabilities as a function of subsidiary or extra-task requirements. Similarly, studies have been executed which reveal that the subjective perception of a physical work load will influence the release of adrenaline with the amount released in some measure attributable to the emotional stress developed during heavy work performance, Franken-haeuser *et al.* (1967, 1968).

Ruff (1963) considered stress in terms of system inputs and outputs and the compensatory mechanisms mobilized by the stressed organism. This holistic approach seems to fit into an examination of the total energy requirements of a system as the determinant of the measurements to be taken. If man is a single-channel data processing system and the imposition of overloads produces physiological manifestations which indicate an increased arousal state and energy cost, then the total energy requirements imposed upon the man must be monitored to prevent stress formation with its consequence of failure. There have been numerous studies of stress in which performance and physiological phenomena have been measured and it seems clear that stress produces an excitation level which increases the total energy requirements demanded of the operator, Benetata *et al.* (1967), Benson and Rolfe (1966), Davis and Edwards (1965), Gautier *et al.* (1965), Kalsbeek and Ettema (1964), Kiriakov (1964), Kozarovitsky (1964),

Hashimoto (1964). It is also clear that stress can adversely affect operator performance (Bondarev *et al.* (1968), Benetata *et al.* (1967), Alluisi and Morgan (1968)), and of primary importance to ergonomics/human factors practitioners is the design of tasks to minimize operator stress loading.

5. Conclusion

If man is accepted as a limited energy input system and excessive demands for energy constitute stress it becomes evident that measurements should be applied which, in fact, examine the total energy demands made upon human operators, both physical and mental. In this period of increasing time compressed operations, e.g. supersonic transport, high-speed ground transportation systems, etc., the consequences of operator failure can be catastrophic. It is the responsibility of practitioners to assess accurately the total demand placed upon the human so that appropriate activity programming, work–rest cycle assignment and man–machine trade-offs can be effected. The criticality of the human operations performance with respect to the consequences of proficiency failure should be the determinant of the measurement techniques applied. There can be no mind–body dualism controversy when the total reliability and safety of highly critical systems is at stake.

References

ADAMS J.A., and CREAMER L.R., 1962, Data processing capabilities of the human operator. *Journal of Engineering Psychology*, **1**, 150–158.
ALLUISI E.A., and MORGAN B.B., Jr., 1968, Effects of practice and work load on the performance of a code transformation task. *Dec. NASA CR*-1261, 73 pp.
ANDREWS R.B., 1967, Estimation of values of energy expenditure rate from observed values of heartrate. *Human Factors*, **9**, 581–586.
ASTRAND P., CUDDY T.E., SALTEN B., and STENBERG J., 1964, Cardiac output during submaximal and maximal work. *Journal of Applied Physiology*, **19**, 268–274.
BAHRICK H.P., NOBLE M., and FITTS P.M., 1954, Extra-task performance as a measure of learning a primary task. *Journal of Experimental Psychology*, **48**, 298–302.
BENETATA G. et al., 1967, Investigations into fatigue in engine-drivers working on diesel engines of the Rumanian railways. Prelim. Rept. Abst. no. 35445, *Ergonomics*, **10**, 79–80.
BENSON A.J., HUDDLESTON H.F., and ROLFE J.M., 1965, A psychophysiological study of compensatory tracking on a digital display. *Human Factors*, **7**, 457–472.
BENSON A.J., and ROLFE J.M., 1969, A psychophysiological evaluation of a visual display. *Ergonomics Research Society Annual Conference*, Abstract.
BINK B., 1962, The physical working capacity in relation to working time and age. *Ergonomics*, **5**, 25–28.
BONDAREV E.V., GURVICK T.T., DZHAMGAROV V.A. et al., 1968, Changes in the rate of information processing by flight personnel during extended flights. *Joint Publications Research Service, Washington, D.C.*, June.
BROUHA L., and MAXFIELD M.E., 1962, Practical evaluation of strain in muscular work and heat exposure by heart rate recovery. *Ergonomics*, **5**, 87–92.
CHAPANIS A., 1969, The growth of ergonomics in the Communist world. *Human Factors Society Bulletin. XII*, 1–3.
CHAPANIS A., GARNER W.R., and MORGAN C.T., 1949, *Applied Experimental Psychology* (New York: John Wiley & Sons).
DATTA S.R., and RAMANATHAN N.L., 1969, Energy expenditure in work predicted from heart rate and pulmonary ventilation. *Journal of Applied Physiology*, **26**, 297–302.
DAVIS G.R., and EDWARDS A.E., 1965, Physiological adaptation to stress: the influence of massed and spaced practice. *Proceedings of the 73rd Annual Convention of the American Psychology Association*, 157–158.
EDWARDS E., 1963, The integration of spaced signals. *Ergonomics*, **6**, 143–152.
FITTS P.M., 1947, *Psychological Research in Equipment Design* (Washington: U.S. Government Printing Office).
FRANKENHAEUSER M., FRÖBERG J., HAGDAHL R., RISSLER A., BJORKVALL C., and WOLFF B., 1967, Physiological, behavioral and subjective indices of habituation to psychological stress. *Physiology and Behavior*, **2**, 229–237.
FRANKENHAEUSER M., POST B., NORDHEDEN B., and SJÖBERG H., 1968, Physiological and subjective reactions to different physical work loads. *Reports from the Psychological Laboratories, University of Stockholm*, no. 254, July, 7 pp.
GAUTIER H., HANOTEL H., and HUGELIN A., 1965, Level of vigilance and type of respiratory movements in waking man. *Journal of Physiology*, Paris, **57**, 1–246.
GIBSON D., and HALL M.K., 1966, Cardiovascular change and mental task gradient. *Psychonomic Science*, **6**, 245–246.
HASHIMOTO D., 1964, Estimation of the driver's work load in high speed electric car operation on the new Tokaido line in Japan. *Proceedings of the 2nd Internal Congress on Ergonomics*, 463–469.
HILGENDORF L., 1966, Information input and response time. *Ergonomics*, **9**, 31–37.
HICK W.E., 1952, On the rate of gain of information. *Quarterly Journal of Experimental Psychology*, **4**, 11–26.

HILDEBRANDT G., and CUNTZE H., 1965, Pneumometric investigations of the bronchomotor reaction to work. *Internationale Zeitschrift für angewandte Physiologie*, **30,** 247–268.

HENRY F., and FARMER D., 1938, Functional tests:II. The reliability of the pulse-ratio test. *American Association for Health and Physical Education Research Quarterly*, **9,** 81–87.

KALSBEEK J.W.H., and ETTEMA J.H., 1964, Physiological and psychological evaluation of distraction stress. *Proceedings of the 2nd International Congress on Ergonomics*, 443–447.

KIRIAKOV K., 1964, Some electroencephalographic criteria of fatigue after mental work. *Aerospace Medical Abstracts*, **35,** 1258.

KOZAROVITSKY L.B., 1964, Dynamics of skin-galvanic reactions in control panel operators during their work of regulating aircraft traffic. *Aerospace Medical Abstracts*, **35,** 1235–1236.

LEHMANN G., 1958, Physiological measurements as a basis of work organization in industry. *Ergonomics*, **1,** 328–344.

LEHMANN G., 1963, *Praktische Orkertphysiologie* (Stuttgart: G.T. Verlag).

MALHOTRA M.S., RAMASWAMY S.S., RAY S.N., *et al.*, 1962, Minute ventilation as a measure of energy expenditure during exercise. *Journal of Applied Physiology*, **17,** 775–779.

MALHOTRA M.S., SENGUPTA J., and RAI R.M., 1963, Pulse count as a measure of energy expenditure. *Journal of Applied Physiology*, **18,** 994–996.

MARITZ J.S., MORRISON J.F., PETER J., STRYDOM P.B., and WYNDHAM C.H., 1961, A practical method of estimating an individual's maximal oxygen uptake. *Ergonomics*, **4,** 97–122.

McNULTY J.A., and NOSEWORTHY W.J., 1966, Physiological response specificity, arousal and task performance. *Perceptual and Motor Skills*, **23,** 987–996.

MÜLLER E.A., 1962, Occupational work capacity. *Ergonomics*, **5,** 445–452.

NAIMARK A., WASSERMANN K., and McILROY M., 1964, Continuous measurement of ventilatory exchange ratio during exercise. *Journal of Applied Physiology*, **19,** 644–652.

OPTON E.M., 1964, Electroencephalographic correlates of performance lapses on an attention task in young and old men. *Duke U., L.C. Mic.* 64-11, 736, 121 pp.

POULSEN E., and ASMUSSEN E., 1962, Energy requirements of practical jobs from pulse increase and ergometer test. *Ergonomics*, **5,** 33–36.

POULTON E.C., BARBOUR A.B., and WHITTINGHAM H.E. (eds.), 1962, *Human Problems of Supersonic and Hypersonic Flight* (London).

RUFF G.E., 1963, *Psychological and Psychophysiological Indices of Stress* (London: Free Press of Glencoe), 33–59.

SHARKEY B.J., McDONALD J.F., and CORBRIDGE L.B., 1966, Pulse rate and pulmonary ventilation as predictors of human energy cost. *Ergonomics*, **9,** 223–227.

SNOOK S.H., and IRVINE C.H., 1969, Psychophysical studies of physiological fatigue criteria. *Human Factors*, **11,** No. 3, 291–299.

STREIMER I., and SPRINGER W.E., 1963, Work and force producing capabilities of man. *Boeing Document* D2-90245, 54 pp.

STREIMER I., TURNER D.P.W., and VOLKMER K., 1967, The effect of tractionless simulator mass upon manual work. *Journal of Astronomical Sciences*, Sept.–Oct.

STREIMER I., TURNER D.P.W., and VOLKMER K., 1968, Task accomplishment times in underwater work. *Journal of Marine Technology*, **2,** 22–26.

STREIMER I., TURNER D.P.W., and VOLKMER K., 1969, Study of work producing characteristics of underwater operations. *North American-Rockwell, Space division Document* 69-20, *Office of Naval Research Cont.* no. N00014-67-03636. *Research Task*, P002, Feb., 41 pp.

SZAFRAN J., 1966, Limitations and reliability of human operators of control systems to process information. *Aerospace Medicine*, **37,** 239–246.

TICHAUER E.R., 1963, Operation of machine tools at high altitudes. *Ergonomics*, **6,** 51–73.

TUTTLE W.W., and DICKENSON R.E., 1938, A simplification of the pulse-ratio technique for rating physical efficiency and present condition. *American Association for Health & Physical Educational Research Quarterly*, **9,** 73–80.

WIENER N., 1948, *Cybernetics* (New York: Technological Press of M.I.T. and John Wiley & Son).

Dr. Bonjer has degrees in physiology and medicine.
Since 1953 he has been head of the Department of
Occupational Medicine in the Netherlands Institute for
Preventive Medicine T.N.O. at Leiden.

Temporal factors
and physiological load

F. H. Bonjer

1.   Introduction

Some industrial tasks require energy expenditure where there is no problem in performance for half an hour, but where undue fatigue is produced should the work be continued for 8 or 9 hours. A similar situation is found in some perceptual tasks.   It is feasible to monitor a radar screen for 10 or even 20 minutes; but it is difficult to continue such a task effectively for eight hours.

Obviously, the limits to the performance of any human operator are, amongst others, set by his personal abilities and by the time during which he is required to sustain his efforts.

It follows that a judgement about the acceptability of any human task must be based on relevant data about the requirements of the task, including the working time, and on information about the operator's capacity, taking into account its rate of reduction as a function of working time.

2.   Requirements

Unfortunately, to describe the requirements of a human task in such a way that these requirements can be matched directly with the available capabilities is possible for only certain forms of dynamic muscular work. It is possible to measure how much energy is expended per minute when pulling in the net of a fishing-boat, and also the number of minutes taken by such an operation. Indeed, work physiology provided us with techniques for this type of investigation fifty years ago.

However, where static muscular activity is involved, it is more difficult to describe the requirements of the task. The product of the exerted force and time does not describe the work done in the sense acceptable to the physicist; it can, however, be used as a kind of quantification.

In the case of perceptual tasks or task components we can indicate the smallest angle of vision needed to detect details necessary for an efficient accomplishment of the task or the lowest intensity of acoustic signals at various frequencies that should be detected under specified conditions of background noise during a specified period of time.

Where information has to be processed, calculations or decisions made or where other forms of mental load are included in the task, it is even more difficult to describe the requirements of the task in terms that can also be used to describe the operator's capabilities.

3.   Capabilities

When the operator's capabilities are tested, this should normally be done under optimal conditions and over a short period of time.

Maximum energy expenditure (i.e. the maximum oxygen uptake or aerobic power) can be measured in a high level exercise test of short duration. The maximum force that can be exerted by certain limbs or parts of the body can be measured in less than a second if a suitable measuring device is available.

Visual acuity and auditory threshold can be tested under equally favourable conditions and these tests are normally so short that maximum effort can be expended without signs of fatigue.

The capabilities for mental tasks are more difficult to measure. There are no standardized procedures because what should be measured is still a topic for discussion.

4. Time factors

The influence of the passage of time on the level of performance has been studied in the field of dynamic and static muscular work to such an extent that laws have emerged. Studies of dynamic muscular work in racing animals (Kennelly 1906) and in the laboratory for human physiology (Grosze Lordemann and Müller 1937, Bink 1964) revealed a negative relationship between the logarithm of the endurance time (distance) and the level of activity (speed) (Figure 1). The higher the fraction of the maximum physical working capacity used, the less time can it be sustained. Similar results were obtained by Monod (1956) and Rohmert (1968), who studied endurance time for different fractions of the maximum muscular strength.

Figure 1. Endurance times in minutes observed in subjects A, B and C, when bicycling at different loads. (Data from the Netherlands Institute for Preventive Medicine TNO.)

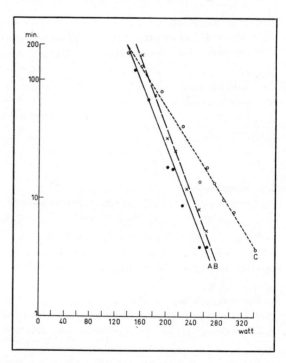

Both groups of experiments had a closer relationship with sports and gymnastics than with industry, the criterion for stopping being exhaustion or a condition near exhaustion.

Such a criterion is difficult to apply to the perceptual and mental components of a task. A detectable deterioration in task performance could be a substitute here, but I have not come across any work in this field.

It can be postulated that the laws that govern the relationship between the level of activity and the maximum duration of the performance also apply to the determination of allowable working times at given levels of activity.

But what do we mean by 'allowable'? Subjective criteria are required at this point. Sometimes the workers themselves state that they can or cannot fulfil a given task under certain conditions for a longer period of time. This may even be the only criterion in perceptual and mental tasks. In other cases the judgement about 'allowability' may be based on long term experience. If there are no signs of undue fatigue such as deterioration of the quality or the quantity of the work, and no signs of interference with the mental and physical state of health, even when the task is carried out for 5 days a week and 50 weeks a year, such a task may be considered as allowable.

Theoretical considerations on the basis of the experiments mentioned and on empirical data obtained from practical industrial situations led Bink *et al.* (1961) to a formula for the assessment of an allowable level of dynamic muscular work ($\dot{A}_t$), as a function of working time (t) and maximal performance ($\dot{A}_4$). It reads as follows:

$$\dot{A}_t = \frac{\log 5700 - \log t}{\log 5700 - \log 4} \times \dot{A}_4.$$

The level of energy expenditure corresponding with maximum oxygen uptake is indicated by $\dot{A}_4$ because the mean value of the endurance time for such an effort amounts to 4 minutes. It may be mentioned here that log 5700–log 480 is almost equal to one, and that log 5700–log 4 equals 3.1. This means that about $\frac{1}{3}$ of the maximum level of energy expenditure proves to be the acceptable level for 480 minutes, that is an 8 hour work period (Bonjer 1968). Figure 2 represents the formula in the form of a diagram.

Figure 2. Diagram showing mean values for an allowable oxygen uptake during working times ranging from 4 to 720 minutes in a subject capable of a maximum oxygen uptake of 3 litres per minute.

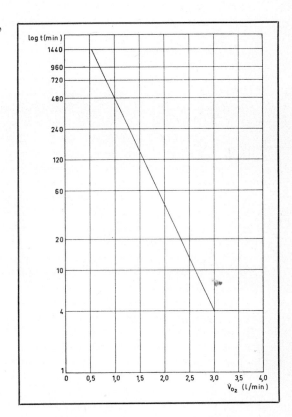

Fortuin (1970) indicates that he has come to a similar conclusion for problems of visual acuity. He points out that details which have to be recognized throughout an 8 hour working day should have a minimal size equal to three times the minimal size that can be recognized with effort during a brief test.

5. Conclusion

We can conclude, therefore, that we have passed the stage of complete ignorance about time factors in operator performance, but that there are still good reasons to pay more attention to this aspect of the functioning of man, both in research and in practice.

References

BINK B., 1964, Additional studies on physical working capacity in relation to working time and age. *Proceedings of the 2nd International Ergonomics Association Congress, Dortmund.* Supplement to *Ergonomics*, 83–6.

BINK B., BONJER F.H., and VAN DER SLUYS H., 1961, Het physiek arbeidsvermogen van de mens. *Tijdschrift voor Efficiency en Documentatie*, **31**, 526–31.

BONJER F.H., 1968, Relationship between working time, physical working capacity and allowable caloric expenditure. In *Muskelarbeit und Muskeltraining*; (Edited by W. ROHMERT) (Stuttgart: Gentner Verlag), 86–98.

FORTUIN G. J., 1970, De mentale belasting bij visuele waarneming. *Tijdschrift voor Sociale Geneeskunde*, **48**, 276–8.

GROSZE-LORDEMANN H., and MÜLLER E.A., 1937, Der Einflusz der Leistung und der Arbeitsgeschwindigkeit auf das Arbeitsmaximum und der Wirkungsgrad beim Radfahren. *Arbeitsphysiologie*, **9**, 454–75.

KENNELLY A.E., 1906, An approximate law of fatigue in the speeds of racing animals. *Proceedings American Academy of Arts and Sciences*, **42**, 273.

MONOD H., 1956, Contribution à l'Etude du Travail Statique. *Institut National de Securité, Paris.*

ROHMERT W., 1968, *Muskelarbeit und Muskeltraining: Schriftenreihe Arbeitsmedizin, Sozialmedizin, Arbeitshygiene* (Stuttgart: Gentner Verlag).

Professor Murrell has a chemistry degree. He has practised ergonomics in the services and in industry, and he has researched into production problems since 1954. He is now head of the Occupational Psychology Department, University of Wales Institute of Science and Technology, Cardiff.

Temporal factors in light work

K. F. H. Murrell

1. Introduction

Heavy physical work can be measured by means of oxygen consumption or heart rate. When these exceed certain levels the subject will become 'fatigued'. The requirements for rest to avoid excessive 'fatigue' can readily be determined. In most modern industry there are large areas of activity in which physical fatigue cannot be detected by these criteria.

Girls engaged on light assembly, for instance, do not show symptoms of physical tiredness of the form just described. Their output fluctuates it is true, and these fluctuations have in the past been described as forming a 'fatigue curve', this curve typically showing a warm-up, a period of reasonably high performance followed by a fall with, perhaps, an end spurt. This use of the word 'fatigue' leads into a curious circular argument. There is a fall in performance; therefore, there must be 'fatigue'. But why has there been a fall in performance? Because the operative is fatigued. All we really know about this performance is that there has been decrement after a certain period of time: it seems reasonable to look at decrements of output and try to determine what is causing them.

2. Experimental work on fatigue

Study in this field goes back to the days of the Industrial Fatigue Research Board set up after World War I (Chambers 1961). These were the first investigators to introduce rest pauses in an attempt to arrest the decrement in performance. One interesting fact emerged from their studies which is most pertinent: quite often when the break was given, performance was better after the break, as might be expected, but also it was better before the break. Now if the decrement were caused by physical tiredness one would have expected to find only the former improvement.

More recent studies are contained in Floyd and Welford (1953). In this book Bartlett suggested that what he called 'psychological fatigue' might be indicated by the appearance of irregularities in the succession of repetitive events. In an extensive laboratory study lasting 7 years, in which the inspection of electrical components was simulated, it was found that after a period of time, irregularities did appear and a technique of studying these was devised applying the principles of quality control to individual cycle times. A period of optimum performance was defined as the 'actile period' (Murrell 1962) and a limit of two standard deviations from the mean during this period was set. Cycles outside these limits were counted and it was found in the experimental task that irregularities began to appear after approximately 70 minutes of work.

It was further found that when subjects were instructed to stop when they felt they wanted to, they would go on sometimes for as long as half an hour beyond the appearance of irregularity; when they took a rest their performance improved, but the effect of the break was comparatively short-lived. However, when rest-pauses were imposed at times determined by the appearance of irregularities, performance tended to improve throughout the working period (Murrell 1965).

3. Vigilance and arousal

In seeking an explanation for this pattern of behaviour I started from the assumption that all continuous tasks have something in common, be they repetitive assembly tasks or inactive monitoring tasks (Murrell 1965, 1967). The former type of task has been very little studied and appears not to have been discussed theoretically. Work on 'blocking' might be relevant, but is usually too short-term for our purpose. Vigilance tasks, on the other hand, have been extensively studied and a series of hypotheses have been developed to explain changes of performance with time. These can act as a useful starting point.

Vigilance hypotheses appear to be of three kinds.

Strategical. The expectancy theory, e.g. Baker (1959), the observing responses of Holland (1958), various statistical approaches of the kind enunciated by Broadbent (e.g. 1963).

Behavioural. Inhibition.

Neurological. Activation or arousal.

Inhibition can be discarded at the outset as being too vague. Effects can equally well be explained on a neurological basis. Neurological and strategical hypotheses are not mutually exclusive; strategical hypotheses are probably specific to vigilance and cannot be used as an explanation of active performance. This leaves only arousal.

Jerison (1967) in a critical appraisal of the arousal theory of vigilance has said 'the activation concept has been so broad it is rather difficult to get experimental results that are beyond its ken'. Lacey (1967) has further pointed out that electro-cortical, autonomic, and behavioural measurements may all give results which appear to be independent; that variable and inconsistent relationships between these functions occur within one individual which appear to be unrelated to arousal patterns; thus a generalized relationship is difficult to specify. Unfortunately this has also been our experience since we have been unable to see any consistent relationships between the different measures which we have been making. On these grounds, then, we might be justified in abandoning attempts to relate our observed behaviour to arousal. But Jerison went on to say 'it is certainly the case that activation or arousal plays a significant role in vigilance performance', and so it has been our view that we should not abandon arousal simply because of difficulties and the present absence of easy success.

Unfortunately, most published results on arousal are rather unhelpful, since they show only results averaged over a group of subjects. Moreover, it is only rarely that any measure other than performance is reported. One experiment in which individual performance is analysed is that of Dardano and Mower (1959) in which time to respond to a change in a CRT display and skin conductance were measured. For twelve Ss skin conductance was negatively correlated to log RT; seven being significant at $p = \cdot 01$. Assuming that skin conductance is an indicator of arousal three of the subjects were able to improve their performance during part of each session by raising their arousal, while four others were apparently able largely to arrest a decrement in performance; but once arousal fell below the initial value the decrement was 10 times greater than it was before. It is worth remembering that this occurred in a situation which should have been non-arousing.

This is an example of how, underlying the averaged results usually published, there are individual differences to which little theoretical attention seems to have been paid. Yet the viability of any explanation of continuous performance must depend on an ability to account for these differences in addition to overall trends (if any). When discussing individual differences in a task Buckner (1963) has proposed several contributory factors including perceived difficulty of signal detection, while Griew and Davies (1963) suggest that there are individual differences in arousability, to which such factors as extraversion/introversion, lability and so on may contribute, which can set a limit on performance. Hebb (1955) has postulated an 'exploratory drive' to explain 'the phenomenon of work for its own sake' but this would seem to be conceived as originating in situations external to the individual. But these proposals still do not explain why some people do well one day and not so well the next; and when all observed phenomena are explained and all conditions are controlled there must be in addition something specific within an individual which will make one person perform well when another in identical circumstances 'could not care less'.

To seek an explanation we must consider the arousal function within the CNS. The mechanism by which the reticular formation delivers non-specific stimulation to the cortex has been well established by both neurophysiological and behavioural studies. In addition, there is activation in the reverse direction from the cortex to the reticular formation. French *et al.* (1955) demonstrated that evoked potentials in the central cephalic brain stem could be detected as a consequence of electrical stimulation of the exposed cortical surfaces in the monkey; they suggested that self-arousal which could sustain wakefulness derived from downward cortical stimulation of the reticular formation. It is only a short step further to propose that this mechanism will also function when an individual 'makes up his mind' to do well in a task. I have called this auto-arousal to avoid confusion with other terms and I would conceive of it coming into maximum play in those situations which are naturally the least arousing. It would be tempting, of course, to equate auto-arousal with motivation, but motivation has so many meanings that this should be avoided. So I would define auto-arousal as 'cortical activation resulting from stimulation of the reticular formation by the cortex, this stimulation being under voluntary control'. I am well aware that this may be just another circular argument since, *prima facie*, we cannot measure voluntary control, I feel that careful exploration could reveal behavioural correlates which are meaningful. It might be said that auto-arousal is low in some Ss, such as sailors, who have no particular desire to do well (except when a bottle of beer depended on the outcome, as related by Buckner (1963)), and this may be a contributory factor to the reported decrements. On the other hand Baker's (1963) paid housewives may have had an entirely different outlook which might help to account for the absence of decrement in some of the runs as well as differences in level of performance.

4.   Auto-arousal and performance

I put forward a tentative theory which could account for the behaviour which we had observed (Murrell 1966). It is in some ways similar to the theory put forward at the same time by Bergum (1966). I proposed that since continuous tasks are basically non-arousing, auto-arousal is necessary to maintain performance. This auto-arousal will push up the arousal level which passes through the zone of good regular performance into the hyper-arousal area and that it is then that the irregularities will occur. In order to maintain homeostasis there must be a damping down of arousal and this will cause the arousal level to fall back into the zone of regular performance (to give an inverted U-shaped curve). The damping down may at the same time cause the extinction of the downward auto-arousal projections so that the level of arousal will continue to fall until it goes below the optimum and irregularity will again occur. This suggestion would account for the appearance of irregularity, the reappearance of regularity and the second phase of irregularity. It is too simple an explanation, which has now been superseded (Murrell 1969), but it provided a useful starting point for reviewing the evidence which was available.

5.   Experiments on auto-arousal

Since it is difficult to use direct neurological techniques in man, behavioural studies accompanied by physiological measurement need to be used. We have started doing this on pilot scale with five subjects, using our repetitive task which is performed for $3\frac{1}{2}$ hours. Our Ss were paid and were with us for a minimum of ten weeks, and they performed continuously both in company and in isolation, they had breaks at predetermined times and at times of their own choosing, they had coffee and no coffee, they had varying quantities of alcohol, caffein and amphetamine and finally they performed with a superimposed vigilance-type task in which a small lamp mounted on headgear so that it was in the periphery of vision was illuminated at irregular intervals. It was extinguished by pressing a Morse key with the non-preferred hand (the main task is one-handed), and the time was measured. After 10 sec it was extinguished automatically if it had not been noticed and this constituted a missed signal.

Skin resistance was measured. We have found in a number of instances that there are inverted U-shaped conductance curves, but the peaks of these curves generally appeared before the onset of irregularity which tended to occur either near the trailing edge of the U or when the conductance was falling. But quite often there were no U-shaped curves at all. There may be a continuous rise in conductance, level conductance, or a continuous fall in conductance after the initial warm-up. But even so the irregularity may still occur. There is some evidence that irregularity may be related to falling conductance or may occur when conductance is low. But if this is so, we would have to explain why performance actually improves over a period when conductance has not risen.

In experiments with the 'vigilance' task superimposed a light came on 12 times in each hour. Different sequences of presentation were used so that after five sessions a signal had occurred once in every minute of an hour. There were relatively few actual missed signals but latency provides a useful measure and as might be expected there were large within-subject and between-subject differences. Nevertheless, in a number of instances there was a sharp increase in latency and misses at about the same time that irregularity made its appearance. Sometimes this would be seen on an individual day ; at other times a relationship could be shown only by grouping the data.

A relationship between latency scores and skin conductance is by no means always easy to establish. In some instances it appears to be non-existent while in others a relationship does appear to exist. For instance, on one day, skin conductance tended to fall after the first $\frac{3}{4}$ hour to reach a low after $1\frac{3}{4}$ hours. In the same way, the number of detections fell from complete success to 100 per cent misses. Just before 2 hours a cup of coffee was taken without stopping work and this was followed by a rise in conductance and an improvement in vigilance performance so that for a half-hour period there was again a 100 per cent detection. This was followed by a fall.

These results indicate that where two different methods of detecting the onset of change of performance are used, they may show a relationship which suggests they have a common cause. One or both of them may be related to changes in skin conductance.

When a period of work commences there is usually a warm-up which shows as an increase in the rate of work, and a reduction in variability. This is generally accompanied by an increase in skin conductance. When 2 Ss are tested together the skin conductance tends to remain high for a period and to be rather variable, possibly due to varying interaction between Ss. When subjects are isolated and remain incommunicado for the whole of the experimental session of $3\frac{1}{2}$ hours the skin conductance tends to fall throughout the period on the first occasion under this condition, and is less variable than when Ss work in pairs. But as the Ss become more accustomed to isolation, the skin conductance pattern fell less steeply and more nearly approached the general shape found when Ss work together: output also increased. In one subject when at the outset there was a steep fall in conductance, output was 1890 and this increased over 4 isolation sessions to 2075, when skin conductance fell much less. Since experimental conditions have not altered in any way, this improvement must have emanated from within the subject herself, so that she was progressively able to compensate for the extremely monotonous condition into which she had been placed. When a subject was put into isolation but with vigilance added skin conductance did not fall steeply as it did when vigilance was absent; in some instances it remained substantially level and production was high. This suggests that the vigilance task is a source of additional arousal which has a similar effect to the auto-arousal which was necessary without vigilance. In one instance the arousal is clearly external but in the other it can only have come from within the subject herself.

Experiments were also carried out using various drugs. When a small dose of alcohol (23·7 ml of whisky containing 40 per cent alcohol) was given the skin conductance level remained high for a period of more than an hour, and output was very high indeed. But with larger doses of double and quadruple this size, there was an immediate fall in skin conductance followed, sometimes, by an increase when the effects of the alcohol had worn off; output was very low.

When the Ss were given a cup of coffee, which contained 1·4 mg of caffeine, ingestion was followed by an increase both in skin conductance and in output. But when 2·5 mg and 5·0 mg were given in capsule form alternately with a placebo there was no obvious effect, except on one day with the 2·5 mg dose. A similar effect was observed when 5·0 mg of dextroamphetamine-sulphate was administered in capsule form as an alternative to a placebo. With most Ss there was no very marked effect, except for a tendency for skin conductance to increase. One subject was different; she had come to believe that the placebo was the amphetamine and there was a marked change in her skin conductance at the time at which she thought the capsule would take effect after the placebo had been administered. She also indicated that when she expected the amphetamine to boost her output she did not need to make much effort herself. This is perhaps reflected in the sharp drop in her conductance before the placebo had 'taken effect'. It was she also, who showed the marked increase in skin conductance after taking the 2·5 mg dose of caffeine, and it transpired afterwards that she again had come to the wrong conclusion that she had been given the 5·0 mg dose.

In general, the Ss in this experiment seem to have performed in a manner determined by their beliefs in the results which would ensue: they knew that a cup of coffee was supposed to be a stimulant but did not know which of the pills contained what, so the pills had little or no effect, except when one S had wrongly made up her mind about their contents. As a result, her performance was influenced in the manner she expected, even though the effects should not have occurred in the way they did. This would explain the anomalous result with the small doses of alcohol. Alcohol is a cerebral depressant but the popular belief is that it acts as a stimulant, and it would seem that a small dose is not sufficiently depressing to overcome the stimulating effect of this belief. Or to put it another way, the belief that alcohol is a stimulant could lead to sufficient auto-arousal to overcome the depressant effect of small doses, but not, of course, be sufficient to counteract large doses. Since in these experiments there were no external determinants of change the anomalous rises in skin conductance and in performance must have been predicated by the Ss themselves; if they knew (or thought they knew) what they were getting their performance was appropriately influenced. This may, therefore, be taken as evidence for the mechanism of auto-arousal.

6. Conclusion

The effects so far described have been seen most clearly in the skin conductance curves. If the skin conductance is a measure of arousal, the behavioural measurements have not, by any means, always followed prediction. One of the difficulties of this kind of work is that from the results obtained it would be possible to produce some which would show that nothing whatsoever has happened; equally as has been described above, results have been obtained which do support the views put forward. That this should be the case need not necessarily disprove the general thesis. It is not part of the proposition that auto-arousal will always occur in a given set of conditions. The very fact that the effects looked for occur only in some of the experiments is further evidence that we have been dealing with an arousal effect which can be manipulated at the will of the subject and over which the experimenter may have only partial control—if, in fact, he has any control at all.

References

BAKER C.H., 1959, Toward a theory of vigilance. *Canadian Journal of Psychology*, **13**, 35–42.

BAKER C.H., 1963, *Vigilance: a Symposium* (New York: McGraw-Hill), p. 45.

BARTLETT F.C., 1953, Psychological criteria of fatigue. In *Symposium on Fatigue* (Edited by W.F. FLOYD and A.T. WELFORD) (London: H.K. Lewis).

BERGUM B., 1966, A taxonomic analysis of continuous performance. *Perceptual and Motor Skills*, **23**, 47–54.

BROADBENT D.E., and GREGORY M., 1963, Vigilance considered as a statistical decision. *British Journal of Psychology*, **45**, 309–323.

BUCKNER D.N., 1963, *Vigilance: a Symposium* (New York: McGraw-Hill), pp. 171–181.

CHAMBERS E.G., 1961, Industrial fatigue. *Occupational Psychology*, **35**, 44–57.

DARDANO J.F., and MOWER I., 1959, Relationship of intermittent noise, intersignal interval and skin conductance to vigilance performance. U.S.A. Ord. Hum. Engng Lab. Tech. Memo 59–7.

FLOYD W.F., and WELFORD A.T. (eds.), 1953, *Symposium on Fatigue* (London: Lewis).

FRENCH J.D., HERNÁNDEZ-PÉON R., and LIVINGSTON R.B., 1955, Projections from cortex to cephalic brain stem (reticular formation) in monkey. *Journal of Neurophysiology*, **18**, 74–95.

GRIEW S., and DAVIES D.R., 1963, Arousability, individual difference and vigilance performance. *Paper given to Bedford Group and reproduced in the printed papers.* P. 51.

HEBB D.O., 1955, Drives and the conceptual nervous system. *Psychological Review*, **62**, 243–253.

HOLLAND J.G., 1958, Human vigilance. *Science*, **61**, 67.

JERISON J., 1967, Activation and long term performance. In *Attention and Performance* (Edited by A.F. SANDERS) (Amsterdam: North-Holland Publishing Company).

LACEY J.I., 1967, Somatic response patterning and stress: some revisions of activation theory. In *Psychological Stress: Issues in Research* (Edited by M.H. APPLEY and R. TRUMBULL) (New York: Appleton-Century-Crofts).

MURRELL K.F.H., 1962, Operator variability and its industrial consequences. *International Journal of Production Research*, **1**, 39–55.

MURRELL K.F.H., 1965, Le concept de fatigue une réalité ou une gêne? *Bull. C.E.R.P.*, **14**, 104–110.

MURRELL K.F.H., 1966, Performance decrement—a tentative explanation. In *Proceedings 2nd Seminar on Continuous Work* (Edited by F.F. LEOPOLD) (Eindhoven: Institute for Perception Research), pp. 122–126.

MURRELL K.F.H., 1967, Performance differences in continuous tasks. *Acta Psychologica*, **27**, 427–435.

MURRELL K.F.H., 1939, Laboratory studies of repetitive work IV: Auto-arousal as a determinant of performance in monotonous tasks. *Acta Psychologica*, **29**, 268–278.

M. Leplat is a psychologist. From 1951 to 1966 he worked on industrial psychology problems at the Centre d'Etudes et Recherches Psychotechniques, and he is now Director of the Industrial Psychology Laboratory of l'Ecole Pratique des Hautes Etudes, Paris. His colleague M. Pailhous, also a psychologist, is a research worker at the Centre National de la Recherche Scientifique.

The analysis and evaluation of mental work

J. Leplat and J. Pailhous

1. Introduction

Mental work, which is essentially non-observable, presents serious difficulties to the analyst who wishes to evaluate it. This kind of work is found increasingly in our society, even where the task requires a relatively low level of skill, and interest in its study is increasing. In this paper we consider various analytical procedures of analysis and evaluation and assess the results obtained. The analysis provides an opportunity to compare the methods of the applied psychologist and the work physiologist and to expose the differences involved. The distinction which we shall propose in terms of load or intellectual mechanisms appears to be quite fundamental; it covers many totally different scientific attitudes; it leads to distinct types of techniques, even if in practice they may be utilised successively at two different periods of an investigation. Such an analysis should contribute to the dialogue with our physiologist colleagues.

2. Practical needs in the matter of evaluation

Numerous actions undertaken in industry demand information relative to evaluation. For developments based on ergonomics this demand is evident: the definition of work conditions, environments, the arrangement and the specification of materials necessitate the use of criteria. At the same time such criteria can be used to orientate actions and to control them. Also, measures of work organisation demand precise data on positions and functions in order to answer such questions as: how much work may be demanded of an individual without impairing his health; how do we balance the work among available personnel; etc.

If needs are numerous they are also multivariate, in relation to the actions to be undertaken, and also in the diversity of mental activity. There is no question of resurrecting the old controversy between manual and intellectual work. In varying degrees all work is intellectual, if only because it is never so elementary and so perfectly planned that it does not require either choice or decision. In mental work one might distinguish a qualitative and an intensive aspect: the former suggests that any mental work may be more or less complex, according to the mechanisms which it involves; the latter characterises the rate at which these mechanisms must operate—the level of mental work may thus be determined by temporary strains entailed by its execution. It will be noticed that, generally, intensive work concerns elementary activities.

Problems arising from mental work in industry have often been avoided by resorting to the results of the job, that is, production. But technological progress has resulted in the multiplication of tasks where the results of the activity of the operator appear after a long delay and where it is not always possible to trace back easily from the result to the action of the operator. This is especially the case in certain jobs in remote control installations. For long periods the operator apparently carries out no action which reveals his intellectual activity and yet this is not absent, as may be ascertained by questioning him on the state of the system under his control. A problem for evaluation therefore arises here, caused by the absence of observable data that may reveal or allow the interpretation of the activity of the operator. The methods of classical descriptions based on observations of movement are inadequate.

Such situations are not the only ones where these problems arise. Numerous jobs can be described where the observable activity, reveals, at least directly, little of the essential part of the work. Inspection tasks and fault-finding are typical examples ; as are telephone switch board operation and card punching.

When faced with the difficulties of classical methods and techniques, it is important to reconsider, for mental work, the questions mentioned at the beginning of this section.

3.   Mental load

The notion of load has been widely used in the evaluation of muscular work and during the past few years it has been used also for mental work. The status of this concept is far from being clear. One must, first of all, distinguish the following.

■ The load, such as can be evaluated by objective physical measures intended for mental tasks. These measures can be expressed, for instance, in terms of information quantity or index of complexity (Fassina 1969).

■ The load for the operator

For a given index of a physical load there corresponds at different moments of training, or for different operators, different loads of work. This load expresses what one might call the degree of mobilisation of the subject, the fraction of his working capacity which he spends on the task.

In the physiology of muscular work, the load is often defined, in an operational manner, by oxygen consumption. It might perhaps be concluded that the physiologist identifies the two variables and that he would be prepared to say that when oxygen consumption becomes k times larger, the load becomes k times greater. If this were true it would be essential to give a precise definition of the load and to examine the relation between this load and its indices. This fundamental problem is best left open, for it would, by itself, require much development, not unlike that which psychology has seen on concepts such as intelligence.

Measures of load generally constitute the global criteria of a given task. They aim at evaluating average cost. These global measures may be of a diverse nature.

One group concerns the evaluation of behaviour. Thus, in the study of air-traffic controllers we chose various criteria of this type (Leplat and Browaeys 1965). One was the average number of conversations exchanged between controller and pilot ; another, the average duration of these conversations. These measures collected for different traffic situations made possible the calculation of the reduction rate of the message-duration for heavy traffic, an expression of work load for the operator, if one admits the hypothesis that 'the work load is greater, when an operation is carried out, by a given subject, in a shorter time'. One can find numerous studies where efficiency criteria have been implicitly considered as measures of work-load. In the laboratory, elementary ergonomics experiments aiming at the comparison of devices and appliances have constantly made use of such criteria (time of execution, errors, volume of transmitted information). These criteria are situated at the interaction level between the operator and his task, hence their specific character.

The global physiological criteria of load are not so prone to this drawback and it is possible to classify jobs as a function of the loads they represent for a given operator.

The same advantage is offered by measuring mental load by means of a secondary task. The modalities of evaluation may be various (Leplat and Sperandio 1967). The application of the method raises numerous theoretical and practical problems. The ratios of the secondary task to the task under evaluation must be very carefully analysed so as to control the effect of interference on the combination of tasks. The application of the method implies the stability of the capacity of the channel : which depends on the attitude of the subject. Kalsbeek (1968) discusses the capacity which the operator is 'willing to spend'. The combination of tasks is, at times, difficult to carry out in the course of work : a few examples of practical use do, however, exist (Brown 1967, Michaut 1968). Here again problems of measurement should be specified. The defined scale is of the ordinal type and the coherence of scales obtained with different tasks should be verified with the secondary tasks on the one hand and the direct measures on the other.

Certain global criteria can be improved. So can those related to the time of execution. It is known that when work is protracted execution time is progressively lengthened: this is regarded as an indication of fatigue (Murrell 1965). If successive portions of execution time are analysed, the dispersion of times constitutes an indication, the variation of which precedes and predicts therefore that of the average time : it is by this criterion that the notion of actile period is defined. Behaviour criteria may also be taken for limited parts of a task. In the same way, physiological criteria may be sampled locally. In other words, the evaluation of load may be made globally or locally according to whether one is interested in the general functioning or in one of its aspects.

In relation to load problems, both psychological and physiological measures constitute an indication of the working of the system : in a somewhat crude manner they indicate the intensity of the working or to what fraction of its maximum potential it does work. However useful it may be this approach is inadequate. It provides us with external information but it tells us nothing of the working mechanisms. However, it can lead to a study of them; to an empirical type of approach which compares two different systems with the aid of global criteria. These different systems are conceived without relation to the information supplied by the criteria. Now, if it is really the modification of the system which ergonomics aims at, one must then obtain information on the mechanisms utilised by the operator.

The same criteria will then no longer be considered as indications of the intensity of functioning but as a means of validating hypotheses on the modalities of this functioning.

4. Intellectual processes

In order to approach intellectual processes we resort essentially to psychological analysis. Neurophysiology also seems to be increasingly interested in these processes but our lack of competence prevents us from dealing with the subject here.

If the study of mental load leads the psychologist to analyse the concomitant variations of inputs and outputs of the black box which constitutes the human operator, the study of intellectual processes leads him to interest himself in what is happening inside that black box. This new point of view entails important changes as much in the objects of investigations as in the approach of the applied psychologist.

Concepts of intellectual mechanisms postulated by the psychologist are purely hypothetical (hypothetical is taken here in the sense of non directly observable). For instance, interiorised operations, mental images, the symbolical function. Organised in a coherent manner, in the logical sense of the term, these concepts constitute, more or less, general theories (Piaget 1947).

For example, consider visual search movements. They are observable and the characteristics of fixation, its relation with foveal vision and with saccade are fairly well known. The relation between stimulus and the pattern of eye movements can be defined. The psychologist therefore knows what the subject sees and what he does not see, the areas where fixation points are concentrated, and the duration and the frequency of these fixations. If he goes no further he will express himself in terms of perceptual load and will conclude, for instance, that the task is too heavily loaded and that the presentation of information must be improved. If the information gathered is considered to reveal signs of the intellectual mechanisms involved, he will consider the fixation points as testifying, in the subject, to the existence of a search programme (Vurpillot 1969). Thus, in a non-problematical task for the subject, it will be said that fixations mean successive hypotheses and choices. The experimenter will therefore be interested in the sequential aspects of fixation points and may conclude, in contrast with the psychologist who is concerned only with load, that if signals are not perceived, this is not necessarily because they are badly presented but rather searched for with an inefficient strategy (with consequences for the training of operators). He may also conclude that fixation points, even in small number, testify to intellectual activity of a high logical level and that, contrary to appearances, the task is too difficult (with possible consequences on the organisation of the task or the arrangement of appliances or devices).

We have developed this example to show that the analysis of intellectual processes does not necessarily lead to new measures but to a new understanding of these measures. The status of responses changes, and these now become observable signs of non-observable mechanisms.

It should not be concluded that any particular type of task leads automatically to any particular type of mechanism : in fact, one must always refer back to the activity of the subject. In the course of training, intellectual tasks which initially are difficult simplify themselves for the operator, who can ultimately ensure their automatic execution. For instance, in a complex supervision task the operator may at first perceive signals as being independent ; he then becomes aware of certain regularities and can introduce a structure into the sequence by which he will be better prepared to organise his responses and even define algorithms. In this latter case he will be able to automatise his responses until they become for him a simple motor sequence occurring with the minimum of perceptual control.

Usually, it is interesting to analyse intellectual processes only in tasks where the operator must solve problems with a certain measure of complexity. Tasks in which the responses have a true status of decision are increasingly numerous. In order to determine the nature of the task it appears to be necessary to resort to observable behaviours which are the most representative of underlying intellectual mechanisms.

According to the distinction made by Piaget (1963) between operations* it may be asked whether the task is of the logical or of the spatial type. According to another conceptual system it may be possible to ask whether the task is random, regular, necessary. The problems which are then set are problems of accountancy between the theoretical systems utilised. Nevertheless, this way of looking at the task allows us to ask relevant questions and to establish protocols of research, which in order to utilise responses of the same nature as more empirical protocols, gives them a none the less particular meaning, which seems to us to be fruitful in relation to applications. Thus, considered from this point of view, an inspection task will lead to the development of probabilistic models with partial regularities, etc., and to a better assessment of the needs of professional training, for instance. From this mode of approach will derive a hierarchy of activities : logical level of the task, co-ordination of operations, and so on. This hierarchy would no longer be based only on the characteristics of responses in the course of execution but, for instance,on the moment when they appear genetically : if the use of inclusion is acquired before that of logical disjunction, it will be possible to say that under certain conditions one operation is easier than another ; and a taxonomy of tasks can then be arrived at.

5. Method of analysis and field of intervention

Whether the psychologist concerns himself with load or with intellectual processes the nature of the measures which he carries out is practically the same : what actually changes is his attitude and especially his objectives.

When he is interested in load, the research worker centres his activity on the worker's physical and mental state. He reasons in terms of discomfort, fatigue, and working conditions : he refers to criteria concerning staff : absenteeism, sickness, turn-over and so on. This outlook is particularly well adapted to work of an intensive type where the worker is stressed by the task : repetitive work being a good example.

In other cases the analysis of intellectual processes might be necessary. For instance, in a traditional accountancy department where the staff deal with simple accounts in a continuous and rapid manner, the psychologist who wanted to study the intellectual mechanisms, to determine economical algorithms in order to make modifications and improve training, would certainly tackle the problem from a very secondary angle. The essential problem in all probability is not apparent but should consist of evaluating the load and fixing for it an acceptable norm.

On the other hand, when the task is of a type involving the solution to a problem, study in terms of load is hardly pertinent because it supplies no means of reducing the perceived

---

(*) For Piaget, the operation (which he distinguishes from the action) has among its characteristics that of being ' interiorised ', which is contrary to the current use of this term in applied psychology.

difficulty. In this case, an analysis of the intellectual processes is essential and has consequences for training, work-design and even selection.

Within engineering psychology, Ochanine (1966) has shown, in an investigation of intellectual mechanisms in remote control tasks, that the mode of presentation of information must be isomorphous with the mental image of the operator and must take into account the operations effected. Without this isomorphism the operator must resort to supplementary coding which is generally a source of error.

The approach in terms of load appears to be more natural and more convenient to the physiologist than to the psychologist. Indeed, work physiology seems directly centred on man, whereas the psychology of work is essentially concerned with situations, that is, with the interaction between man and the technical system. The psychologist must therefore see to it that modifications of the situation, such as occur in the course of experimental reduction, do not transform his problem. The problem could be solved at the time of its definition if there existed a valid taxonomy. This is why the interaction between laboratory and field appears to be easier for physiologists than for psychologists. If in work physiology the field asks the laboratory the valid questions, in applied psychology the field also supplies the valid answers. A certain amount of laboratory practice in psychology may be quite fictitious and the isolation of a variable element or part of the situation—a frequent reason for a return to the laboratory—does not appear to have the same value as in physiology. The execution of a tapping task with sampling of criteria such as E.E.G., E.M.G., etc. in order to investigate the effects of speed is not analogous with the learning of meaningless syllables in order to study memory.

The physiologist very quickly analyses a situation into a pattern of variable elements which he does not hesitate to isolate or to transfer into various contexts. The psychologist is much more cautious, knowing how easily the initial situation can be distorted. While attempting to simulate one situation he runs the risk of creating another which is only apparently related to the original. This is one of the reasons which render so delicate the justification of laboratory investigations in psychology (Chapanis 1967).

These remarks might explain why the fall-outs from work-physiology investigations are often effected in terms of norms (norms of environment for instance); and why this is quite exceptional in psychology.

The applied psychologist is at a disadvantage over the physiologist. Although he might be very successful within a given situation, he comes up against problems when he tries to generalise from his results: it is at this stage that empirical techniques (such as are used in investigations of dials) reveal themselves as inadequate. To give an example: in order to determine the best possible design of a product (e.g. maps, dials, keyboards) a whole scale of these designs might be constructed with preliminary hypotheses. The results are then compared in order to isolate the best design. The technique is simple, its carrying into effect is costly in experimental terms, cheap on the conceptual; its efficacy is often satisfactory. But at the end of the experiment the psychologist has learnt nothing that can be transferred to another situation. It is only when he refers a fine analysis of activities to a theoretical context that generalisations become possible and justified.

These generalisations will not concern the design and will not produce norms of the type 'all maps must be made in such a manner' (a mistake very frequently met with) but they will extend the results to tasks with similar structures. It is thus possible to say that a road-map is closer to a cabling-plan than a geological map, on account of the activities to which it gives rise (Pailhous 1969).

This said, generalisation to other tasks always poses a problem since task structure analysis is accomplished by the postulation of hypothetical constructs. Moreover, in adopting a too systematic attitude in this direction the psychologist runs the risk of putting on blinkers and so masking certain important problems.

Conceptual ergonomics allows us to escape partially from the dangers of generalisation because the new system arises directly from preceding analyses.

To end this section on the methods of evaluation and the fields of intervention we have evoked several 'modes of approach': among these it appears to us that there is much more which complements than opposes. We have stressed that it is often the objective of the research worker and the means placed at his disposal that will decide the choice between one approach and an other. Also, these different modes of approach may frequently constitute parts of the same investigation and thus lead to a fine diagnosis of the problem and its optimum treatment. An illustration is supplied by the aerial navigation control at Orly Airport where an investigation which began in terms of load (Leplat and Browaeys 1965) was continued with investigations of intellectual mechanisms (Leplat and Bisseret 1965). It belongs to the psychologist and the physiologist to set and treat problems at the level where the efforts they devote to their solution stand the greatest chances of being effective.

References

BROWN I.D., 1967, Measurement of control skills, vigilance, and performance on a subsidiary task during 12 hours of car driving. *Ergonomics*, **10**, 665–674.
CHAPANIS A., 1967, The relevance of laboratory studies to practical situations. *Ergonomics*, **10**, 557–578.
FASSINA A., 1969, Un intermédiaire dans le système homme-travail: le dessin technique. Lecture et écriture des schémas explicatifs. *Thèse, Paris-Sorbonne.*
KALSBEEK J.W.H., 1968, Measurement of mental load and of acceptable load: possible applications in industry. *International Journal of Production Research*, **7**, 33–45.
LEPLAT J., and BISSERET A., 1965, Analyse des processus de traîtement de l'information chez le controleur de la navigation aérienne. *Bull. C.E.R.P.*, **14**, 51–68.
LEPLAT J., and BROWAEYS R., 1965, Analyse et mesure de la charge de travail du controleur du trafic aérien. *Bull. C.E.R.P.*, **14**, 69–80.
LEPLAT J., and SPERANDIO J.C., 1967, La mesure de la charge de travail par la technique de la tache ajoutée. *Année Psychologique*, **67**, 255–277.
MICHAUT G., 1968, Etude de la tâche de conduite à l'aide d'une charge de distraction. *Travail Humain*, **1–2**, 95–110.
MURRELL K.F.H., 1965, *Ergonomics* (London: Chapman and Hall).
OCHANINE D., 1966, The operative image of a controlled object in ' man-automatic machine ' systems. 18ème *Congrès International de Psychologie, Moscou*, 48–56.
PAILHOUS J., 1969, Représentation de l'espace urbain et cheminements. *Travail Humain*, **1–2**, 87–139.
PIAGET J., 1947, *La Psychologie de l'Intelligence* (Paris: Collin).
PIAGET J., 1963, *La Construction du Réel chez l'Enfant* (3ème édition) (Neuchatel: Delachaux et Niestlé).
VURPILLOT E., 1969, Activité occulo-motrice et activités cognitives. *Bulletin de Psychologie*, **70**.

Section 1-Man

Epilogue

Any lingering illusion about the possibility of providing simple measures of man must surely have been dispelled by the range of relevant factors exposed in the preceding papers. It must be accepted that any description of man which claims to be comprehensive, even in relation to the simplest of work situations, will be inherently multidimensional and yet almost certainly incomplete.

However, acceptance of this position does not absolve the investigator from his responsibility to restrict the range of measures taken to the minimum commensurate with a reasonable description of the demands of the situation on the man. The currently popular method is to use a vast range of instrumentation and data handling equipment. This is not likely to be successful unless it is used in the context of a high level of sensitivity to the conceptual relationships between the various scores and the relevant descriptors. On the other hand it seems to be equally true that any investigation or experiment in which a single simple measure is taken, e.g. speed of performance or metabolic rate, is unlikely to provide a description of the human activity which can be interpreted usefully in either theoretical or practical terms.

Some generalizations about what the physiologist and the psychologist have been doing appear possible. It seems to be agreed that the physiologists have been orientated towards the concept of human energy with the consequent concentration on metabolic mechanisms. Progress has been made by relying on overall measures of the load on the operator, particularly successfully in relation to dynamic muscular work. It has proved possible to provide reliable measuring techniques and standards of acceptable performance. The methods are most precise when considering questions of maximum levels of activity. Even for these problems, however, subjective judgment must be used to interpret standards in the context of particular work situations. These become more diffuse for questions of long duration activities.

Partly because of the success of these methods in the field of energy expenditure there have been many attempts to transfer the same methods and ideas to problems of information processing. This is illustrated by the concept of mental load as an analogue of the concept of physical load. This is still extant but it has not been very useful. Mental load has at least two independent dimensions: complexity and intensity. The general trend seems to be that for mental activity global measures are less fruitful than micro-studies aimed at identifying more precisely the mechanisms involved. It then emerges that individual differences of method as well as of capacity must be considered. There are changes not only between individuals but within individuals at different times, that is, learning and fatigue effects. We have to abandon that most valuable summarizing abstract: the average man. Although it is still possible to think in terms of subsets of the population studied which behave similarly.

There are clearly some fundamental difficulties about laboratory methods of investigation which have not been fully explored and described. Internally these stem mainly from the uncontrolled variables brought into the situation by the man as an experimental subject: his attitudes, expectancies, motivation and so on can materially affect the results of apparently objective experimentation. In addition output measures are often not comprehensive, usually in that the cost to the subject is not determined. Externally, the extrapolation from laboratory behaviour to performance in real systems is equally loaded with unknown factors.

The current enthusiasm for field studies may have caused us temporarily to overlook some of the problems here which stem mainly from lack of ability to control the situation being studied. In both the field and the laboratory we are entirely dependent on the sensitivity and creativity of the investigator. Experimental design in the statistical sense is no substitute for this. In addition we need improved conceptual models which will facilitate the transition between field and laboratory. We are now at the stage where those borrowed from physiology and psychology are insufficient. We need to develop a philosophy of man–machine systems. It is argued that only physiological measures are reasonably consistent for man ; psychological measures cannot be divorced from situations.

In almost all situations there are measurable demands on both the vegetative and neuronic systems of the man. This naturally raises questions of cause and effect but these do not seem to be fruitful. Man is a complex set of interacting mechanisms with extensive feedback loops within and between systems. He must also be considered as an on-going total system continuously adjusting his patterns of internal and external activities so that the study of cause/effect relationships is essentially fallacious. This is not to say that time is not an important variable. On the contrary it is a dominant variable and the duration of any activity is one of the main parameters.

The concept of stress is a useful illustration of the close interaction between what used to be regarded as the separate domains of the psychologist and the physiologist. It is necessary to distinguish between stress and strain and to recognize that the stressor can only be measured by the strain effects. Stressors can be either physiological, e.g. heat, noise; or psychological, e.g. fear, incentives; but invariably a particular man at work is subject to an inextricable combination. There have been many attempts to identify characteristic patterns of reaction (both physiological and psychological) which will lead to reliable separation and identification but these have not yet been successful.

The difficulties of setting adequate standards for physiological stress highlights the problem of equivalent standards for psychological stress and also the lack of knowledge of combined effects of stressors. Even if we did have a reasonable comprehension of effects of stressors we would still have problems of determining what are reasonable or acceptable stresses and strains which we can expect workers to accept. It can be argued that the worker will, given the right conditions and freedom, select his own optimum level of work but this is still uncertain. Preference levels are probably different from upper and lower acceptable limits and again we know little about inter- and intra-individual differences.

To summarize, it seems that current knowledge has only reached the point where it is possible to state that most well tried methods are not adequate alone. They are too simple, each only reveals one facet of the total picture. To stress the main point once again, the state of man is essentially multidimensional and there are few measures which always correlate highly with other measures or with any one descriptor. Even when we reach the state of making reasonable inferences about descriptors there is the further problem of correlating these with the socio-economic standards which must be determined by the society in which the work is being done.

Section 2 - Techniques

Prologue

In this section, the papers review techniques which are available for obtaining data relevant to human performance and welfare in the man–machine system. We are concerned essentially with measurement, and so the familiar criteria of validity and reliability are appropriate for evaluating the techniques. First, validity: are the measurements related, in a predictive sense, to salient characteristics of the man–machine system? Second, reliability: how stable and repeatable are the measurements obtained by a given technique in a specific situation?

Validity seems to involve two separate aspects. First, as discussed in the prologue to the preceding section, there is the establishment of relationships between possible measuring techniques and the useful descriptors described in that section. The most convenient techniques do not always produce data which are very meaningful in this sense, and the ergonomist may find that his range of familiar measurements needs extending to cater for the important facets of modern man–machine systems. The second aspect of validity concerns the interpretation of the measurements: norms should be available, indicating the range of values expected in a given population. Such norms provide a frame of reference for evaluating observations from a particular study, and are essential to the development of a rigorous and comprehensive discipline.

Reliability of measurement is achieved by minimizing the influence of irrelevant internal and external variables on the measurement process. This implies clear definition of the concepts involved, and then much can be accomplished by increased standardization and refinement of the equipment and procedures used in investigations. The technological nature of ergonomics emphasizes the importance of field studies, and it is these extra-laboratory investigations which tend to produce the less reliable measurements. In general it is true that the mechanical and electrical design of data acquisition and recording equipment is very advanced, but, as we shall see, there are some areas of difficulty still to be overcome, particularly in relation to field-work. The efficient analysis of data may also be difficult, and this applies equally to laboratory and field research in certain areas. Standardization and control of procedures is, of course, much more easily achieved within the laboratory, and the conflict between this desirable aspect of laboratory research and the essential 'realism' of field-work has always been of concern to ergonomists.

The relevance and utility of continuing the distinction between 'physiological' and 'psychological' approaches have already been questioned at several points in this book. It may be convenient in many respects to see the physiological techniques as related to extracted measurements, and the psychological measures as concerning performance. However, the electro-physiological measures, which we shall find discussed at length in the following papers, may well provide a significant common ground for the previously separate interests of the two academic disciplines. Again, any argument that a physiology/psychology dichotomy corresponds to a well-being/efficiency distinction in terms of basic criteria is rather naive; in advanced systems, observation and interviews may be used to evaluate operator strain, and evoked cortical potentials may provide a more sensitive measure of efficiency.

The selection of papers in this section attempts such a review of most of the useful techniques already in use or under development. Wisner provides a broad review of the electrophysiological measures for mental tasks, and this sets the scene for our primary concern with mental work. The evoked potential techniques are discussed in more detail by Groll-Knapp and by Defayolle and his colleagues. The former provides an introduction to the field, and the latter give two striking examples of the flexibility and sensitivity of the techniques. Turning to physical work, Bonjer provides a comprehensive discussion of the well-established procedures of work physiology—perhaps the most reliable of all ergonomics techniques.

The next two papers deal with two specialized techniques which have been subject to much research. Kalsbeek discusses sinus arrhythmia, and its relationships with dual task methods. Rey reviews the extensive research on flicker fusion, and sets out some cautionary guides for its use in practice. There is then a paper by Borg, which attempts to draw together our two underlying disciplines in a discussion of the 'psychophysiology' of heavy physical work.

The paper by Edwards attempts a comprehensive classification of all the measurement techniques discussed in this section, and this may help to advance the integration of physiological and psychological approaches. Rolfe provides an extensive review of the secondary task technique: its theoretical foundation, a variety of applications, and some difficulties.

The last two papers are primarily concerned with large-scale man–machine systems. Rabideau reviews the techniques and problems of observation of system personnel, particularly under field conditions. Chiles underlines the need for studies involving more realistic tasks, by emphasizing the challenge of complex operator performance typical of advanced systems.

Professor Wisner has physiology and psychology
degrees. He has worked in research institutes and with
the Renault automobile company, as head of the
laboratory of applied physiology and biomechanics. He is
now Professor of Work Physiology and Ergonomics at
the Conservatoire National des Arts et Metiers, Paris.

Electrophysiological measures for tasks of low energy expenditure

A. Wisner

1. Introduction

It has been suggested that we should be able to measure mental activity with the same technique used for muscular activity, namely oxygen consumption. However, this possibility seems remote at present.

Having no good measure of the energy aspect of brain activity, researchers have recorded and evaluated the electrical activity of the brain and some physiological functions which are more or less faithful images of central activity. In early studies technical difficulties restricted interest to certain variables and experimental conditions, and the resulting interpretations are now of little importance. Many people now prefer to measure both in laboratory experiments and in the field and the classification of work situations in terms of a single scale of arousal is no longer accepted. As well as estimating the cost of high or low extremes of activation, we must also determine the orientation of attention towards different aspects of the work situation. Only thus can physiological measures be of any use in the study of man–machine systems.

2. Measurements in the work situation

Ever since Ferre measured the variations of skin resistance for the first time in 1886, many research workers have tried to use changes of the biological state as measures of mental work. The research has been limited because of equipment difficulties which have permitted measures to be taken only before and after work. Further, the tasks were often extremely short, over-simplified and thus irrelevant to real work situations.

2.1. *Measurements before and after work*

Techniques of measurement before and after work are currently used in industry when the physical load (e.g. muscular work or heat exposure) is important and persistent enough to have a significant effect on such parameters as heart rate or body temperature. Even when the task is interrupted, the effects persist, and so significant comparisons can be made with the levels at rest. (See, for example, Brouha 1961, Lundgren 1959.)

Where mental load is being considered, such measures have very limited relevance. The very fact of changing the task seriously affects brain activity, which is more sensitive to a positive or negative change than to the actual level of stimulation (Bloch 1966). Although Duffy (1962) asserted that 'a review of studies of a number of physiological functions has demonstrated that measures of these functions show, in most instances, consistent variations with changes in the apparent demands of the stimulus situation', we must qualify this statement. The *apparent demands* are not only related to the characteristics of the tasks but also to the way they are perceived and accepted. This is true both for the work task and for the task chosen as the test to be undergone before and after work. *Changes in the stimulus situation* also cause modifications in arousal, affecting different physiological variables. It is not surprising, in these circumstances, that the results of studies where we compare the measures taken before and after work are rather disappointing.

The flicker fusion frequency test (F.F.F.) has however given some interesting results (Grandjean 1967), but this is usually with intense and extended tasks. Rey and Rey (1965) showed the prevailing effect of the visual characteristics of work on F.F.F. changes.

We must also stress the findings of Lille *et al.* (1968) which show the persistence of amplitude decrease of evoked cortical potentials (E.C.P.) for some minutes after the execution of an intense but brief mental task (Figure 1). These results can be compared with industrial observations by Manigault *et al.* (1969) who report the disappearance of α waves from people working under heavy mental load (dispatching, precise settings, and so on) during short pauses (6 to 7 minutes); and their reappearance only during longer pauses (15 minutes).

Figure 1. Visual evoked potentials obtained from the same subject during and after three mental auditory tasks of different intensity. Results obtained within the same experimental session (after Lille *et al.* 1968).

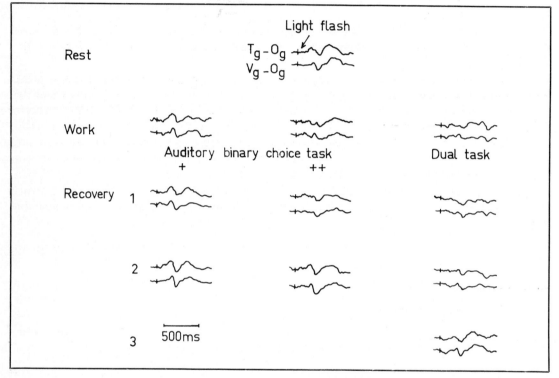

Angiboust *et al.* (1966) estimated the alertness level of strategic reserve pilots by testing their E.E.G., not in the actual stand-by situation, but by leaving them resting in the dark for some minutes.

2.2. *Measurements in the real work situation*

The highly controlled laboratory study will continue to be necessary for establishing fundamental laws, but the essence of ergonomics is to relate clear but perhaps partial results from the laboratory to the rich but complicated observations of the real situation. The works of Gantchev, Haider, Kalsbeek, Pottier, Pternitis and Tarrière illustrate this approach very well.

2.2.1. *Measurement techniques*

Improvements in equipment have ensured that recordings in the real work situation, even for lengthy studies, are quite common.

The first advances were in pick-up devices. They are now easily tolerated for a long time: 1–2 days for E.E.G., 10–20 days for E.K.G. and rectum thermistance with no deterioration of the contact between the electrodes and the skin.

Radio transmission is readily available. The subject carries only a small, light transmitter and he can stand at the very back of a shop, at the top of a building under construction, or on the moon.

Magnetic tape recording brings the advantages of permanent storage of large amounts of data, with high fidelity, using an easily portable apparatus. Even now, analysis is often best undertaken by direct visual investigation of waveforms, as with E.E.G. and arousal patterns. However, where appropriate, computer analysis is highly efficient, and it can allow on-line processing of observations as the subject performs.

2.2.2. *The difficulties of interpretation*

In spite of the technological revolution in measurement techniques, research is still limited by our methods of interpreting data. For instance, there is very little research describing the E.E.G. of the normal, active, alert man ; while there has been wide E.E.G. coverage of such situations as vigilance, sleep, resting in the dark and pathological conditions. So the results of E.E.G. measurements taken on the working subject are still difficult to interpret. We know very little of the effects of various situations, or of factors such as motivation and distraction.

Gastaut and Bert (1954) noted 'a large burst of alpha waves occurred with each relaxing of attention, as when, at the cinema, there were episodes arousing no special interest'. With such observations it is possible to begin to explore the orientation of attention, as will be seen later. The effect of distraction upon the physiological indices has also been noted; Bitterman and Soloway (1946): 'During work on a clerical test, it (the pulse rate) was higher during work than during rest, and higher when the individual was working under distraction than when he was working under conditions of quiet—this in spite of the fact that there was no decrease in the efficiency of the performance under distraction'. Such independence of performance and arousal measures forms the subject of studies about the physiological cost of mental work, to be discussed below.

From the published results, it cannot be concluded that all the physiological measures are equally easy to obtain and interpret in all work situations. For instance, surface E.M.G. is easily recordable, but can only be analysed during an experiment in which conditions remain very similar. Again, the recording of eye-movements is done using very complicated techniques and the analysis of the results is difficult or impossible except in precise, steady situations such as assembling electronic or fine mechanical devices, piloting aeroplanes or driving cars, where posture and direction of vision are constrained.

3. Arousal levels

3.1. *Activation theory*

The numerous attempts to relate the characteristics of a situation to the physiological response of the system are characterized by the work by Cannon (1915), showing clearly the energizing mobilization, and by Freeman's attempt to generalize the notion of 'Energetics of Human Behavior' (1948), related by Lindsley (1950) to the new discoveries of Moruzzi and Magoun (1949) about the physiology of the brain reticular formation. Duffy's book, *Activation and Behavior* (1962) provides a good background for our present discussion. Bloch's account in *Traite de Psychologie Experimentale* by Fraisse and Piaget (1966) has a more critical orientation towards this general theory.

In Duffy (*op. cit.*) are found the fundamental elements of activation theory: 'the level of activation of the organism may be defined, then, as the extent of release of potential energy, stored in the tissues of the organism, as this is shown in activity or response. . . .

'When the intensity of behavior, or the degree of activation of the organism, is the subject of investigation, it is observed that a large number of measures of autonomic functioning, of skeletal–muscle functioning, and of cortical functioning vary with considerable consistency in one direction with increased stimulation of the organism and in the opposite direction with decreased stimulation of the organism. . . .

'These changes are not specific to sleep, or to "emotion", or to any other particular condition. On the contrary, they may be found, not only during sleep on the one hand, and intense excitement on the other hand, but also during such intermediate conditions as waking relaxation, work on easy tasks, or work on more difficult tasks. They apparently vary in a continuum. . .

'Changes in the level of activation may be produced by any type of stimulus, physical or symbolic. They may be brought about by drugs, by hormones, by the chemical products of fatigue, by simple sensory stimuli, or by complex situational stimuli, present, past, or anticipated'.

3.2. *Arousal in varied working situations*

The above quotations from Duffy help to clarify the notion of activation (or *arousal*), which has aided discussions in this area.

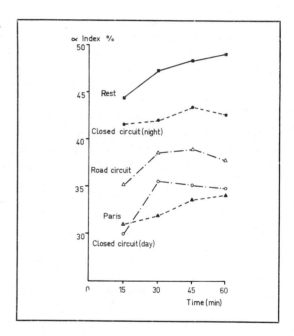

Figure 2. Evolution of α activity of the E.E.G. during an hour of experiment. Graphs of means for eight subjects at rest and in four different driving conditions. The results are expressed in relative value (%) of the α activity of the subject at rest with closed eyes (after Pin *et al.* 1969).

Some experiments show an influence of the work-load corresponding exactly to the theory of arousal. Bujas and Petz (1954) showed the quick decrease of the α index at the beginning of an addition session, which then remained low during an hour's work. Michaud *et al.* (1964) showed that, with monotonous car driving (on a closed circuit), heart rate and respiration frequency decreased progressively during the experiment (3 hours). In addition, the number of steering wheel movements was reduced. Pin *et al.* (1969) (Figure 2) showed that some varied physiological parameters (the α index of E.E.G., heart rate, instantaneous variations of heart rate and electrodermal reactions frequency) were located on very different levels according to the driving conditions. They then classified them by the degree of arousal in the following order: awake at rest in a stationary car, night driving on a closed circuit, road circuit, driving in Paris. They concluded with rules related to the discipline of car driving (Lecret and Pottier 1970). These studies can be compared with Hulbert's studies (1957) which established the correlation between frequency of psychogalvanic responses and density of car-traffic and Kondo's (1961) demonstration of the influence of urban traffic on arousal level.

Hartemann *et al.* (*op. cit.*) (Figure 3), recorded E.E.G. and heart rate during whole days of working at jobs of high mental load in the car industry. They established a complete and permanent disappearance of the α waves of the E.E.G. and a continuous increase of heart rate during the day. Although there is a marked influence of work load on physiological variables, there is no clear E.E.G. trend during the day as heart rate increases. Heart rate increases not only with intellectual load but also with thermal environment, digestive activity and stimulants (alcohol, coffee, tobacco). Nycthemeral variations of physiological activity level can also be evoked. However, the experiment of Pin *et al.* (*op. cit.*) shows in the light load situation that car driving for 6 hours causes a constant decrease of heart rate during the same part of the nycthemer as was noted in the observations of Hartemann *et al.*

Figure 3. Evolution of the heart rate and of the spectral density in the α band frequencies of the E.E.G. during a work session and during two rest sessions — female sewing machine operator in automobile industry (after Manigault *et al.* 1969).

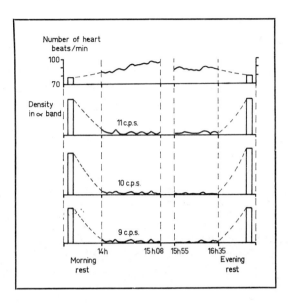

3.3. *Limitations of arousal theory*

A simple theory of arousal is made suspect by cases of divergence between two physiological variables. Bloch (1965) showed divergences between E.E.G. and E.M.G.; Parrot and Baumstimmler (1968) found few relationships between heart rate, rectal temperature and frequency of blinking (this last being highly correlated with performance).

Sleep studies, also, present a complex picture of trends in physiological parameters, at variance with a hypothetical arousal scale. During the paradoxical period E.E.G. is characteristic of a light sleep (first period), heart rate and respiratory frequency are high, eyes move, but the immobility of the sleeper is complete and the E.M.G. is almost absent (see particularly Kleitman 1967).

Thus, it is unwise to classify the varied states of human activity according to a continuum where intense intellectual work is near the state which follows an accident in which death could occur. We must always measure several parameters belonging to varied categories, for instance E.E.G., E.M.G. and heart rate. In many cases, divergence of the measures may be as useful as convergence. Also, any comparison task used in estimating mental load must be of the same nature. An important example of this is the suppression of the E.E.G. α rhythm; this is not related to general mental overload, but to the visual processes. Angiboust *et al.* report (personal communication) that difficult mental calculations or intense auditory tasks performed in the dark show much α rhythm.

4.   The estimation of mental workload

4.1. *Arousal and mental load*

A fundamental difference appears between the effects of mental load and the effects of physical load. Physical activity under light load is easy and is progressively impaired by increasing load. In contrast, for mental load there is an optimum. Task performance is related to arousal level, as expressed by physiological variables, by an inverted U function (Figure 4). The peak occurs at lower arousal levels when the task is difficult or new.

Figure 4. Relation between the results of experimental conditioning and the variations of vigilance caused by drugs on the rat. Performance is related to the level of excitation of the reticular activating system by an inverted U curve (after Cardo 1961).

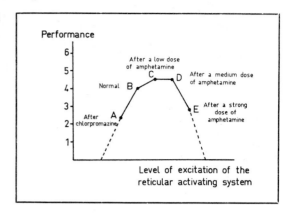

Figure 5 presents an idealized version of the graph, emphasizing three phases. The left upward phase corresponds to low arousal levels, where increasing arousal is closely related to increasing performance. The right downward phase corresponds to the overload levels, with high arousal causing progressive deterioration of performance. The central phase is more important; it is a phase of steady performance, bounded on the left by some indications of vigilance decrease, and on the right by some signs of physiological overload. We suggest that our physiological estimation of mental load should be concentrated on these two areas, where performance is maintained but where there are physiological signs of instability. Thus, we can distinguish means of studying overload and underload respectively.

Figure 5. Schematic variations of the performance realized and of the physiological variables related to arousal in terms of the work load. A satisfactory performance level of the required task can occur with physiological signs of mental underload or overload.

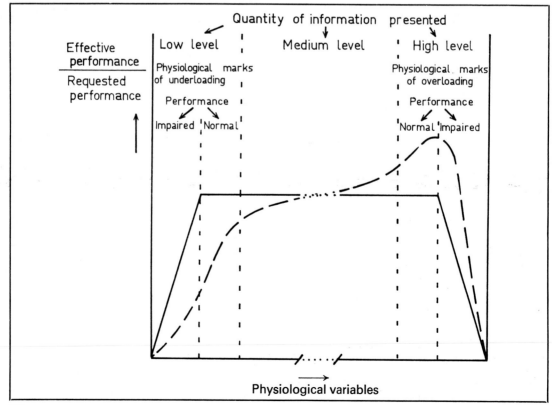

4.2. *The physiological signs of overload without deterioration of performance*

Studies show that alarm signals exist which can detect overload situations in industry: Haider and Popper (1964) studied heart rate among workers in the electronics industry; Manigault *et al.* (1969) measured E.E.G. and heart rate among the operators in the car industry.

Yet, it is not possible to estimate mental load if there are no results obtained in standard situations and some research must be undertaken in the laboratory and the results used in real situations. So Kalsbeek, after noticing that sinus arrhythmia decreases not only during physical but also mental strain, took care to adjust the variations of sinus arrhythmia on a task of variable and measurable difficulty; the response to binary choice at a paced rhythm (Kalsbeek and Ettema 1968, Kalsbeek and Sykes 1967). Auffret *et al.* (1967) showed in the real life task of landing an aircraft the difference in pilot load according to the use of one or another design of cockpit. Here, the measure of sinus arrhythmia is much superior to the single measure of heart rate (Williams *et al.* 1947, Ruffell-Smith 1968).

Laville's research (Laville 1968, Laville and Wisner 1966) is on the variations of physiological data during a precise task performed at high rate (Figure 6). He studied performance (quantity, quality and sometimes response-time), arousal level (heart rate), and visual attention—expressed through varied parameters (distance between eye and task, E.M.G. of the nape muscles and blinking frequency). E.M.G. is used not as a measure of arousal or dexterity, but as evidence of the rigid posture required for the task. This use of postural E.M.G. has become easier since the fundamental work on muscular activity during static work (Scherrer and Monod 1960).

Figure 6. Influence of the work load and of the duration of the experiment on varied physiological variables, performance remaining approximately constant. Average of two subjects over 24 experiments (distance between eye and task and number of blinkings), average of 13 subjects over 26 experiments (integrated E.M.G.) (after Laville 1968).

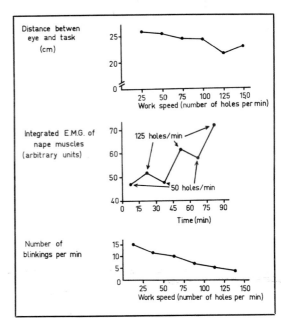

Over a two hours' experiment, we notice the steadiness of the performance, the decrease of arousal expressed by the decrease of heart rate, a slight reduction of the distance between eye and task and a very important increase of the electrical activity of the nape muscles which then varies inversely with the arousal level. (The increase in E.M.G. can be explained by the static work realized by the nape muscles under the weight of the head.)

As the speed of the task increases, the distance between eye and task becomes smaller, E.M.G. increases and frequency of blinking decreases. Heart rate is little affected.

In a second group of experiments (Figure 7), Laville and Teiger (1970) showed the effect of increasing the quantity of information for each element. The higher the quantity of information by time unit, the more the performance decreases, but the physiological signs of overload always appear before performance deteriorates. The utility of these overload measures is being assessed currently in the electronics industry.

Figure 7. Variations of the value of integrated E.M.G. in terms of the number of signals/min. and in terms of the complexity of the task. The task is responding
(1) in a regular and predictable order (ordered signals)
(2) in response to signals randomized only in their position (random signals)
(3) according to instructions taking into account the preceding signal (paired signals). Averages for 21 experiments on 3 subjects (after Laville and Teiger 1970).

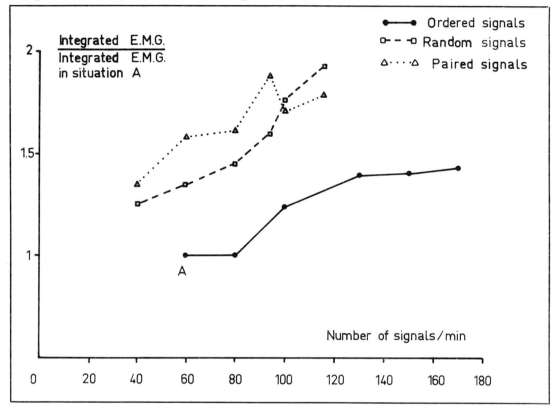

Gantchev *et al.* (1968) are working along the same lines by establishing a relation between heart rate and quantity of information treated and by then checking this in the workshop.

The laboratory research of Lille *et al.* (1969) shows the progressive influence of varied levels of mental work upon the amplitude of evoked cortical potentials (E.C.P.). In spite of great practical difficulties, this is a very important way of estimating high level mental loads. In work situations where the load is reasonable, however, fluctuations of attention do not permit a convenient use of E.C.P. (Pternitis 1969).

4.3. *The physiological signs of underload without deterioration of performance*

The monitoring of a system by a man is becoming more common and important. Car-driving is one example, and we know that 15–20 per cent of accidents occur on a straight road with no crossing or obstacle (Tarriere and Wisner 1960). There have been many attempts to relate 'classical' vigilance studies (Mackworth 1950) to such problems in process plant control rooms, but more recently Leplat (1968) has suggested a better classification of such work situations.

The study of concomitant physiological phenomena is also interesting. For Lindsley (1950) 'The α rhythm is a manifestation of a resting state, physiologically, an inattentive waking state and psychologically, a relaxed and sensory free peripheral state'. Pin *et al.* (*op. cit.*) showed that the less stimulating driving situations are accompanied by the most abundant E.E.G. α rhythms. Haider (1967) showed in the laboratory that signals in a vigilance task are more likely to be missed when the E.E.G., which is observed for 3 seconds before the signal, has a lower frequency. For the detected signals, the longer reaction times correspond to the lower frequencies of E.E.G. during the 3 seconds before the signal. Haider's experiments, like those of Tarriere and Hartemann (1968) and other authors, stress the highly variable receptivity state of the central nervous system, the periods of intense attention being expressed by a low probability of drowsiness figures on the E.E.G. and the relaxing periods by a high probability of such signs. In monotonous tasks where the density of response is quite high, a careful investigation of the performance is enough to detect unfavourable periods, but in the important cases considered (car-driving, monitoring in control rooms), the specific signals are infrequent and stereotyped behaviour is usual. Here, study of the usual actions of the operator cannot predict the adequacy of the response to an unexpected situation, whereas physiological measures may perhaps do this. Two technological devices can efficiently be compared from the point of view of the possible quality of monitoring. Defayolle uses the amplitude of auditory E.C.P. as a means of estimating the quality of the radar watch in a real situation; and he can then recommend one of the work time schedules proposed for the watchers (Defayolle 1969, Defayolle and Dinand 1968).

In the same way, Angiboust *et al.* (1966) use frequency analysis of the E.E.G.s to evaluate the long term effect of different sleeping periods on the quality of vigilance by strategic command pilots.

Some twenty years ago physiological 'alertness monitors' were proposed, using frequency analysis of the E.E.G. or forehead muscle E.M.G. (Kennedy and Travis 1947). However, in practice, important lapses of attention may be of very short duration indeed, and development of such devices has been discontinued.

5. Task orientation

As well as measuring arousal level, it is equally important to determine the direction of the operator's attention (Broadbent 1961, Bloch 1966). E.E.G. and eye-movement measures may be of use here.

5.1. *The spontaneous E.E.G.*

We have already noted above that α rhythms are copious in the E.E.G. of a subject who accomplishes a non-visual task, however difficult, yet they disappear if the work needs even moderate visual perception activity. So the study of occipital α rhythms is a first approach to the study of orientation towards a task.

5.2. *Evoked cortical potentials*

The investigation of E.C.P. brings much more precise results. Many authors show that the amplitude and sometimes the latency of E.C.P. change considerably with the attention given to the signal, the cortical response to which is recorded (Wilkinson 1967). The response to the A sensory modality diminishes when a task using the B sensory modality is made more difficult (Lille *et al.* 1967).

We also find this reduction of the cortical evoked response for the same sensory modality if the subject has to distinguish two categories of similar signals, where the instructions lead him to become interested in some signals rather than others. So high amplitude E.C.P. for low intensity but meaningful signals can be observed, as can low amplitude evoked responses to non-significant signals even if their intensity is high (White 1968).

When the subject does not understand the instructions, the wanted signals in a complicated set do not cause any E.C.P. No E.C.P. are observed among mental defectives when conditioning does not occur (Lelord *et al.* 1969).

These few examples are enough to show the very great importance of E.C.P. for the study of attention orientation. As White (1968) says: 'they seem to be connected with the activities of information processing in the brain'. Unfortunately, current technical difficulties exclude E.C.P. studies from practical situations.

5.3. *Measurement of eye movements*

Eye movement research has been carried out in laboratories for some sixty years. Our knowledge of the typical saccadic movement is fairly complete, and the *organization* of visual search has received much attention. The subject appears to base his visual operations on a model which is made up of information from numerous sources. Yarbus (1967) showed that the same painted or photographed scenery was differently scanned according to the information the subject was looking for: meaning of the scenery, age of the participants, characteristics of the furniture, and so on. Free scanning of a non-structured space is also done according to a strategy which leaves out some areas (the middle and edges) unless strict instructions are given (Ford *et al.* 1959). These observations have been made in the laboratory, with the visual material and the head fixed, and with measurement of eye movement by E.O.G. With similar arrangements, Gabersek (1962) showed that subjects' educational level and familiarity with the text influenced reading performance (Figures 8 and 9).

Figure 8. Scopogramm (E.O.G. recording) of the same line of a French text for subjects of varied educational levels. The fixation spans increase with educational level while the number of verifications, the reading time for a line and the duration of fixations decrease. The illiterate subject does not take enough information to allow any reading (after Gabersek 1962).

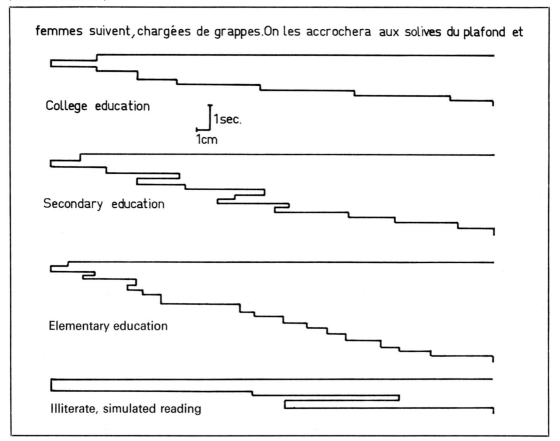

The recording of eye movements during a real work situation is now possible but requires much technical work: either by E.O.G. measures of horizontal and vertical components of ocular electric field and spotting the position by means of an XY recorder (Shackel 1966); or by collecting a corneal reflected light-ray. (The light source is attached to a helmet worn by the subject, to allow for head movement); and simultaneous and superimposed recording of the explored area and of eye movements (Mackworth and Mackworth 1958). For this purpose the ciné camera must be fixed upon the head which has the disadvantage of weight, or it must be connected by glass-fibre optics to a much lighter objective fixed on the forehead.

Figure 9. Scopogramm (E.O.G.) recording of the same line of a French text for the same subject during first and second readings. The increased familiarity with the text has an effect similar to the higher educational level (see Figure 8) (after Gabersek 1962).

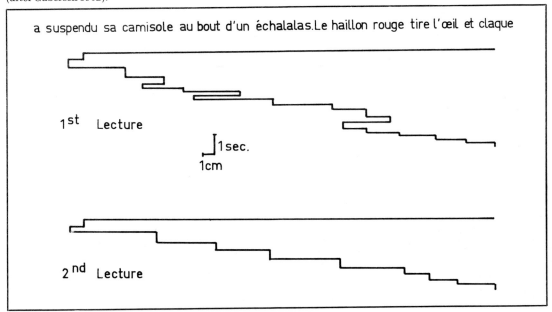

So researchers now have efficient and handy apparatus which allows them to understand visual scanning, to explain the mistakes and the omissions, to understand training processes and to prepare more efficient devices. There is the classical study of Jones *et al.* (1949) on the optimal layout of aircraft instrument panels, the work of Ford and White (1959) on improving aim by showing the subject the difference between the point to aim for and the point he looks at. Thomas (1968) reviews a succession of interesting examples: study of landscape scanning during car driving, of radioscopic investigation by radiologists and students, of inspecting pictures by normal and ill patients.

In spite of the technical achievements, only limited use of eye movement techniques has been made by researchers. Perhaps the technique is *too* powerful, providing more information than we can possibly interpret.

6. Conclusion

It is obviously naive to try to reduce the extraordinary complexity of relations between man and machine to a pure physiological phenomenology. All the psychological exploratory techniques available are also necessary. Nevertheless, working difficulties of man–machine systems are sometimes revealed earlier by an impairment of the operator than by a change in the behaviour of the system. Moreover, physiological measures encourage us to consider more particularly the overloading of the human operator and the effects on his health. Finally, if we are looking for the cause of trouble in the functioning of the system, we cannot neglect the information given by man's perceptual and motor links. The careful observation of an exploratory modality, or of execution difficulties might explain and solve previously intractable problems.

All of these considerations encourage close collaboration in the future between psychologists and neurophysiologists in the study of man–machine systems.

References

ANGIBOUST R., GALBAN P., GOUARS M., et VEDEL R., 1966, Evolution des aspects E.E.G. au cours d'une tâche de surveillance visuelle effectuée de nuit. *Revue de Medecine Aeronautique*, **5**, 13–18.

AUFFRET R., SERIS H., BERTHOZ A., et FATRAS B., 1967, Evaluation d'une tâche perceptivep ar la mesure de la variabilité du rythme cardiaque. Application à une tâche de pilotage. *Travail Humain*, **30**, 309–310.

BITTERMAN M.E., and SOLOWAY E., 1946, The relation between frequency of blinking and effort expended in mental work. *Journal of Experimental Psychology*, **36**, 134–136.

BLOCH V., 1965, *Le contrôle central de l'activité électrodermale* (Paris: Masson).

BLOCH V., 1966, Les niveaux de vigilance et l'attention. In FRAISSE P., et PIAGET J., *Traité de Psychologie expérimentale III* (Paris: Presses Universitaires de France), pp. 79–122.

BROADBENT D.E., 1961, Human arousal and efficiency in performing vigilance tasks. *Discovery*, 314–318.

BROUHA L., 1961, *Physiology in Industry* (London: Pergamon).

BUJAS Z., et PETZ B., 1954, Les modifications des ondes alpha au cours du travail mental prolongé. *Travail Humain*, **17**, 201–206.

CANNON J., 1936, *Bodily Changes in Pain, Hunger, Fear and Rage* (New York: Appleton).

CARDO J., 1961, *Rapports entre le Niveau de Vigilance et le Conditionnement chez l'Animal* (Paris: Masson).

DAVIS R.C., 1956, Electromyographic factors in aircraft control. The relation of muscular tension to performance. *U.S.A.F. School of Aviation Medicine Report 55.122.*

DEFAYOLLE M., 1969, Application de la méthode des potentiels évoqués aux tâches de veille. *Centre de Recherches du Service de Sante des Armees. Division de Psychologie, Lyon.*

DEFAYOLLE M., et DINAND J.P., 1968, Potentiels évoqués auditifs au cours d'une tâche de vigilance. *Comptes Rendus Societe Biologie*, **162**, 709–714.

DUFFY E., 1962, *Activation and Behavior* (New York: Wiley).

FERRE C., 1900, *Sensation et Mouvement: Etudes Expérimentales de Psychomécanique* (Paris: Germer-Bailliere).

FORD A., and WHITE C.T., 1959, The effectiveness of the eye as a servocontrol mechanism. *U.S. Navy Electronics Laboratory Report 634.*

FORD A., WHITE C.T., and LICHTENSTEIN M., 1959, Analysis of eye movements during free search. *Journal of the Optical Society of America*, **49**, 287–292.

FREEMAN C.L., 1948, *The Energetics of Human Behavior* (Ithaca: Cornell University Press).

GABERSEK V., 1962, Applications possibles de l'électro-oculographie en ophtalmologie. *Annales d'Oculistique*, **45**, 298–335.

GANTCHEV G., DANEV S., et KOITCHEVA V., 1968, La corrélation entre la fréquence cardiaque et la quantité de l'information. *Actes du Troisieme Congres de la Societe d'Ergonomie de Langue Francaise* (Bruxelles: Presses Universitaires de Bruxelles), 352–356.

GASTAUT H., and BERT J., 1954, E.E.G. changes during cinematographic presentation. *E.E.G. ClinicalNeurophysiology*, **6**, 433–444.

GRANDJEAN E., 1967, *Physiologische Arbeits Gestaltung* (Thun: Ott), pp. 119–122.

HAIDER M., 1963, Experimentelle Untersuchangen über Daueraufmerksamkeit und cerebrale Vigilanz bei einförmigen Tätigkeiten. *Zeitschrift für Experimentelle und Angewandte Physiologie*, **10**, 1–18.

HAIDER M., 1967, Vigilance, attention, expectation and cortical evoked potentials. In *Attention and Performance* (Edited by A.F. SANDERS) (Amsterdam: North-Holland Publishing Company), pp. 246–252.

HAIDER M., und POPPER L., 1964, Arbeitsbeanspruchung im modernen Betrieb. *Arbeitsgemeinschaft zum Studium von Arbeits Belastungen, Vienne.*

HARTEMANN F., MANIGAULT B., and TARRIERE C., 1970, An endeavour for evaluating the nervous load at working stations in line production. *IVth Bedford Group Meeting Report.* In press.

HULBERT S.F., 1957, Drivers C.S.R. in traffic. *Perceptual and Motor Skills*, **7**, 305–315.

JAVAL E., 1905, *Physiologie de la Lecture et de l'Écriture* (Paris: Alcan).

JONES R.E., MILTON J.L., and FITTS P.M., 1949, Eye fixations of aircraft pilots: IV. Frequency, duration and sequence of fixations during routine instrument flight. *U.S.A.F., A.F. Technical Report 5.975.*

KALSBEEK J.W.H., et ETTEMA J.H., 1968, L'arythmie sinusale comme mesure de la charge mentale. *Actes du III Congres de la Societe d'Ergonomie de Langue Francaise* (Bruxelles: Presses Universitaires de Bruxelles).

KALSBEEK J.W.H., and SYKES R.N., 1967, Objective measurement of mental load. In *Attention and Performance* (Edited by A.F. SANDERS) (Amsterdam: North-Holland Publishing Company).

KENNEDY J.L., and TRAVIS R.C., 1947, Prediction of speed of performance by muscle action potentials. *Science*, **105**, 410–411.

KLEITMAN N., 1967, *Sleep and Wakefulness* (Chicago: University of Chicago Press).

KONDO T., 1961, Studies on physical and mental reactions of motor drivers to the change of traffic conditions. *Journal of Sciences of Labour*, **37**, 195–210.

LAVILLE A., 1968, Cadences de travail et posture. *Travail Humain*, **31**, 73–94.

LAVILLE A., et TEIGER C., 1970, Performance et variables psychophysiologiques dans diverses tâches répétitives. In press for *Travail Humain*, **33.**

LAVILLE A., et WISNER A., 1966, Effets physiologiques du travail précis et rapide. *Actes du XV Congres International de Medecine du Travail*, A IV 3.

LAVOISIER A.L., 1865, Mémoires et rapports sur divers sujets de chimie et de physique pures ou appliquées à l'histoire naturelle générale et à l'hygiène publique. *III, Imprimerie Nationale.*

LECRET F., et POTTIER M., 1970, Effet des pauses sur l'activité de conduite de longue durée. In press for *Travail Humain*, **33.**

LELORD G., GOUSSET A., et HEININ-RIBEYROLLES D., 1969, *L'Intérêt de l'Étude du Conditionnement des Activités Évoquées en Pathologie Mentale* (Paris: Masson).

LEPLAT J., 1968, *Attention et Incertitude dans les Travaux de Surveillance et d'Inspection* (Paris: Dunod).

LILLE F., POTTIER M., et SCHERRER J., 1968, Influence chez l'homme des niveaux d'activité mentale sur les potentiels évoqués. *Revue Neurologique*, **118**, 476–480.

LINDSLEY D.B., 1950, Emotion and the electroencephalogramm. In *Feelings and Emotion* (Edited by M.I. REYMERT) (New York: McGraw-Hill), pp. 238–246.

LUNDGREN N., 1959, The practical use of physiological research methods in work study. *Forest Research Institute of Sweden. Department of Operational Efficiency*, Rep. 6.

MACKWORTH N.H., 1950, Research on the measurement of human performance. *Medical Research Council Special Report* 268 (London: H.M.S.O.).

MACKWORTH J.F., and MACKWORTH N.H., 1958, Eye fixations recorded on changing visual scenes by the television eye marker. *Journal of the Optical Society of America*, **48**, 439–445.

MANIGAULT B., VALENTIN M., et TARRIERE C., 1969, L'évaluation de la charge nerveuse. *Conditions de Travail, Regie Renault*, **8**, 5–23.

MICHAUD G., POTTIER M., ROCHE M., et WISNER A., 1964, Etude psychophysiologique de la conduite automobile. *Travail Humain*, **34**, 193–219.

MORUZZI C., and MAGOUN H.W., 1949, Brain stem reticular formation and activation of the E.E.G. *E.E.G., Clinical Neurophysiology*, **1**, 455–473.

PAILLARD J., 1966, L'utilisation des indices physiologiques en psychologie. In FRAISSE P., et PIAGET J., *Traité de Psychologie Expérimentale III* (Paris: Presses Universitaires de France), pp. 1–78.

PARROT J., et BAUMSTIMMLER Y., 1968, Evolution du clignement palpébral au cours de l'exécution prolongée d'une tâche d'attention visuelle soutenue. *Actes du III Congres de la Societe d'Ergonomie de Langue Francaise* (Bruxelles: Presses Universitaires de Bruxelles).

PIN M.C., LECRET F., et POTTIER M., 1969, Les niveaux d'activation lors de différentes situations de conduite. *Bulletin de l'Organisme National du de Securite Routiere*, **19**, 1–11.

PTERNITIS C., 1969, Etude des modifications du potentiel évoqué visuel moyen au cours d'une tâche expérimentale de surveillance et après un travail industriel de surveillance de 8 heures. *Centre d'Etudes et de Recherches Minieres. Mazingarbe (59–France)*.

REY P., and REY J.P., 1965, Effect of an intermittent light stimulation of the critical fusion frequency. *Ergonomics*, **8**, 173–180.

RUFFELL-SMITH P., 1968, Personal communication.

SHACKEL B., 1960, Electro-oculography: the electrical recording of eye position. *Proceedings of the Third International Conference on Medical Electronics*, pp. 322–335.

SCHERRER J., et MONOD H., 1960, Le travail musculaire local et la fatigue chez l'homme. *Journal de Physiologie*, **52**, 419–501.

TARRIERE C., et HARTEMANN F., 1968, Importance du niveau de vigilance physiologique dans l'exécution d'une tâche de surveillance. *Travail Humain*, **31**, 125–156.

TARRIERE C., et WISNER A., 1960, L'épreuve de vigilance. *Psychologie Francaise*, **5**, 261–283.

THOMAS E.L., 1968, Movements of the eye. *Scientific American*, **219**, 88–95.

WHITE C.T., 1968, Some aspects of evoked cortical responses. *H.F.S. Bulletin*, **11**, 1–3.

WILKINSON R.T., 1967, Evoked response and reaction time. In *Attention and Performance* (Edited by A.F. SANDERS) (Amsterdam: North-Holland Publishing Company), pp. 235–245.

WILLIAMS A.C., MACMILLAN J.W., and JENKINS J.G., 1947, Preliminary experimental investigations of tension as a determinant of performance, in flight training. *C.A.A. Division of Research Report* No. 54.

YARBUS A.L., 1967, *Eye Movements and Vision* (Plenum Press).

Dr. Groll-Knapp is a psychologist, with special interest in neuropsychology. Since 1960, she has researched into problems of attention and mental fatigue, in the department of Environmental and Social Hygiene, Hygiene Institute, University of Vienna.

Evoked potentials and behaviour

E. Groll-Knapp

1.   Introduction

The electrical activity that can be recorded from the surface of the human head, the so-called electroencephalogram (EEG) has been used, apart from its clinical application, in psychophysiological studies. This ongoing or spontaneous rhythmic activity of the brain may be used among other things as an indicator of the arousal state of the organism but at the same time it masks the more specific brain responses. Such EEG responses evoked by different stimuli constitute an important group of phenomena for the neurophysiologist as well as for the psychologist. The method of acquisition of such so-called evoked potentials and our knowledge about their interpretation have reached a state of development where it seems justified to add this method to the other physiological methods which are used to find better strategies for man–machine design. In this paper I will briefly summarize the common techniques and then give some examples of application in vigilance studies, attention and expectation studies; studies on basic time relations in sensory motor interactions and investigations on communication problems especially concerning the gnostic discrimination ability.

2.   Methods

Evoked potentials are faint signals with an amplitude about ten times lower than the ongoing background EEG activity. However they are of a fairly constant pattern and their occurrence is time related to the stimuli. These properties may be exploited by various techniques to improve the signal to noise ratio.

2.1. *Superimposition technique*

The simplest way of increasing the signal to noise ratio involves the superimposition of several records. This method was first described by Dawson (1947). For example, EEG signals may be made to produce vertical deflections of an oscilloscope beam. The beam itself is triggered by each stimulus and repeatedly sweeps horizontally across the photographic plate. This results in a composite record with a central modal pattern where individual traces have repeatedly coincided. This method is technically relatively simple. It gives the picture of an 'average response' in a not very precise way and the improvement of the signal to noise ratio may be inadequate for some purposes. On the other hand this method gives some impression about the variability of the phenomenon.

2.2. *Summation techniques*

Various digital computers of average transients are available commercially. The principal manner of increasing the signal to noise ratio is simply to have the computer add up a large number of responses. The summation of signals may be performed independently in several channels. The computer may be triggered directly by the stimuli in on-line programmes or the trigger may be recorded on a multichannel magnetic tape together with the brain potential changes picked up by several electrodes. In the averaging process only those potential changes that occur with a constant time relation to the trigger are summed up; those unrelated in time cancel out even though in each single record they may be of higher amplitude. The small special purpose computers give only a final output of the averaged potential and there is no information about other characteristics: for instance variability. For such purposes an analogue to digital conversion of the EEG has to be undertaken and then the more flexible general computer used. The slow potential changes which appear in expectancy situations or as motor readiness potentials before voluntary movements need some special methodology as for instance long time constant or d.c. amplifiers and non-polarizing electrodes.

3. Interpretation

Evoked potentials as recorded from the human scalp consist of a complex series of waves. Specific early 'primary components' (De Becker and Desmedt 1964) have only been demonstrated in very rare special cases with shocks to the peripheral nerves and special scalp electrode locations. In other situations for example with auditory stimuli the first 50 msec very often show artefacts possibly of muscular origin. The worker in this field has therefore to be especially careful in his interpretation. He should be very clear in the description of the stimulus he uses and also in the description of the components of the evoked potential on which he concentrates his analysis. The main components of the evoked potentials appear within the first two hundred msec. They seem to vary with a number of stimulus and organismic variables. After this main component some rhythmic after-discharges may appear. Under special conditions a slow late potential change may follow the evoked potential or may arise between two or more stimuli. This phenomenon depends very much on the experimental situation and has to be interpreted accordingly.

One advantage of the brain potential studies over other physiological methods seems to be that we are dealing with a central rather than a peripheral component. One might therefore expect a wide range of psychological variables to be reflected in brain potential changes. The evoked potentials and the slow potential changes may be influenced by several stimulus and organismic factors. Some of such factors are: sense modality, intensity of stimulus, rate of stimulus presentation, regularity of stimulus presentation, background electrical activity at the time of stimulation, arousal state of the organism, meaning of the stimulus for the organism, attention and expectation. Only a few of these variables will be treated in the following examples.

4. Some brain potential studies related to vigilance, attention, expectation and sensory motor interaction

One direction of research which I want to exemplify is the use of some appropriate response variables as an indicator of what the subject is doing and to combine them with simultaneous brain potential studies. Such an approach has proved useful in many situations in which, for example, reaction time responses or a discrimination task has been required of the subject and the speed or accuracy of response was used as some indication of variables like attention, vigilance or expectation (Davis 1967, Sutton *et al.* 1968).

One approach is to try to correlate vigilance performance and the amplitude of the evoked potential. In such a study (Haider *et al.* 1964) it could be shown that the amplitude and the latency of visual evoked responses correlated with fluctuations in attentiveness during a prolonged visual vigilance task. Non-signal flashes occurred at a rate of one every three seconds. Randomized signal-stimuli were approximately 0·2 log units dimmer and were presented ten times during each five minute period. As vigilance performance fluctuated and waned during the course of the task, amplitude and latency of the prominent peak of evoked potentials to non-signal stimuli showed corresponding variations. Amplitude was reduced and latency increased during periods of lowered signal detection efficiency (Figure 1). To focus upon a more specific aspect of attention, averaged evoked responses to signal stimuli were separately computed for signals which were correctly detected and for those which were not detected, thus contrasting attentive and non-attentive conditions (Figure 2).

Figure 1. Computer-averaged evoked potentials for 100 non-signal stimuli presented during successive 5-minute periods of a visual vigilance task, together with the percentages of a randomly interspersed signal stimuli correctly detected during the same time periods.
Recordings: occipital to vertex reference; negativity upward.

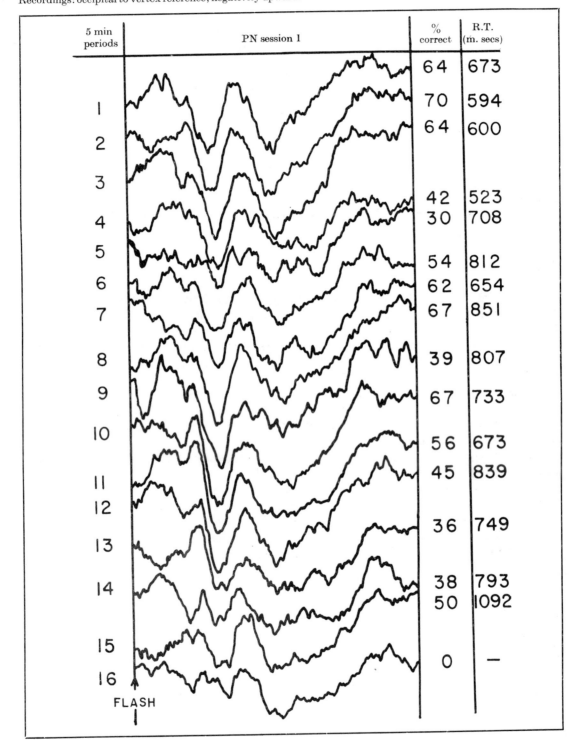

5 min periods	PN session 1	% correct	R.T. (m. secs)
1		64	673
		70	594
2		64	600
3			
4		42	523
		30	708
5			
6		54	812
7		62	654
		67	851
8		39	807
9		67	733
10		56	673
11		45	839
12			
13		36	749
14		38	793
		50	1092
15			
16		0	—

FLASH

The evoked potentials to signals which a subject failed to detect during an experiment were typically reduced in amplitude as compared with those of an equal number of signals which were correctly detected. These differences between detected and missed signals were equally

77

Figure 2. Computer-averaged evoked potentials for equal numbers of detected and missed signal stimuli. Recordings: occipital to vertex reference, negativity upwards, analysis time 500 msec.

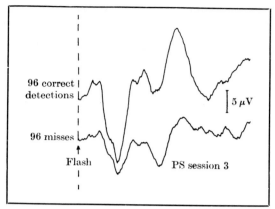

marked during early and late stages of the task, thus showing that the differences were not dependent upon the general decline of vigilance.

Obviously, vigilance tasks are of a monotonous and soporific kind. This may lead to long term fluctuations and reduction of cerebral vigilance or, more broadly speaking, of arousal or activation level as well as long term fluctuations and reduction of vigilance performance. These may be of a sleep-like nature or even really passing into sleep. In such instances we would expect that later components of evoked responses show more pronounced changes (Wilkinson 1967).

Besides such broad activation viewpoints, it may be argued, that expectancies about signal rate and signal occurrence may be important factors in vigilance performance. In one of such studies it could be shown that the evoked reactions to neutral stimuli immediately following the correctly detected signal stimuli are smaller than the evoked potentials to the neutral stimuli, preceding the signals (Figure 3). Such results are consistent with predictions from an expectancy theory viewpoint. In any case it seems to show that evoked responses could be used as a measure of 'observing behaviour' in continuous signal detection tasks, whereas performance criteria are only helpful in determining the actual 'detection behaviour'. Expectation, anticipation and motor readiness are essential to all kinds of motor learning and to skilled performance.

It can be shown that brain potentials evoked by unexpected stimuli show a late wave complex with peak latencies of 250–350 msec and relatively large amplitudes (Haider *et al.* 1968). These 'orienting potentials' are evoked by unexpected changes of the occurrence of stimuli in time as well as by unexpected changes of sense modalities. If the deviation of the time model of the stimulus is larger, the amplitude of the orienting potential is larger too (Figure 4). Orienting potentials in such situations are especially prominent in cases in which a motor reaction to the unexpected stimulus has to be performed.

Figure 3. Computer-averaged evoked potentials for equal numbers of signal stimuli after long and short intersignal-intervals and of missed signals during an acoustic vigilance task.
Recordings: vertex-mastoid reference, negativity upwards.

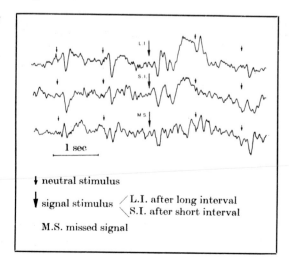

Figure 4. Computer-averaged evoked potentials for (*a*) regular clicks (*b*) unpredictable change of stimulus 200 msec after the beginning of analysis with this irregular signal (*c*) unpredictable change of stimulus 370 msec after the beginning of analysis with this irregular signal. Recordings: vertex-mastoid reference, negativity upwards, number of Summation—50.

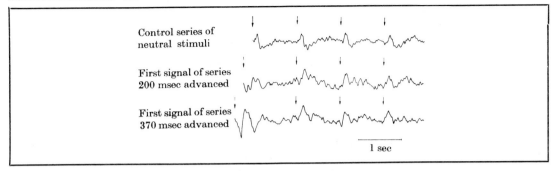

During the last few years much neurophysiological research has been concerned with slow brain potentials related to these processes. An 'expectancy wave' has been described between a conditional or warning stimulus and an indicative or imperative signal (Walter *et al.* 1964). The relationship between attentiveness, expectancy, contingent significance, operant response and the amplitude of this slow potential change suggests that it may be a kind of electrophysiological sign of cortical priming. A comparable phenomenon in the cortex preceding voluntary movements has been described as 'motor readiness potential' (Kornhuber and Deecke 1965), or in response to unexpected stimuli as 'orienting potential'. Together with the neurosurgeon Ganglberger (1969) we were able to demonstrate for the first time the thalamic involvement in the origin of expectancy and motor readiness potentials in man (Figure 5). The results fit well into a modern concept of central regulation of the motor system and its close sensory-motor interaction at different levels.

Figure 5. Computer-averaged evoked potentials for 30 combinations of two clicks followed by a motor reaction. First line: bipolar cortical derivation, second line: bipolar thalamic derivation, third line: time of motor reaction, negativity upwards.

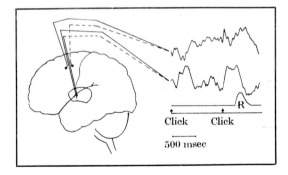

Looking to the future it may be that brain potential studies will help us to solve communication problems. The gnostic discrimination may be disturbed, for instance in noisy working conditions. Many psychological studies have dealt with communication difficulties under noise. In a recent study an attempt has been made to correlate electrophysiological data with the understanding of spoken words (Burian *et al.* 1969): nonsense syllables and meaningful words were differently reinforced by light flashes and separately analysed at two computer channels. By introducing an electronic remote control selector switch it was possible to use one set of electrodes and amplifiers for both computer channels thus reducing artefacts due to individual amplifier or electrode drifts. It was found that although the expectancy wave varies in amplitude and latency it could be used as a criterion for the understanding of the difference between both word groups. Below the 'gnostic threshold' no differences between the two situations occurred. In this fashion one could attempt to obtain a kind of 'objective speech discrimination threshold' (Figure 6).

Figure 6. 75 dB intensity above the gnostic threshold the subject is able to discriminate between meaningless and meaningful words. Expectancy wave only for meaningful words.
25 dB intensity below the gnostic threshold, there is no discrimination between meaningless and meaningful words. Expectancy wave occurs in both situations.

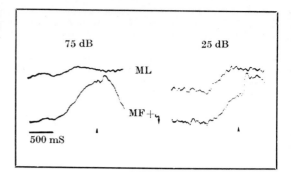

Brain potential studies in relation to psychological phenomena and problems are interesting and provocative. But the studies need rather complicated technical equipment, a thorough neurophysiological methodology and a precise and systematic experimental design. So far our knowledge in this fascinating field of investigation is incomplete and the practical applicability obviously restricted.

References

BURIAN K., GESTRING G.F., and HAIDER M., 1969, Objective speech audiometry. *Int. Audiology*, **8**, 378–390.

DAVIS R., 1967, Intermittency and selective attention. In *Attention and Performance* (Edited by A. F. SANDERS) (Amsterdam: North-Holland Publishing Company), pp. 57–63.

DAWSON G.D., 1947, A summation technique for the detection of small evoked potentials. *EEG clin. Neurophysiol.*, **6**, 65–84.

DeBECKER J., et DESMEDT J.E., 1964, Les potentiels évoqués cérébraux et les potentiels de nerf sensible chez l'homme. *Acta Neurol. Psychiat. Belg.*, **64**, 1212–1248.

GANGLBERGER J. A., GESTRING G. F., GROLL-KNAPP E., GUTTMANN G., und HAIDER M., 1969, Stereotaktische Hirnoperationen und neuropsychologische Forschung. *Bericht über den 26 Kongress der Deutschen Gesellschaft für Psychologie* (Göttingen: Hogrefe).

HAIDER M., SPONG P., and LINDSLEY O.B., 1964, Attention, vigilance and cortical evoked potentials in humans. *Science*, **145**, 180–182.

HAIDER M., GROLL-KNAPP E., und STUDYNKA G., 1968, Orientierungs- und Bereitschaftspotentiale bei unerwarteten Reizen. *Exp. Brain Research*, **5**, 45–54.

KORNHUBER H.H., und DEECKE L., 1965, Hirnpotentialänderungen bei Willkürbewegungen und passiven Bewegungen des Menschen, Bereitschaftspotential und reafferente Potentiale. *Pflügers Archiv.*, **284**, 1–17.

SUTTON S., TUETING P., ZUBIN J., and JOHN E.R., 1965, Information delivery and the sensory evoked potential. *Science*, **148**, 395–397.

WALTER W.G., COOPER R., ALDRIDGE U.J., McCALLUM W.C., and WINTER A.L., 1964, Contingent negative variation: an electric sign of sensorimotor association and expectancy in the human brain. *Nature*, **203**, 380–384.

WILKINSON R.T., 1967, Evoked response and reaction time. In *Attention and Performance* (Edited by A. F. SANDERS) (Amsterdam: North-Holland Publishing Company), pp. 235–245.

Dr. Defayolle has degrees in medicine and psychology. He is head of the Psychology Division of the Centre de Recherches du Service de Santé des Armées, Lyon. His major interest is the psychophysiology of individual differences. Dr. Dinand holds medical and psychology degrees, and Mlle. Gentil is a psychologist.

Averaged evoked potentials in relation to attitude, mental load and intelligence

M. Defayolle, J. P. Dinand and M. T. Gentil

1. Introduction

1.1. *The concept of mental load*

Quantitative assessment of modern industrial tasks requires measurement of the 'mental load' of the operator. However, a rigorous measure of information content of a task is feasible only in very simple situations, and even then this would refer only to the *objective* complexity of the task. The *subjective* complexity, which is the basis of mental load, depends on the operator's abilities and previous experience relevant to the situation, and, as such, shows large differences between individuals. Indeed, as with the more common approaches to physical work-load, we are forced to estimate mental load on the basis of indirect measurements.

Various measures of the state of arousal, such as variations in heart rate and rhythm, in skin conductance, and in E.E.G. activity, have been suggested as indicators of mental load. At the very least, we can argue that arousal level varies with the intensity of mental strain the subject accepts, and that it partly reflects the working of some informational processor. This is the *quantitative aspect* of mental load. But no less important to explore is the *qualitative aspect*, that is the nature of processed information. Processing of information, to be effective, implies that mental load is mainly devoted to pertinent information. The ability to filter out non-pertinent information is a necessary condition for an optimum use of the processor, working at a given level of intensity.

Thus, the notion of mental load seems to involve two components: on the one hand, the level of excitation of some hypothetical mental processor, on the other hand the selective value of the attention focused on the task. Both condition the efficiency of performance. Mental load corresponds with processing of information which we could represent by a vector, the length of which would be the quantity of 'mental energy' invested in the task, and the direction of which would correspond with the nature of the informational content. We assume that the energy invested in a task corresponds with the subjective complexity of the task.

1.2. *The method of dual tasks*

Such a model accounts well for the results obtained by the method of dual tasks, for indirect measurement of mental load. Two independent tasks (A and B), may be plotted on two orthogonal axes. Their respective values (a_1 and b_1) indicate the mental load involved in the achievement of each task separately. The vector $\vec{V}$, drawn from the origin, indicates the mental load involved in the two tasks performed simultaneously. The model is represented in Figure 1.

The projection of $\vec{V}$ on B, for a given complexity (a_1) of A, indicates the effective subjective complexity (b_1) of B. This is smaller as a increases and the angle ϕ becomes more acute. The length L of $\vec{V}$ corresponds with the notion of channel capacity. It is supposed to be constant for a short period of time. The angle ϕ corresponds with the direction of attention towards A or B. If one wants to assess the subjective complexity (a) involved in a non-quantifiable task A, it is possible to measure the output variable B, which is calculable.

Generally the task B will be a simple one, for ease of quantification (barring test, reaction time, detection time, calculation, tapping . . .). Thus, we assume that the task B has a small and stable complexity, and that the variations of its performance reflect the fluctuations of the mental load applied to the other task A. With these conditions, and with channel capacity constant during a short time period, each change in the subjective complexity of the main task A, will cause an opposite variation in the performance of the accessory task B. This vectorial model allows us not only to assess mental load, but also to account for the subject's attitude towards both tasks, when they can be regarded as common and constant. Then it is possible to distinguish explicitly induced attitudes, that is, deliberate attitudes (when the subject is instructed to focus attention on one task) from implicit or involuntarily attitudes which are produced by unconscious biases on the part of the subject.

Figure 1. Vectorial model of double task. Angle ϕ is always superior to 0°; the corresponding value to it is called residual.

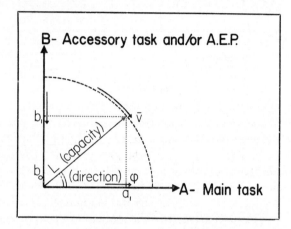

1.3. *Evoked potentials as an accessory task*

This vector model of mental load can be extended to similar situations where averaged evoked potentials (A.E.P.) are recorded. It is well known that evoked potentials have greater amplitude when the subject is presented with a single detection task, as compared with two simultaneous tasks. Referring again to Figure 1, if B is the inherent mental load in the stimulus information, and A is the mental load of the main task, the projection of the vector $\bar{V}$ on B shows the amplitude of A.E.P. as a function of the direction of attention towards A or B.

If such instructions are given to attend to task A, the vector $\bar{V}$ will tend to be orthogonal with regard to B, as the task A is increasing in complexity. The amplitude of A.E.P. will decrease, as has been found by experiment. So we assume that, on the one hand, the accepted mental load for A, and, on the other hand, the electrical evoked response for B, determine the angle ϕ. In other respects evoked potential may be considered as the response of a system. The system is in some state, and this influences its transmittance characteristics, among which L and ϕ are the most important components.

1.4. *Attitude and mental load*

Thus, the unit of our measuring technique depends on the angle ϕ, which itself is determined largely by the subject's attitude towards the stimulations and/or the accessory task. Kalsbeek has emphasized how the subject's attitude can influence the results—the subject may choose to ignore the accessory task, or combine the two tasks in one optimal strategy. Again, Broadbent (1958) has distinguished two separate aspects of a task—'load' and 'speed'. This is related to two modes of performing two separate tasks: if the sum of the complexities of the tasks is less than or equal to channel capacity at any given time, the tasks can be performed simultaneously, and otherwise attention must be shared between them (cf. 'time-sharing' of computers). With time-sharing, the problem is not only how, but also when, to achieve the accessory task without disturbing the main one: the efficient use of the channel depends on the movableness of 'programmes', that is, the capacity for fast-changing from one task to another. This adaptability of self-programming has been studied by psychotechnicians, who have tried to show up some factor of 'fluency' which could make different attitudes alternate quickly during problem solving. More recently, Welford, then Broadbent, assessed the lost time during the changing from one attention mode to another. This 'shifting' delay is about one or two seconds during which the channel is not available.

We shall present now two experiments in which we used evoked potentials as an indirect indicator for mental load and attitude induced by an accessory task. Firstly we shall examine the common technical data for both experiments.

2. Technical data

The left temporovertex derivation was chosen as giving the most utilizable activity. Silver electrodes were used, stuck to the scalp by collodion. Signals were pre-amplified up to about one volt, by an E.C.E.M. pre-amplifier, and then recorded on magnetic tape by an AMPEX SP 300. Recorded signals could be inspected 'on-line' by injecting them again into the E.C.E.M. post-amplifier and plotter. Evoked potentials were extracted by averaging with an analyser 'A.R.T. 1000' equipped with a digital output driving an I.B.M. 024 card puncher. Several parameters of evoked potentials were studied, but only two will be considered here, amplitudes A_1' and A_3' of the two main positive components (1 and 3) of the potential; each one was measured from the top of the negative component (2) between them (Figure 2).

Component 1 is at about 70 msec and component 3 at about 100 msec after the stimulus.

Figure 2. Definition of the two components of A.E.P.

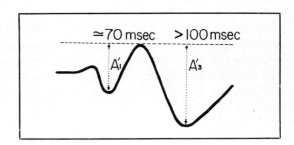

3. Experimental methods and results

3.1. *First experiment*

3.1.1. *Methodology*

This study dealt with the variations of auditory evoked potentials, as a function of the circadian rhythm, among subjects doing simultaneous tasks. Only the dual tasks effect will be considered here, the others having been published elsewhere. Ten subjects (young adult males) participated in the experiment. Each one was tested six times at different hours in the nycthemere, the hours being arranged in a latin square plan to avoid learning effects. Each watch lasted 90 minutes and it was shared between two-minute sequences during which the subjects had to achieve one or two tasks, according to one of the patterns: 00, 0+, +0 or ++. The first of the tasks (A) consisted of auditory watch keeping during which the subject had to detect a modulated sound emerging from background noise. The second task (B) was to detect breaks in a succession of clicks 1 ± 0.2 seconds apart, that is intervals greater than two seconds. The two tasks were each recorded on a separate track of a tape recorder. The performance in each task and the amplitude of auditory evoked potentials (A.E.P.) were treated statistically by analysis of variance, according to a plan with repeated measures (6 hours × 3 times on the task × 4 instructions).

3.1.2. *Results*

Effects of circadian rhythm have been reported elsewhere. We shall recall only that the amplitude of evoked potentials varied as the subjects' rhythm of vigilance: and consider mainly the effect of the deliberate ('explicit') or involuntary ('implicit') attitude of the subject towards the tasks.

Implicit attitude. This effect appears if we compute correlations between the performances in both tasks, for each subject. These correlations were revealed to be heterogeneous; subjects can be divided into two distinct groups, in each of which the correlations are homogeneous. In the first group (six subjects) the correlations between tasks 1 and 2 were positive. The whole combined correlation is $+ \cdot 366$. This positive relationship indicates a trend for succeeding or for failing simultaneously at both of the tasks. It seems to correspond with a diffuse attention. In the second group of four subjects, the correlations between tasks were negative, and their combination gave the whole correlation of $- \cdot 413$. It means that subjects of this group made a choice for one or the other task, so that success in one was coupled with failure in the other. We assume it corresponds to focused attention. Thus these two groups seem to be defined by their implicit attitude towards the dual task situation. Let us examine what happens for evoked potentials in each group. We sought some correlation between the amplitudes (A_1' and A_3') and the performances in each of the tasks (A or B). Recall that the accessory task B constituted at the same time stimulations for producing A.E.P. For the subjects with focused attention, the correlations were found as follows:

	A_1'	A_3'
Task A	$- \cdot 268$	$- \cdot 107$
Task B	$+ \cdot 296$	$+ \cdot 040$

The positive correlation of A_1' with task B shows clearly the increase of amplitude when the performance is good in the task B, i.e. when attention is directed towards the stimuli which produce potentials. Conversely we see the negative correlation with task A, the flattening when the results are good with the task A, that is, when attention is focused on another stimulation. By contrast, for the subjects with diffuse attention, the correlations have the same sign in both tasks:

	A_1'	A_3'
Task A	$+ \cdot 215$	$+ \cdot 234$
Task B	$+ \cdot 221$	$+ \cdot 169$

The evoked potentials of these non-focused subjects vary as their global efficiency, whatever the direction of attention may be.

Explicit attitude. Here we can easily distinguish A.E.P. as they are produced in a definite state of focused attention, according to one of the four possible instructions:

	Task A (Signal detection)	Task B (Breaks in clicks)
1	+	+
2	+	0
3	0	+
4	0	0

The normalized means for A_1' and A_3' in the four situations are plotted on Figure 3. Amplitudes of the A.E.P. are differentiated by the direction of attention: they are maximum in the situation $0+$, where the instruction is to take into account the stimulations only. They are minimum in the situation $+0$, where subjects had to detect the signal, neglecting the stimulations. The two other situations (00 and $++$) equally share attention towards the two tasks, at different levels: they do not differentiate clearly the amplitudes. It seems that an analysis based on a lattice model could show up this effect better. The four instructions can be represented as a two-dimensional graph (Figure 4).

Figure 3. Influence of focalization on amplitude of A.E.P.

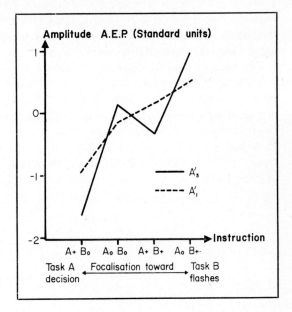

Figure 4. Lattice representation of double task.

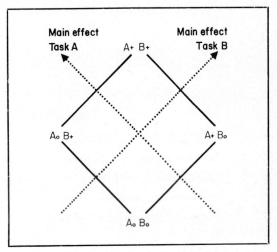

The main effects (in the analysis of variance sense) of the tasks A or B correspond to the two stippled median lines, but the four situations may be considered as two different types of partial order. The first one is represented by a projection on the vertical diagonal, of three points: $(++)(+0, 0+)(00)$. It corresponds with a dimension of task complexity. The other one is represented by a projection on the horizontal diagonal: $(+0)(++, 00)(0+)$ and corresponds with the direction of selective attention. These two dimensions may be subjected to analysis of variance, if we applied orthogonal coefficients such as to show off the wanted contrasts. The analysis of variance in these conditions reveals that both dimensions have a significant effect on amplitude, for both A_1' and A_3'. In particular, the direction of attention accounts for 90 per cent of the variance of amplitude.

3.2. *Second experiment*

The aim of this study was to determine if evoked potentials would reflect the mental load in a quantifiable manner according to the levels of task complexity.

3.2.1. *Methodology*

Sixteen young male adult subjects participated in the experiment. Previous psychological tests had evaluated their intellectual level and emotional stability. The experimental task included decision-making with three complexity levels, and an accessory task of flash-detection with two levels. The successive instructions gave priority to one or/and the other task, by a latin square plan for controlling possible order effects.

The task of decision-making consisted of eight series of problems. Each problem-solving implied a definite number of logical relations. Each series involved 14 slides shown on a 95×67 cm screen. Each slide (Figure 5) presented several forms of cars which differed from one another by some characteristics: M: the make: Citroën (M_1) or Renault (M_2) or Peugeot (M_3); C: the colour: black (C_1) or white (C_2); D: the direction: to the right (D_1) or to left (D_2); P: the power: low (P_1) or high (P_2). The characteristic of power distinguished only '2 C.V.' from 'D.S.' among Citroëns. Eight instructions combined these characteristics, using relations of intersection and equivalence in logical equations.

Figure 5. Task of decision making (example of item).

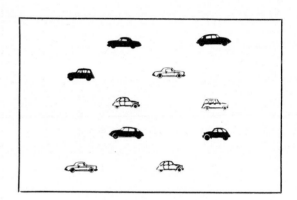

	Proposition	Equation
1	Three Citroëns are going to the right.	$M_1 \cap D_1 = 3$
2	There are as many Citroëns going to the left as D.S.	$M_1 \cap D_2 = M_1 \cap P_2$
3	There are as many little white cars as D.S.	$P_1 \cap C_2 = M_1 \cap P_2$
4	There are as many little black cars as Renaults.	$P_1 \cap C_1 = M_2$
5	There are three little white cars.	$P_1 \cap C_2 = 3$
6	There are an equal number of little white cars, of Citroëns going to the right and of Peugeots.	$P_1 \cap C_2 = M_1 \cap D_1 = M_3$
7	There are as many Citroëns going to the right as Renaults.	$M_1 \cap D_1 = M_2$
8	There are as many little cars going to the left as black Citroëns, or as Renaults.	$P_1 \cap D_2 = M_1 \cap C_1 = M_2$

There are two relations implied for the propositions 2 and 3, and three relations for 6 and 8. The four propositions implying only two relations are combined according to two other dimensions. The first one is the nature of the class of equivalence, which may be a cardinal number (1 and 5) or a qualitative attribute (4 and 7). The other dimension concerned the implication of Citroëns (1 and 7) or not (4 and 5). For each proposition, 14 successive slides were shown to the subject for his unpaced response, by pressing a green knob for 'true' or a red knob for 'false'. Each response (correct or not) made the next slide come on. So, the objective complexity of the task was well known. The subjective complexity was estimated by a psycho-metrical technique. For this purpose, 64 subjects took the test. In each series, the decision times and the number of errors were measured. These two variables are highly correlated ($r = 0.84$). Their balanced sum gives a synthetic value for the assessment of subjective complexity of each series. The order of subjective complexity follows the objective one, except series 7. Moreover, we see that comparisons with a cardinal number are easier than with an attribute. We distinguished three levels of objective complexity in the analysis of variance. They are (0) no problem to solve (slides are shown without instructions); (1) two logical relations (or less); (2) more than two logical relations.

The *accessory task* consisted in detecting red flashes among white flashes which produced Visual Evoked Potentials (V.E.P.). Intervals between flashes were from 850 to 1250 msec. The ratio of red flashes to white flashes was 10 per cent. The subject identified the red flashes vocally. The stimuli were presented in the peripheral field of vision, in the right temporal sector. The flash-lamp was soundless and delivered flashes of 0·2 joule at 50 cm from the subject's eyes.

3.2.2. *Results*

Three main sources of variation show a significant effect on the amplitudes A_1' and A_3' of evoked potentials.

Effect of the complexity of the decision task. The three levels of complexity cause significant differences between values of A_3'. Figure 6 shows the percentage of flattening of the A_3' component, plotted against complexity, for the eight propositions. It can be seen that the main contribution to the significant effect of complexity is from the differences between the lower levels of complexity. For the higher levels of complexity, the effect on the amplitude A_3' is constant, perhaps because many subjects relinquish the decision task when it becomes too hard.

Figure 6. Effect of subjective complexity on flattening of A.E.P. Numbers correspond to number of items. Position of task is arbitrarily defined.

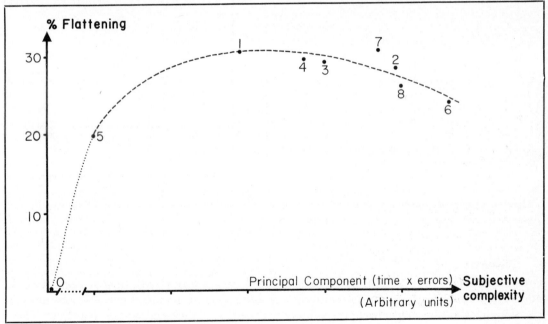

Effect of the subject's attitudes. As we did in the first experiment, we could verify that A_1' and still more A_3' increased when the instructions focused attention upon the flashes: the mean amplitude of the late positive wave then passes from 5·74 to 8·44 microvolts, and this difference is highly significant. Thus this effect of explicit attitude towards the task B is proved as before. Besides, as the experimental plan controlled the effect of task order, we could establish that, when an experiment began with a sequence where the subject did watch the flashes, evoked potentials were always greater, even for the whole experiment. Here too, the difference is highly significant: visual evoked potentials for subjects who began with watching flashes had mean amplitude of 8·14 microvolts. For subjects who began without watching flashes, mean amplitude was 5·74 microvolts. This suggests an implicit attitude connected with adhering to the first instruction. When the subjects were questioned after the experiment they reported no awareness of this attitude.

Effect of the subject's level of intelligence. No simple relationship is observed between subject's intelligence and amplitude of V.E.P. However, an interaction between intellectual level and task complexity shows a significant effect on amplitude. The explicit changes in orientation of attention make the evoked potential amplitude fluctuate more or less according to subjects. The degree of fluctuation can be quantified by computing a ratio as for assessment of mental load. This 'coefficient of evoked potential plasticity' is obtained by computing $(A_F - A_T)/\overline{A}$, where A_F, A_T and $\overline{A}$ represent the values of the amplitude A_3' in the situations 'watching flashes only' (for A_F), 'watching decision task only' (for A_T), and the mean value of A_3' for $\overline{A}$. This coefficient corresponds, in our lattice model, with the relative numerical value of the diagonal representative of the task organization. The rank correlation between the coefficient of plasticity and the intelligence quotient, estimated by a G factor test (D.48), is $\rho = \cdot 79$. Figure 7 shows this relationship.

Figure 7. Correlation between IQ and plasticity of V.E.P.

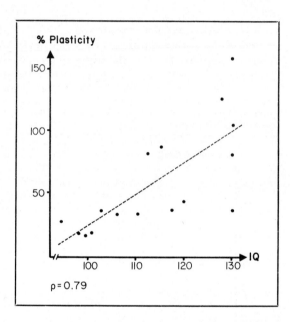

Other indices of plasticity may be used, such as the standard deviation of intra-subject measures, and these give analogous results.

4. Discussion

These experiments reveal first that the attitude explicitly induced by the instructions given to subjects plays a great part in determining the amplitude of averaged evoked potential. In addition, involuntary attitudes of the subject may either reinforce or oppose this effect. These effects might explain, partly, some divergences between conclusions of different researchers. The classical results are those of Haider (1967) and Sheatz (1969) which show an increase of amplitude when attention is focused on synchronizing signals; also Garcia-Austt (1964) observed a flattening of evoked potential when a distractive task was imposed on subjects. However, Satterfield (1965) and Spong (1965) find exceptions to these data. Callaway (1966) imputes most of these divergences to the ambiguity of the notion of attention, and emphasizes the need to distinguish between the active attention that implies an effective participation of the subject in decision tasks, and the passive attention, required for simple numbering tasks, for example. Groves and Eason (1969) also demonstrate that evoked potentials increase as the degree of attention is focused upon stimuli, only when the subject considers these as important ones. These conclusions accord with Sutton (1967), who suggests that the most important parameter is the information carried by the stimuli. According to that point of view, evoked potential would be an index for uncertainty or novelty as Pribram says. All these data are linked with the notion of individual attitude towards the proposed tasks. High motivation of the subject, great novelty in the task, can enhance the situation. A non-pertinent stimulus may become more important when the subject tries not to be troubled by it, rather than when he shows a simple passive attitude. Thus we have observed, in an unpublished experiment, that, when subjects were sleep-deprived, flash-evoked potentials increased though the main task was difficult enough. When we finally questioned the subjects, they said they found the stimuli distracting, and they had to strain to filter them out.

On the question of attitudes explicitly induced by instructions, we shall examine two main points: the notion of plasticity of these attitudes, and the possibility of using evoked potentials as an accessory task for the assessment of mental load. The relationship we found between plasticity of evoked potentials and G factor suggests a dynamic exploration of mental processes that correspond to the characteristics of intelligence. Also, the notion of fluency is another aspect of this plasticity. Indeed, mobility of attitude is a necessary condition for the achievement of intellectual tasks that necessitate several different successive strategies. Callaway (1969) proposes an interpretation which is at first sight opposite to ours: he thinks the most intelligent subjects have the most stable evoked potentials in respect of amplitude. As a matter of fact, both results are complementary. For us, intelligence goes with a great variability according to different situations, and for Callaway it goes with a great stability within one given situation. But in fact, speed of mobilization and strength of concentration are two conditions of mental efficiency. We propose as a hypothesis a combination of these two aspects as represented in Figure 8.

Figure 8. Hypothetical representation of combination of variability and stability, as an assessment of intelligence.

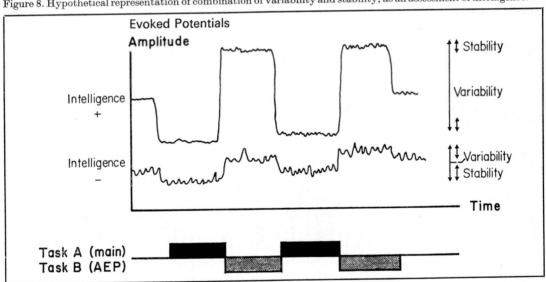

According to this assumption, the 'availability of attitudes' might be measured as a ratio between intra-situation and inter-situation variances. This index might be correlated with G factor and with fluency. We plan a future study with two samples of subjects, one with brain disorders, and the other normal.

The use of evoked potentials as a measure of mental load has been implicitly applied by Lehmann. Taking Welford's model, he interprets the flattening of A.E.P., during the achievement of a task, as a sign of saturation of the information channel. However, the measurement of mental load by evoked potentials seems to be restricted to small loads (as in the present experiment). Indeed we saw that the amplitude of the large positive late component ceased to diminish when the content of the problems exceeded two logical operators. Very likely this result is related to the unpaced presentation of slides, which leads the subjects to use continually their information channel, except during the changing of slides. Then the decrease of A.E.P. reflects the sum of elementary mental loads. It has long been accepted that the capacity of the information channel, and of the short term memory which is associated with it, is about seven objects or relations (Miller 1956). This broadly corresponds with the number we gave, assuming that the subject must keep in mind previous comparisons and treat one object, one relation of belonging, and one operator of verity (true or false). Within this elementary mental load, the vector model can be directly applied; the spare channel capacity is shared between problem solving and signal detection. Beyond this limit, the elementary mental load cannot grow and the whole complexity of situation must be divided between some sequences. Recently Trumbo (1969) has shown that a secondary task leads to a deterioration of a main one, in a linear way between 1 and 3 alternatives. Then the main performance stays constant for 4 or 5 alternatives. Thus, mental load has another dimension, the duration of channel occupation, which constitutes the third dimension of our model. This concerns the assigned time to this task or that. When the main task is achieved unpaced it occupies the channel constantly, which depends on the attention it demands according to the type of instructions. If these imply both tasks simultaneously, the share devoted to the main problem is reduced; less information is processed during each time-unit, and performance deteriorates. By contrast, when the main task demands a shorter time for completion, the problem is to take samples of information. The amplitude of potential will depend on the probability of its occurring during a rest-period, and also on the plasticity of the system for changing the orientation of its attention. Figure 9 shows a hypothetical representation of the effects of these various possibilities: (a) no task to achieve, (b) tasks with smaller complexity than the channel-capacity, (c) continuous task with equal complexity and (d) with greater complexity than the channel capacity, (e) non-continuous task. The last case raises again the problem of availability of attention which is a necessary condition for the achievement of a heterogeneous task-series.

Figure 9. Hypothetical representation of effect of mental load on amplitude of A.E.P. Thick line represents the occupation of channel by main task and arrows its mean values.

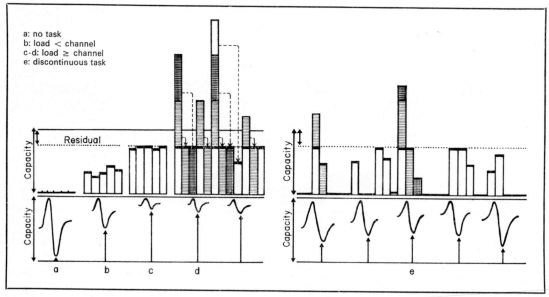

5. Conclusion

What conclusions do we have concerning the efficiency of the method of evoked potentials, applied to ergonomics research? We have seen its contribution to measuring mental load, and the abilities of operators. However, we would not suggest that it possesses an absolute objectivity although it is a physiological measure. Indeed, we saw that very subjective phenomena, such as attitudes, may greatly affect evoked potentials. Only the simultaneous study of performance, of subjective feelings, and of evoked potentials has been able to show this effect. Evoked potentials have the status of a performance measure in an accessory task, and therefore they are as difficult to interpret. The originality and the interest of the method consist in the fact that it enables us to assess the operator's transmittance without requiring a measure of some effector system (Callaway 1969). Thus, performance measures, operator interviews and electrophysiological records are complementary methods.

References

BROADBENT D.E., 1958, *Perception and Communication* (London: Pergamon Press).

CALLAWAY E., 1966, Averaged evoked responses in psychiatry. *J. Nerv. Ment. Dis.*, **143**, 80–94.

CALLAWAY E., 1969, Evoked potentials for assessment of intellectual potentials. *Colloque sur le Traitement des Paramètres Électrobiologiques* (Toulon).

DAVIS H., 1964, Enhancement of evoked cortical potentials in human related to a task requiring a decision. *Science*, **145**, 182–183.

DINAND J.P., and DEFAYOLLE M., 1969, Utilisation des potentiels évoqués pour l'estimation de la charge mentale. *Colloque sur le Traitement des Paramètres Électrobiologiques* (Toulon). In 1969, *Agressologie*, **10**, Numéro Spécial.

DONCHIN E., and COHEN L., 1967, Averaged evoked potentials and intra-modilaty selective attention. *E.C.G. Clin. Neurophysiol.*, **22**, 537–546.

EASON R.G., AIKEN L.R., WHITE C.T., and LICHTENSTEIN L., 1964, Activation and behavior: visually evoked cortical potentials in man as indicants of activation level. *Perceptual and Motor Skills*, **19**, 875–895.

GARCIA-AUSTT E., BOGACZ J., and VANZULLI A., 1964, Effects of attention and inattention upon visual evoked response. *E.E.G. Clin. Neurophysiol.*, **17**, 136–143.

GROVES M., and EASON R.G., 1969, Effects of attention and activation on the visual evoked cortical potential and reaction time. *Psychophysiology*, **5**, 394–398.

HAIDER M., 1967, Vigilance attention, expectation and cortical evoked potentials. In *Attention and Performance* (Edited by A.F. SANDERS), 246–252.

KALSBEEK J.W.H., 1967, Objective measurement of mental load. *Acta Psychol.*, **27**, 253–261.

LEHMANN D., BEELER G.W., and FENDER D.M., 1967, E.E.G. response to light flashes during the observation of stabilized and normal retinal images. *E.E.G. Clin. Neurophysiol.*, **22**, 136–142.

PRIBRAM K.H., 1969, The neurophysiology of remembering. *Scientific American*, **220**, 73–86.

SATTERFIELD J.H., 1965, Evoked cortical enhancement and attention in man. A study of responses to auditory and shock stimuli. *E.E.G. Clin. Neurophysiol.*, **19**, 670–675.

SHEATZ G.C., and CHAPMAN R.M., 1969, Task relevance and auditory evoked responses. *E.E.G. Clin. Neurophysiol.*, **26**, 468–475.

SPONG P., HAIDER M., and LINDSLEY D.B., 1965, Selective attentiveness and cortical evoked responses to visual and auditory stimuli. *Science*, **148**, 395–397.

WELFORD A.T., 1967, Single channel operation in the brain. In *Attention and Performance* (Edited by A.F. SANDERS), 5–22.

Dr. Bonjer has degrees in physiology and medicine. Since 1953 he has been head of the Department of Occupational Medicine in the Netherlands Institute for Preventive Medicine TNO at Leiden.

The contribution of work physiology to the evaluation of man - machine systems

F. H. Bonjer

1. Introduction

Validation of man–machine systems, either existing or in some phase of design, should take account of, amongst others, three considerations.

The *first* category includes questions such as: 'Is this component of the task more suitable for a human operator or for a machine?' Such questions can be answered only if sufficient human factor and technical knowledge is available. Many branches of biological and behavioural sciences have contributed to our knowledge of the capabilities, limitations and vital needs of man. This paper surveys what work physiology has to offer when describing the human body as an 'energy converter'.

The *second* category of questions deals with the adaptation of the machine to the worker. The possibility of investigating the efficiency of tools and machine operations by means of work physiology will be discussed in relation to a general treatise on the concepts of the 'energy converter'. Problems that can be solved better by functional anthropometry or experimental psychology are ignored.

The *third* category of considerations is covered by environmental physiology and could be illustrated by the question: 'To what kind of an environment is the human operator exposed and to what extent are his capabilities reduced by a hostile environment?' Although many of these questions can be answered on the basis of methods, measurements and calculations similar to the methods of work physiology, problems of the environment will not be discussed in this paper.

2. The human body as an energy converter

If the human body is considered as a system in which chemical energy can be converted into mechanical energy, the output is muscular work. Muscular work, however, is not held in high regard in Western countries and it is a general philosophy in industrialized countries, that it should be eliminated by mechanization and automation.

The British physiologist Barcroft (1934) expressed other views: 'The condition of exercise is not a mere variant of the condition of rest, it is the essence of the machine'. The human engine is not created for idling, but for performing. It functions better, the more regularly and intensively it is used. No occupational disease will develop from appropriate physical activity, that is as long as the pattern of movements and the body posture are 'natural' and prolonged repetitive work is avoided.

Unlike the machine, the human body, within certain limits, repairs itself, and there are strong indications that physical activity can postpone and mitigate the pathological processes that threaten ageing populations which have a high standard of life.

It may be concluded that a high level of metabolism or a high heart rate do not necessarily indicate a failure in the design of the work. High intensities of bodily functions can be completely physiologically acceptable and pathological effects only arise in cases of overloading that may be preceded by equally high or even higher intensities of the same functions.

This means that the work designer, the human engineer or ergonomist, must know more about the laws that govern the capabilities and limitations of the human body as an energy converter and that he must also have some understanding of the vital needs of the biological organism. The maintenance of machines is an accepted feature in industry, but the physical fitness of the worker is not yet the concern of the employer.

The maximum extent to which the chemical energy can be converted into mechanical energy is often described as the physical working capacity. This term compares very well with the term horse power used to describe the performance of a petrol engine. Both the performance of a petrol engine and that of a human subject can be measured by means of a special device, which measures the amount of work produced per unit of time (the power) and which usually converts mechanical energy into heat. Such a test gives more than the cubic capacity of the petrol engine or the body weight of the subject.

More elaborate studies have yielded two laws.

■ The longer the body has to work, the lower is its power to convert chemical energy into mechanical energy. This aspect has been treated elsewhere in this book in the paper 'Temporal factors and physiological load'.
■ The older the organism is, the less the power per kg body weight.

It is not feasible, however, to measure the amount of chemical energy withdrawn from the stores in the body or the mechanical work performed, whenever physical work needs to be studied. It is for this reason that physiologists prefer to study body functions that accompany the conversion of chemical energy into mechanical energy without measuring the input or the output.

Such body functions are oxygen uptake from the surrounding atmosphere, including pulmonary ventilation and diffusion, and oxygen transport to the active tissues by means of the circulatory system. This is readily explained by the fact that the conversion of chemical into mechanical energy is an oxidative process, that can take place only if there is an adequate supply of oxygen to the tissues. As there is no capability for the storage of oxygen in the human body, the uptake and transportation of oxygen are really good indicators of the intensity of the process.

The rate of conversion is limited by the capacity of the body to take up and to convey oxygen and therefore many authors define the physical working capacity as the aerobic capacity, but the newer term 'aerobic power' is preferable, because it is really power that is being described.

The costs of 'fuel' for the human body are much higher than those of a similar amount of chemical energy in the form of petrol or coal. Moreover, it must be stressed that the mechanical efficiency—this is the ratio between the produced mechanical and the invested chemical energy —is lower in any human operation than was the case in the steam engines of the oldest railways.

We may conclude that physical activity is much more a normal function, and even a vital need, than a strain on the human body, if due attention is paid to the natural laws. The use of human labour will be satisfactory when there is variety in the things to be done and if the requirements change frequently.

On the other hand, the human body is an unsatisfactory and expensive source for mechanical energy and should never be used in uncomplicated and repetitive systems.

3.   Methods of measurement

The methods described in this paper only refer to the study of the human organism seen as an 'energy converter'. In principle these methods could be directed towards the input of the system, chemical energy (food) intake and oxygen uptake: and to the output, mechanical energy, carbon dioxide and heat. For reasons explained above and because carbon dioxide expiration is governed by processes other than metabolism, oxygen uptake is the most feasible yardstick of the intensity of the process. As oxygen uptake can be translated into terms of metabolism and this can be expressed in kcal/min, this approach is known as indirect calorimetry. Direct calorimetry (measurement of the produced heat) is not feasible in a practical situation, and, moreover, is without meaning unless the mechanical work can be measured at the same time.

In principle there are two methods for the measurement of oxygen uptake. The subject may be connected to a closed system containing oxygen and the oxygen uptake calculated from the reduction of the amount of oxygen caused by the respiration of the subject. Closed systems, however, are heavy and bulky and for that reason only applicable in the laboratory. Moreover, the subject must stay at the same location in the room during the measurement.

Much better possibilities are offered by the so-called open systems. The subject inspires fresh air and the expired air is measured and analysed for its oxygen content. A classic tool in work physiology is the Douglas bag (Douglas 1911), a large rubber or plastic bag worn on the subject's back to collect the expired air during a short period (Figure 1). Measurement of the volume and gas analysis follow afterwards.

Figure 1. Illustration of the use of the Douglas bag.

Newer developments are the portable gasmeter of Kofranyi and Michaelis (1941) and the integrating motor pneumotachograph (I.M.P.) designed by Wolff (1958) (Figure 2).

Both instruments combine the collection and volume measurement of expired air and also collect a sample of expired air for gas analysis after the experiment.

Prototypes of instruments do exist already in which the volume measurement and the gas analysis are performed continuously on the subject's back (Figure 3). By means of a cable or telemetry a continuous record of the oxygen uptake can be obtained. In this way heavy loads of short duration and complex situations can be studied (Bleeker and Hoogendoorn 1969, Bonjer 1969).

Apart from oxygen uptake other related physiological functions can be measured. The ventilatory minute volume can be used as a yardstick for oxygen consumption, because ventilation of the lungs is essential for it. The same reasoning applies to heart rate, but it is obvious that these two functions will be good substitutes for oxygen uptake only, if—in the first case—the extraction of oxygen from the air, and—in the second case—the stroke volume of the heart and extraction of oxygen from the blood by tissues are constant. Unfortunately, there is no guarantee for this.

Figure 2. Illustration of the use of the I.M.P.

Figure 3. Portable instrument for continuous measurement of the flow of expired air and gas analysis.

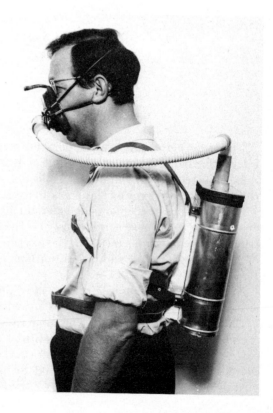

Heart rate, however, can be studied throughout a working day by telemetry or using miniature tape recorders. Therefore, it is attractive to combine a continuous heart rate recording with intermittent measurements of the energy expenditure. The heart rate reveals all changes in the workload and can be calibrated to a certain extent by repeated simultaneous measurements of oxygen uptake. This has less meaning, however, if changes occur in the working posture or the environmental factors and if heart rate is influenced by fatigue, apprehension or cigarette smoking.

More details about techniques and the methods for the measurement of energy expenditure throughout the whole working day or even 24 hour periods have been published elsewhere (Bonjer 1965, Bonjer in press). Body temperature and sweat rate also reflect the intensity of energy conversion or metabolism, but they are crude measures.

4. Applications

In the validation of man–machine systems knowledge of the human body as an energy converter can be applied as follows.

4.1. *Distribution of components of the task between man and machine*

Man should never be used as a source for mechanical energy or electricity. Motors and generators are superior. Man should not be used for simple, repetitive movements, nor to exert large forces. These should be taken over by mechanization. Sometimes such unsuitable activities are hidden in a complex work situation. They may be detected by continuous recordings of oxygen uptake and/or heart rate.

This can be illustrated by the results of an investigation into the workload of manufacturers of truck tyres. Oxygen consumption was measured by the aid of Douglas bags. This means that expired air was collected over a period of several minutes. The measures showed that this work was heavy, having an average energy expenditure of 5·5 kcal/min. Even more important, continuous recording of the heart rate revealed peaks of over 140 beats per minute, always occurring at the same action, whereas the average heart rate was about 127 beats per minute.

Figure 4. Usual operation of a truck tyre manufacturer.

The particular action was exerting a horizontal force on a handle bar in order to move another cylinder of rubber around a motor-driven drum and preceding layers of rubber, when balancing on one foot and controlling the machine with the other (Figure 4). It was clear that this component of the task should be transferred from the man to the machine. The necessary mechanization was provided according to our advice (Figure 5).

Figure 5. Pneumatic instrument providing a force for the operation shown in Figure 4.

4.2. *Adaptation of tools, machines and work methods to the man*

The highest efficiency of the man–machine system can be attained if tools, machines and work methods are adapted to workers in general or perhaps even to the individual worker. Sometimes it is difficult to decide on the best body posture: standing, sitting, or kneeling.

Paving, for instance, is done in the Netherlands in any of these three mentioned working postures according to the region. The regional habits can partly be explained by differences in weight of the bricks used, partly by local convention. When national training courses for brick laying were established, a decision had to be made on the preferred method. Physiological tests (Figure 2) indicated that kneeling was the optimal posture for the most frequently used types of bricks (Bonjer 1959).

In other cases it is preferable to provide such facilities that the body posture can be changed from time to time. Road traffic control cabins, for instance, were designed in such a way. Both observation of the traffic and work at the control panel could be carried out equally well in a seated or a standing position.

Handgrips should be adapted to the forces to be exerted, both in form and location. In a study of the optimal handgrips for building blocks of nearly 29 kilograms, 30 different models were moulded out of a clay and tested on a dummy of the same weight. The maximum lifting power expressed in kilograms and the minimum oxygen consumption and heart rate during a standard handling task were taken as criteria to determine the preferred model which was accepted by industry as a standard (Bobbert 1960). In a study with dustbin loaders (Figure 1) different types of loading systems were encountered, using different loading heights. A frame fitted with a specimen of the loading system, that had proved to require the least energy expenditure during comparative physiological studies, was set at different heights. Oxygen uptake and heart rate were used as criteria for optimal operation (Figure 6).

In summary, comparative studies of different work methods are often useful to allow decisions based on physiological functions.

Figure 6. Oxygen uptake and heart rate as encountered in a standard task carried out at a standard rate, but at different heights.

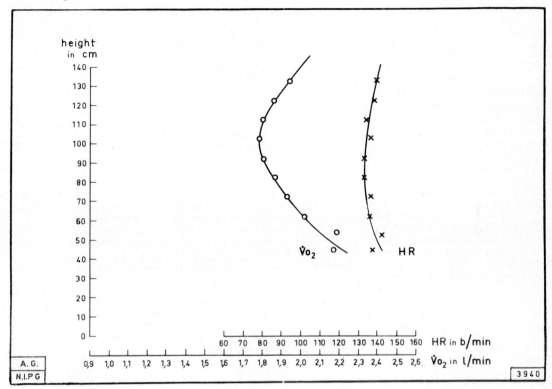

4.3. *Maximum allowable energy expenditure*

If due attention has been paid to the above aspects, in other words if the man is used in a proper and efficient way, it still has to be decided whether his physical working capacity can meet the requirements of the job. If he can, there is no reason for intervention: he might even have the benefit of increased physical fitness. If not, there is a danger of overloading, which may result in fatigue, loss of efficiency and ill health.

To decide whether the available physical working capacity meets the requirements of the work, one needs to know the aerobic power of the worker, the actual working time per day and the average energy expenditure on the job.

The available physical working capacity is a fraction of the aerobic power. The longer the working time, the smaller the fraction. Its value for a working time of 500 minutes is 0·33.

The following table derived from the study of truck tyre manufacturers may explain how the available working capacity can be compared with the demands of the job.

The apparent overloading in case I resulted from an inappropriate estimation of the workload by the time study engineer, who had allotted too little time and so unduly paced the production. He changed his opinion as soon as he knew the result of the study.

Case VI revealed the insufficient working capacity of the worker. Although he was given one of the two least tiring jobs, he was not able to do it without undue fatigue. A lighter job elsewhere was recommended for him.

The degree of loading and its application to problems of maximum allowable physical work has been described earlier in more detail (Bonjer 1962, Bonjer 1968).

	Case I	Case II	Case III	Case IV	Case V	Case VI
Energy required (M in kcal/min)	5·55	3·72	5·19	4·13	4·72	3·79
Avail. phys. working cap. (A in kcal/min)	5·43	5·14	5·43	5·55	5·14	3·72
Degree of loading (M/A)	1·02	0·72	0·96	0·74	0·83	1·02

References

BARCROFT J., 1934, *Features in the Architecture of Physiological Function* (London: Cambridge University Press).

BLEEKER J., and HOOGENDOORN M., 1969, A portable apparatus for continuous measurement of oxygen consumption during work. *Acta Physiol. Pharmacol. Neerl.*, **15**, 30–31.

BOBBERT A.C., 1960, Optimal form and dimensions of hand-grips for handling certain concrete building blocks. *Ergonomics*, **3**, 141–147.

BONJER F.H., 1958–59, The effects of aptitude, fitness, physical working capacity, skill and motivation on the amount and quality of work. *Ergonomics*, **2**, 254–261.

BONJER F.H., 1962, Actual energy expenditure in relation to physical working capacity. *Ergonomics*, **5**, 467–470.

BONJER F.H. (Editor), 1965, *Fysiologische Methoden voor het Vaststellen van Belasting en Belastbaarheid* (*Physiological Methods for the Assessment of Work Load and Working Capacity*) (Assen: van Gorcum).

BONJER F.H., 1968, Relationship between working time, physical working capacity and allowable caloric expenditure. In *Muskelarbeit und Muskeltraining*, hrsg. W. Rohmert, pp. 86–98. *Proc. Internat. Kolloquium*, Darmstadt, 1968. *Schriftenreihe Arbeitsmedizin, Sozialmedizin, Arbeitshygiene*, Bd. 22 (Stuttgart: A W. Gentner Verlag).

BONJER F.H., 1969, Application of a portable apparatus for continuous registration of oxygen intake. *Acta Physiol. Pharmacol. Neerl.*, **15**, 32.

BONJER F.H., In press, Energy expenditure. In *Encyclopaedia of Occupational Safety and Health* (Geneva: I.L.O.).

DOUGLAS C.G., 1911, A method for determining the total respiratory exchange in man. *J. Physiol.*, **42**, XVII.

KOFRANYI E., and MICHAELIS H.G., 1941, Ein tragbarer Apparat zur Bestimmung des Gasstofwechsels. *Arbeitsphysiol.*, **11**, 148–150.

WOLFF H.S., 1958, The integrating motor pneumotachograph; a new instrument for the measurement of energy expenditure by indirect calorimetry. *Quart. J. Exp. Physiol.*, **43**, 270–283.

Dr. Kalsbeek is a psychologist. He is head of the
Laboratory of Ergonomic Psychology of the Netherlands National Health Organisation TNO, situated in
Amsterdam. His research interests have consistently
included the measurement of mental load.

Sinus arrhythmia and the dual task method in measuring mental load

J. W. H. Kalsbeek

1. The phenomenon

The heart rate pattern of normal healthy subjects sitting at rest is irregular. Momentary irregularity of up to ten or fifteen beats per minute can occur. In the medical literature this phenomenon is generally referred to as sinus and respiratory arrhythmia. If one concentrates one's attention on a perceptual motor task the irregularity of the heart rate pattern tends to disappear as a function of the number of signals per minute one has to deal with. The mean heart rate, however, changes little if at all.

2. Sinus arrhythmia and mental load

2.1. *Apparatus and scoring method*

For recording sinus arrhythmia, a non-integrating cardio-tachometer (recently developed by the T.N.O. Institute for Medical Physics in Utrecht) has been used which measures the time between two successive R tops of heart cycles, and from this produces a direct pen recording in beats per minute: if a heart beat follows the one preceding it after a short interval, a high rate is recorded; if the time between successive R tops is longer a lower rate is shown. Continuous recording in this way results in a wavy line if the pattern is irregular, but in a smoother one if the rate is more regular.

Figure 1 shows graphs obtained with this cardio-tachometer. Dynamic work raises the level of heart rate and diminishes the irregularity of the rest pattern. Static load (e.g. when holding a weight with an outstretched arm) has the same effect. Increasing the number of signals per minute again diminishes the irregularity of the rest pattern but without affecting the level of heart rate (Kalsbeek and Ettema 1963, 1965).

In order to score the fluctuations of the recording the following procedure has been adopted (Kalsbeek and Ettema 1963, 1965): for every period of observation a smooth line was drawn by hand on the record to indicate the mean rate, and further lines were drawn three, six and nine beats per minute on either side of this mean. The number of times per minute that the deviation of the heart rate exceeded each of these levels was counted and the sum of all three counts—for three, six and nine was used as an *irregularity score* (Figure 2). This score took account of magnitude as well as frequency of variation in rate as fluctuations exceeding nine beats per minute were counted three times whereas those between nine and six were counted only twice, and those between six and three only once.

2.2. *Laboratory experiments*

Table 1 shows data obtained from a group of healthy subjects within the age-range 20 to 29 years. The experimental tasks were auditory binary choice tasks during 3 minutes at rates corresponding to 60 and 90 per cent of the maximum speed at which each subject could perform without error. All differences between rest and experimental condition were significant. The irregularity score of the 90 per cent group was significantly lower than that for the 60 per cent group, so it may be concluded that an increase of mental load consisting of the number of binary choices per minute is reflected by a decrease in the scored irregularity of the heart rate pattern.

Table 1. Effect on heart rate and frequency pattern (sinus arrhythmia) for different perceptual conditions. The figures given are percentages of resting values with no task and are averages. Standard deviations are given between brackets.

	Rest	Auditory signals		Spanish text	
		60% individual maximum	90% individual maximum	hearing	listening
Condition	R	III	IV	V	VI
Number of subjects		11	11	13	13
Heart rate	100	101 (6·1)	104 (7·9)	101 (3·4)	101 (6·5)
Sinus arrhythmia	100	68 (17·1)	45 (24·1)	86·6 (26)	50 (22·8)

Another experiment studied the suppression of sinus arrhythmia as an objective measure of whether a subject really listened to, or only heard, what was read to him. The first part of a Spanish text was read in a loud voice to a subject sitting at rest, who did not know the language. The subject was not instructed to listen or to pay attention. The second part was then read, when the subject had to count the times the article 'los' was used. In this way an attempt was made to introduce, experimentally, the difference between hearing and listening. Table 1 shows

Figure 1. Cardiotachogram of a subject. a and d: sitting at rest (s.a. score 13). b: walking a treadmill (6 km/hr). c: holding right arm outstretched (6th and 7th minute). e: tone-task, 40 signals/minute (s.a. score 1·5). f: tone-task, 70 signals/minute (s.a. score 0·5).

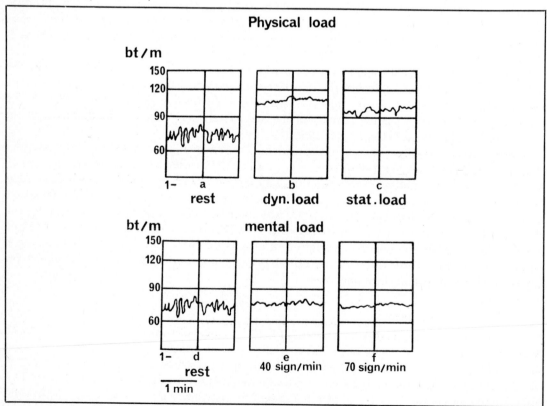

the results. The 'hearing' condition differed significantly from rest at $p < 0.01$ and from listening at $p < 0.005$ (Kalsbeek and Ettema 1965).

Figure 3 shows an attempt to scale different tasks according to the progressive suppression of sinus arrhythmia. According to this scale listening carefully would constitute the same mental

Figure 2. Method of scoring sinus arrhythmia.

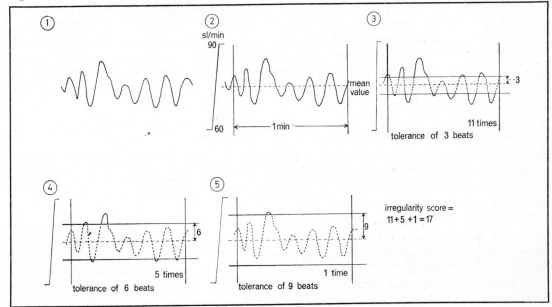

Figure 3. Experimental tasks arranged according to the suppression of sinus arrhythmia.

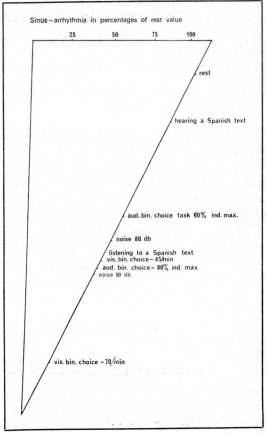

load as answering about 40 binary choices per minute (Kalsbeek and Ettema 1965a). Mulder and Hoogstraten (1969) compared four different resting conditions as regarding their reliability and usefulness as conditions of reference. No overall significant difference was demonstrable between the resting conditions, although one of the conditions (tapping on the table keeping in pace with the 20/min beat of a metronome) showed a variability score of the heart rhythm even lower than the one found during a very loaded task condition.

This puzzling outcome led Ruyssenaars (1970) to design an experiment which as its main objective had a closer look at the adequacy of the binary choice task as a validation condition. Physiological measurements were taken under seven conditions of increasing choice, motor load and under seven conditions of motor load without choice-making activity. It was found that the effect of the motor monitoring part of the binary choice task was not negligible, a result which makes the interpretation of observed changes in physiological measurements no easier.

2.3. *Studies under working conditions*

Recording sinus arrhythmia in working conditions on magnetic tape with and without telemetry techniques proved to be feasible with telephonists, car-drivers, pilots, workers on the assembly line, etc. Sometimes, however, it is difficult to obtain recordings at the work place without electrical interference.

Krol and Opmeer (Kalsbeek 1968a, Krol and Opmeer 1970a) found a significant difference between different phases of a flying task with junior pilots. Sinus arrhythmia scores decreased respectively with level flight, holding, take-off and approach. They also found a marked increase in the same order of the mean heart rate. Increasing heart rate causes some difficulty. With laboratory tasks there was only a small increase in the mean heart rate, if any. In practice, however, there is often quite an important increase in the mean heart rate due to dynamic and static workload, tension and emotion, or climatic conditions. Unfortunately as can be seen in Figure 1 changes in the mean heart rate have a depressing influence on the irregularity of the heart rate pattern. So finding a depressed sinus arrhythmia value in practice, one does not know which part might be due to the mental workload, and which part to other kinds of influence.

To tackle this problem studies have to be made with combinations of different kinds of workload.

3. Mental workload combined with physical workload

Opmeer (1969) studied the effect of heart rate, increased by physical load, on the differentiating power of both sinus arrhythmia and respiratory rate. By giving different groups of subjects different dynamic loads (watts) on a bicycle-ergometer subjects' heart rates increased to 105 beats/m, 123 beats/m, 135 beats/m and 151 beats/m, while one group which did not cycle showed a mean heart rate of 66 beats/m.

The subject was given a headset and on hearing a high tone produced by a binary choice generator he had to push the right-hand button on the handle-bar, while on hearing a low tone he had to push the left-hand button. In this way he could be given a binary choice task on top of the cycling task.

The number of tones the subject had to respond to was given at three levels of difficulty:

no tones at all;

50 per cent of the subject's maximum capacity;

80 per cent of the subject's maximum capacity.

These phases were separated by periods of rest and the order was randomized.

The maximum capacity was defined as the maximum number of tones/min the subject could respond to during 2 minutes without making more than 6 errors. This maximum capacity was measured before each session. After an adaptation session, each subject went 3 times through the same procedure.

An analysis of variance showed that in the group which did not cycle sinus arrhythmia and respiration rate differentiated both equally well between the 3 levels of mental load ($p = 0 \cdot 01$). Taking into account all the results at the varying levels of heart rate, respiration rate came out a better predictor of mental load than sinus arrhythmia. From Figures 4 and 5 it can be seen

Figure 4. The mean *sinus arrhythmia* score (S/N) of 5×3 subjects, under different levels of mental load.
Differentiated for the introduced 5 levels of physical load:
H1 = the subject did not cycle
H2 = cycling up to a mean of 110 beats/min. heart rate
H3 = cycling up to a mean of 125 beats/min. heart rate
H4 = cycling up to a mean of 140 beats/min. heart rate
H5 = cycling up to a mean of 155 beats/min. heart rate
(Opmeer 1969).

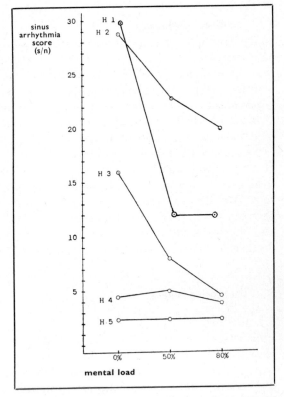

Figure 5. The mean *breathing rate* of 5×3 subjects, 3 sessions each, under different levels of mental load.
Differentiated for the introduced 5 levels of physical load:
H1 = the subject did not cycle
H2 = cycling up to a mean of 110 beats/min. heart rate
H3 = cycling up to a mean of 125 beats/min. heart rate
H4 = cycling up to a mean of 140 beats/min. heart rate
H5 = cycling up to a mean of 155 beats/min. heart rate
(Opmeer 1969).

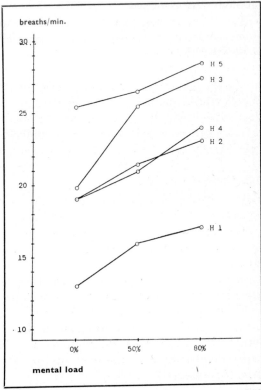

105

that respiratory rate still differentiated between levels of mental load, even at heart rates above 130 beats/min.

Thus one can conclude that at heart rates increased (above 130 beats/m) by energetic load, respiratory rate provides a better criterion of mental load than sinus arrhythmia.

4. Emotional workload

In developing physiological measuring methods of emotional load there is the difficulty that emotional load is not easy to vary systematically. In an experiment with parachute-jumpers (Krol and Opmeer 1969), both sinus arrhythmia and heart rate differentiated between levels of emotional stress. These levels were:

rest at the end of the day, when jumping had stopped;

15 minutes before the subject went into the plane;

in the plane, from take-off until the last 2 minutes before the jump;

the last 2 minutes before the jump.

Heart rate was slightly better, though both showed differences at $p = 0.01$. Between the phases 'rest' and '15 minutes before the jump' only heart rate showed a significant difference. Respiratory rate showed strong inter- and intra-individual differences but did not differentiate between the levels. So the conclusion must be that with emotional stress (information handling and physical load being at a minimum) as the independent variable, the differentiating power declines respectively with heart rate, sinus arrhythmia and respiratory rate.

5. The scoring method

5.1. *Criteria*

Following the initial handscoring method of Kalsbeek and Ettema different scoring methods were put forward of which, until now, the S/N score (see below) is the most frequently used.

An extensive study was done by Mulder and Mulder (1969) comparing different scoring methods in order to get a computer program. Criteria for the evaluation of different scoring methods were as follows.

- The differentiating power between laboratory conditions in which the number of stimuli per minute in a binary choice task is systematically varied.
- Sensitivity of the method regarding the influence of the value found during the preceding rest (law of initial value).
- Making the correlation with heart frequency as low as possible.

Which scoring method is best, depends on the criteria used. Which criteria have to be chosen depends on the *conditions* of mental workload. It is, for instance, conceivable that conditions of 'overload' demand scoring procedures differing from those for conditions of 'underload'. In any case it is necessary for the scoring program to have built-in procedures to correct for artefacts.

5.2. *Computer analysis*

For the quantitative analysis of variations in heart rate, these times have to be measured accurately. This variation, sinus arrhythmia, is currently expressed as a quantity per minute.

For the calculation the formula S/N is used where S is the sum of the positive difference between successive intervals and N the number of times a certain level of negative differences is exceeded.

In a computer programme written in Algol-60 this quantity can be calculated for a series of experiments, each experiment running a different number of minutes.

Many precautions against artefacts have to be taken and limits are set within which the time-intervals are measured beforehand with a multi-channel analyser giving the consecutive intervals on paper tape. Programmes are available permitting the analysis on-line of sinus arrhythmia with different methods of calculation simultaneously on a small computer-oriented system (Blom 1969).

6. The validity problem

6.1. *Scaling tasks with regard to mental load*

Some investigators however have not found significant scores on recording sinus arrhythmia during obviously different conditions of mental workload. So one can ask: how far can sinus arrhythmia differentiate between industrial tasks?

Sinus arrhythmia has been proved to differentiate between laboratory conditions in which the intensity of mental load was systematically varied. To validate sinus arrhythmia as a parameter of mental load in practice, the industrial tasks to be compared have to be chosen very carefully.

Tasks differing in difficulty or in complexity do not necessarily differ in mental load. Skill and quality of performance have to be taken into account. Tasks differing in the amount of information per minute to be handled do not necessarily differ in mental load either. The coding systems have to be taken into account. Tasks obviously differing in many aspects but falling largely within the mental work capacity cannot be expected either to give rise to distinguishable physiological reactions.

Mental load, systematically varied in the laboratory, consists mainly in a load for the central decision making mechanism, the sensory and the motor components being extremely simple. In practice however a decision is generally based on an integration of sensory input, the parts of which are presented to the central system one after another. During this time the decision system is not employed. In practice, also generally, the motor component is not a simple movement, but consists of a programme of movements to be carried out. Moreover the stimuli to be answered in practice are generally not presented in a randomized way but allow anticipation. In other words the central decision mechanism is supposed to be engaged only at moments of performance uncertainty calling for central control which could not have been anticipated. So we have to choose industrial tasks which differ regarding their use of the central decision making mechanisms as based on job descriptions and on tests of skill and experience of the subjects.

To summarize: in order to answer the question whether sinus arrhythmia can differentiate between industrial tasks one must be clear beforehand that there is actually something to be differentiated.

6.2. *Information handling activity per time as a constant*

But why should we look for a refined scaling method to differentiate between industrial tasks? Maybe the underlying philosophy is wrong. Maybe, disregarding the exigencies of the tasks, human beings will tend to maintain a level of mental activity, within certain limits as a norm in a cybernetic system, depending on the state of wakefulness (Kalsbeek 1963, 1967). The system will tend to apply strategies of shedding load in some cases or in other cases to search for a load ɪf this norm is threatened. The task-induced information is not the only one handled by the information processing system. There is a constant flow of more general situational information and a constant flow of self-generated information by the system. When the intensity of this latter flow changes, the intensity, or the fluctuations in intensity, of the first flow can be counterbalanced.

If task-induced information handling activity is taking the place which otherwise would be taken by more general situational information or by self-generated information the overall load remains the same. Therefore changing physiological parameters are not to be expected.

Only in cases of constraint are physiological parameters likely to differ. This constraint can be inherent in the task such as in security jobs and in cases of ambition, competition or fear. The constraint can also arise from within, when there is a great inner urge to generate information, e.g. in cases of pathological worries or personal problems. Finally the constraint can arise from the situation: noise, distraction, etc.

According to this model a detailed scaling of tasks regarding the mental load involved in their completion cannot be expected from recording physiological parameters.

Generally such a detailed scaling is desirable, for salary purposes. However physiological parameters are more useful to indicate to what extent the mental work-load is *acceptable*. If during the whole working period an important suppression of sinus arrhythmia were found in a group of workers doing the same kind of work, this would mean that the job needs to be redesigned.

Indicators of acceptable load would be comparable to values of maximum allowable concentrations (M.A.C. values) in toxicology.

This point of view has also implications for the scoring method. Until now the sinus arrhythmia score is averaged over the whole working period. A more sensitive scoring method according to this view would be scoring, say, the number of minutes per hour the irregularity score is less than a certain percentage of the preceding rest value.

In other words the number of peak loads would provide a better criterion than averaging the variability of the heart rate pattern scored per minute over the whole working period. This method is currently being tried out.

7. Sinus arrhythmia and pathology

It can be assumed that physiological systems will generally not function at their maximal capacity. A certain bandwidth of reserve capacity will therefore be available for cases of constraint. So according to this model one could say that physiological parameters are indicating the amount of reserve capacity available but not occupied. A complete suppression of sinus arrhythmia would mean that there is no reserve capacity left unoccupied.

A complete suppression, however, has been shown only for a short duration (about 3 minutes) in our experiments. After 3 minutes sinus arrhythmia reappears and subjects tend to make errors. So it can be assumed that the reserve capacity serves in the first place to cope with *peak loads* due to a sudden increase of information. Thus reserve capacity could also be called emergency capacity (Kalsbeek and Ettema 1964). Recovery time after a peak load is relatively long. According to an experiment of Lille *et al.* (1968) it took about 12 minutes after a demanding task of 3 minutes for EEG variables to return to their initial value.

Continuously occupying someone's reserve capacity would lead to exhaustion. On the other hand performance will break down if on top there comes a sudden peak load.

If a man feels, over a long period, urged to treat an inner flow of self-generated information of unusual intensity his working capacity will be very low as is the case in psycho-neurosis. If he nevertheless tries to keep up his normal performance he has to use his reserve capacity and risks a nervous breakdown if there is a sudden peak load.

In close collaboration with Offerhaus (1969) of the Jelgersma psychiatric clinic at Leiden an experiment was done in which sinus arrhythmia was recorded with groups of patients and non-patients. These groups can be significantly differentiated regarding the suppression of sinus arrhythmia during rest conditions. Patients show rest values equal to values only found with non-patients during quite demanding mental tasks. It is as if in patients the intensity of internal preoccupation is abnormally high.

How far the phenomenon of the regularization of the heart rhythm means a load, or even a danger, for the cardiac system itself is not yet known.

8. The dual task method

8.1. *The traditional way*

To determine to what extent the performance of a task occupies the available capacity, the subject is asked to perform a secondary task at the same time, which serves to absorb the mental capacity not used by the main task.

Bornemann (1942), Poulton (1950), Brown (1962), Haider (1962), Michon (1966), Rolfe (1966, 1969) and others have used this method. By measuring the 'spare capacity' an indication is obtained of the mental load implied in the performance of a job. In the use of the dual task method, the main task which had to be performed perfectly was a normal one, the mental load involved in the performance of which had to be measured. The mental spare capacity was measured by adding a simple repetitive task which was easy to score, such as simple arithmetical problems. The more problems the subject was capable of solving while performing the main task properly, the less this task was assumed to occupy mental capacity, and vice versa.

A difficulty however arises from the fact that the degree of activity is not without significance for the central nervous system. Above, attention was drawn to the concept of reserve capacity: and experiments with distraction stress described below show that the psychological state of the subjects depends on the degree of mental activity (cf. Kalsbeek 1964).

8.2. *The method of distraction stress*

In experiments using distraction stress the main task to which preference was given was a simple repetitive one, the binary choice task (Kalsbeek 1962, 1965; Schouten *et al.* 1962; Kalsbeek 1968 b).

In such experiments the binary choice task can be regarded as a stress condition in the performance of a secondary task. This way of applying the dual task method has been called 'the method of distraction stress'. In experiments using distraction stress normal task performance can be systematically broken down by progressively occupying the subject's channel capacity by increasing the rate of the binary choices. Figures 6 and 7 give examples of the effect on normal behaviour of progressively reducing the channel capacity left to control performance. Figure 6 shows the effect of distraction stress on performing a maze-test. The left-hand side of the figure shows the performance of a subject with only a few binary choices per minute as distraction stress. On the right is the performance of the same subject, under increased distraction stress, showing his incapacity to look ahead under these conditions.

Figure 6. Two instances of maze tracking as a second task by one subject.

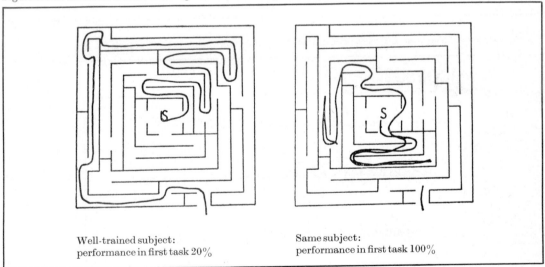

Well-trained subject:
performance in first task 20%

Same subject:
performance in first task 100%

Figures 7a and 7b show the effect of increasing distraction stress on normal handwriting. The subject was asked to write on a freely chosen theme. Handwriting became irregular, the letters grew larger, sentences became shorter, the content simpler and more infantile. In the end the subject was capable only of writing the alphabet or repeating one and the same letter. Under decreasing distraction stress, however, the same steps, but in reverse order, were obtained. Between the increasing and decreasing conditions there were no rest-pauses, which proved that the disintegration was not due to fatigue or to panic. The disintegration of handwriting on a freely chosen subject shows, at the same time, the breaking down of controlled movement and of thinking and reasoning. It is amazing that the complete disintegration of both can be

Figure 7a. Specimen of spontaneous handwriting as a second task with increasing performance in the first task.

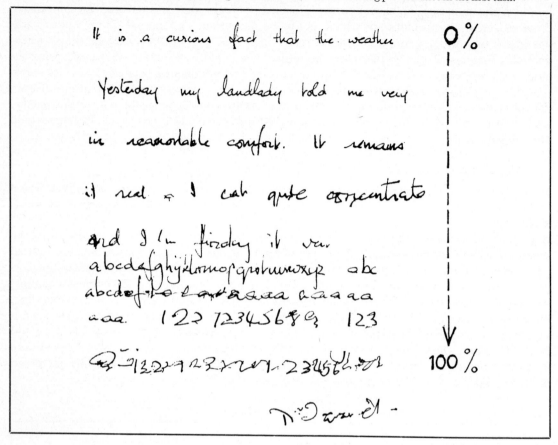

Figure 7b. Specimen of spontaneous handwriting with decreasing performance in the first task.

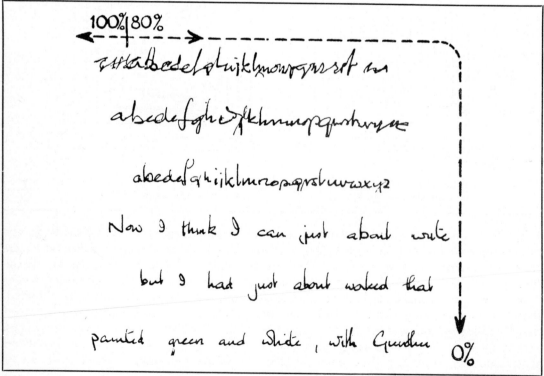

brought about experimentally simply by increasing the rate of binary choices in the distraction task. Thus, the deterioration of performance of complicated tasks, provoked by distraction stress, is comparable to the deterioration caused by other sources of stress which do not lend themselves so easily to experimentation, such as fatigue, lack of oxygen, ageing, etc. By increasing the load in the stressing task the system is forced to reduce progressively the load involved in the performance of the secondary task. By this method it is possible to investigate what strategies the system adopts spontaneously in order to reduce the demands put on it by the performance of a task. In general these strategies can be described as a successive shifting to simpler levels of organized behaviour. The characteristics of each level, if known, could be used to diagnose the level. The extent to which the full hierarchy of levels is operational could thus be used as a yardstick of mental load. On the other hand, knowing which kind of errors could be expected and in what order of probability, tasks could be so arranged as to make them foolproof against those anticipated errors that may endanger the production process. However, capacity may not be available either because there is no spare capacity or because the brain system keeps too much capacity in reserve. Kalsbeek and Ettema (1964) conducted an experiment to demonstrate this: the subject was given the writing task with increasing distraction stress as above, but on one occasion the subject was told to give of his utmost and on another to take care not to get too tired in view of later experiments. Table 2 shows that identical levels of deterioration were obtained, but in the motivating or *plus* condition as an effect of 30 and 40 binary choices per minute and in the less motivating or *minus* condition only by 10 or 20. Recording the irregularity of the heart beat pattern at the same time can give the kind of results obtained. Because the bandwidth of reserve capacity can vary according to the level of motivation the total capacity spent in a dual task situation could be called the 'willing to spend' capacity. It seems difficult to make assumptions about maximum mental capacity based on the study of maximum performance only.

Table 2. Identical levels of deterioration in writing performance as an effect of different levels of distraction stress according to fluctuations in the 'willing-to-spend' capacity. Scores of sinus arrhythmia reflecting these fluctuations.

Writing performance	+ (increased motivation)		— (reduced motivation)	
	Number of distracting binary choices	Sinus arrhythmia in percentage of rest value	Number of distracting binary choices	Sinus arrhythmia in percentage of rest value
Sentences	20	45	10	78
Repetition: whole alphabet	30	28	15	55
Repetition: parts of alphabet	40	17	20	55
Agrafismes and lines	70	35	30	40

8.3. *Deterioration of performance as a measure of mental fatigue*

Decrement in performance due to prolonged activity is often called fatigue. In the model described above it is better to say: prolonged activity results in reducing the capacity.

As far as decrement in performance is a linear function of reduced capacity the relative mental load remains the same. On the other hand if decrement in performance is greater than the reduction of capacity, decrement in performance can be looked at as a strategy to *avoid* fatigue.

So, only taking into account the decrement in performance nothing can be said about the reduction of capacity nor about the onset of fatigue. The subjective need for effort even to accomplish a poor performance is generally called fatigue. Following our model this could also be caused by an unusual intensity of the flow of self-generated information. The latter case is generally referred to as psychological or neurotic fatigue.

8.4. *The method of distraction stress as applied to practice*

Krol and Opmeer (1968, 1970b) studied the process of breaking down the flying task by distraction stress in a flight simulator. They also found that the reading of integrated instruments was more resistant to distraction stress than the reading of non-integrated instruments. Together with the National Aerospace Laboratory we made a comparative study of two different ways of presenting motor instruments using this method (*De Ingenieur* 1969). Performance was the same for both methods of presentation without distraction stress but could be differentiated under distraction stress (Kalsbeek 1968 b). In general it can be said that distraction stress can be used to test the resistance of a task to overload which can occur in emergency; by a drop in mental capacity (e.g. illness); or by an increase of the intensity of self-generated information (internal preoccupation). The method of distraction stress could also be used for selection and training, testing the subject's resistance to overload.

Few systematic studies are as yet available. The method of distraction stress is thus not put forward as a technique of measuring mental load. For this purpose the traditional application of the dual task method is better, although it would be naïve to think that the measuring task is just added to the measured task as was argued above. An elegant way of applying the traditional dual task method is based on the tapping task of Michon (1966). The measuring task consists of the subject having to say 'pom-pom' in a regular way. This technique was applied by Michaut (1967) to compare different cars and proved to interfere very little with the measured task.

References

BLOM J.L., 1969, A method for the analysis of cardiac time series. *Report TNO–Laboratory of Ergonomic Psychology.*

BORNEMANN E., 1942, Untersuchungen über den Grad der geistigen Beanspruchung. *Arbeitsphysiologie*, **12**, 142.

BROWN I.D., 1961, Measuring the spare 'mental capacity' of car drivers by a subsidiary task. *Ergonomics*, **4**, 35–40.

De Ingenieur, 1969, Fifty years aeronautical research in the Netherlands, **14**, 64.

HAIDER M., 1962, Experimentelle Untersuchungen über geistiger Beanspruchung durch Arbeitsleistungen. *Zeitschrift für Angewandte Physiologie einschlieszlich Arbeits Physiologie*, **19**, 241–251.

KALSBEEK J.W.H., 1962, Loss of information in information generating processes as a measure of distraction-stress induced by the method of subsidiary tasks. XXII International Congress of Physiological Sciences Excerpta Medica, *International Congress Series* No. 48, p. 1104.

KALSBEEK J.W.H., 1963, A model for studying repetitive and monotonous work. *1st Seminar on Continuous Work* (England, Bedford), 48–50.

KALSBEEK J.W.H., 1964, On the measurement of deterioration in performance caused by distraction stress. *Ergonomics*, **7**, 187–195.

KALSBEEK J.W.H., 1965, Mesure objective de la surcharge mentale; nouvelles applications de la méthode des doubles tâches. *Le Travail Humain*, **28**, 121–132.

KALSBEEK J.W.H., 1967, *Mentale Belasting, Theoretische en Experimentele Exploraties ter Ontwikkeling van Meetmethoden* (N.V. te Assen: Van Gorcum & Comp.).

KALSBEEK J.W.H., 1968 a, Objective measurement of mental workload; possible applications to the flight task. *AGARD–NATO, XVIth Avionics Panel Symposium* (Amsterdam). AGARD Conference Proceedings, **55**, 4.

KALSBEEK J. W. H., 1968 b, Measurement of mental work load and of acceptable load: possible applications in industry. *The International Journal of Production Research*, **7**, 33–45.

KALSBEEK J.W.H., and ETTEMA J.H., 1963, Scored regularity of the heart rate pattern and the measurement of perceptual or mental load. *Ergonomics*, **6**, 306.

KALSBEEK J.W.H., and ETTEMA J.H., 1964, Physiological and psychological evaluation of distraction stress. *Proceedings of 2nd International Ergonomics Association Congress, Dortmund.*

KALSBEEK J.W.H., and ETTEMA J.H., 1965a, Onderdrukking van de onregelmatigheid van de hartslag als een objectieve maat voor de mentale inzet. *Tydschrift Sociale Geneeskunde*, **43**, 644.

KALSBEEK J.W.H., and ETTEMA J.H., 1965b, Sinus arrhythmia and the measurement of mental load. *Communication at the London Conference of the British Psychological Society, December.*

KROL J.P., and OPMEER C.H.J.M., 1968, Sinusaritmie en Dubbeltaak bij vliegprestatie. *Report TNO–Laboratory of Ergonomic Psychology.*

KROL J.P., and OPMEER C.H.J.M., 1969, Sinus arrhythmia, heart rate and respiratory rate with parachute jumpers. *Report TNO–Laboratory of Ergonomic Psychology.*

KROL J.P., and OPMEER C.H.J.M., 1970a, Physiological parameters during different phases of flight in a flight simulator. *Report TNO–Laboratory of Ergonomic Psychology.*

KROL J.P., and OPMEER C.H.J.M., 1970b, The influence of a distraction task upon physiological parameters and flight performance during the approach phase of a simulated flight. *Report TNO–Laboratory of Ergonomic Psychology.*

LILLE F., POTTIER M., and SCHERRER J., 1968. Influence chez l'homme des niveaux d'activité mentale sur les potentiels évoqués, *Rev. Neurol.*, **118**, 476–480.

MICHON J.A., 1966, Tapping regularity as a measure of perceptual motor load. *Ergonomics*, **9**, 5.

MICHAUT G., et PIN M.C., 1966, Effects sur quelques variables psychophysiologiques de la conduite automobile urbaine. *Le Travail Humain*, XXIXe Année, nos. 1–2, 147.

MULDER G., and HOOGSTRATEN J., 1969, De sinusaritmie als meeinstrument voor mentale belasting. Een evaluerend onderzoek. *Report Laboratory of Ergonomic Psychology–TNO*.

MULDER L., and MULDER G., 1969, Computer-verwerking van sinusaritmie-registraties. *Report Laboratory of Ergonomic Psychology–TNO*.

OFFERHAUS R.E., 1969, Mentale belasting bij psychiatrische patiënten. *Referaat no. 23b. Jelgersmakliniek-Oegstgeest*.

OPMEER C.H.J.M., 1969, Sinusaritmie als maat van mentale belasting bij verschillende niveau's van de hart-frekwentie. *Report Laboratory of Ergonomic Psychology–TNO*.

POULTON E.C. 1958,Measuring the order of difficulty of visual motor tasks. *Ergonomics*, 1, 234–239.

ROLFE J.M., 1966, A study of setting performance on a digital display, *Royal Air Force IAM Report* 365.

ROLFE J.M., 1969, An evaluation of the effectiveness of a secondary task, *RAF IAM Report* 473.

RUYSSENAARS N.J.M.G., 1970, Sinus arrhythmia as mental load, sinus arrhythmia and neuroticism. *Report Laboratory of Ergonomic Psychology–TNO*.

SCHOUTEN J.F., KALSBEEK J.W.H., and LEOPOLD F.F., 1962, On the evaluation of perceptual and mental load. *Ergonomics*, 5, 251.

Professor Rey holds degrees in medicine and public
health. She has worked in European industry and
research centres, and in the U.S.A. Her current work at
the Institute of Physiology, University of Geneva,
Switzerland includes applied studies and research on
flicker fusion and the visual system.

The interpretation of changes in critical fusion frequency

Paule Rey

1. Introduction

Our intention is, in this paper, to discuss in what conditions, in laboratory experiments or in field studies, a decrease in CFF can be postulated to be due to mental fatigue. It is well known that, when the frequency of an intermittent light is increased to a certain value, our eye is no longer able to perceive flicker; the frequency of this light at which fusion just occurs is usually called critical fusion frequency or CFF. Since the time Simonson and Enzer suggested that CFF could be used as an indicator of mental fatigue, numerous studies have been devoted to this problem (Simonson and Enzer 1941, Schmidtke 1965, Grandjean 1967). As pointed out by Schmidtke the usefulness of the decrease in CFF as a test for mental fatigue hinges on the question as to whether the variability of this function has its source in the receptors, or whether it is of central origin (Schmidtke *op. cit.*). This author showed that, measured in the same conditions, CFF could be lowered even in purely mental tasks, that is tasks which did not involve a visual participation.

However, we have to remember that CFF is primarily a visual performance and that it reflects the activity of the visual system, from receptors to cortex. Therefore, we believe that a more valuable approach to our knowledge than before-and-after-work measurements depends upon a complete investigation of all the factors which may influence CFF apart from fatigue, in order to avoid attributing to it what is due to other variables. Moreover, CFF is not sufficient to describe man's perception of flicker. De Lange has shown that fusion can be experienced, at any frequency of a sinusoidally modulated light, provided that the modulation amplitude is set, for each frequency, at an appropriate level; plotting, on a double log scale, frequency versus

Figure 1. De Lange curves recorded in normal subjects with sinusoidally modulated light, under different conditions of the intermittent stimulation. BR = background reduced intensity; CR = concomitant brightness reduction of test and surrounding fields; REF = references; EF = eccentric fixation; SR = intensity reduction of the test field.

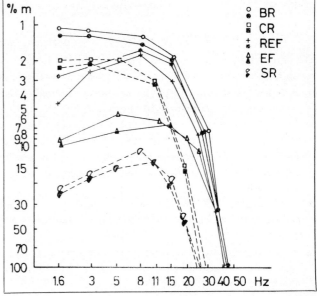

modulation amplitude, this author described an entire curve which is called after his name. The point at which the curve cuts the x-axis is the well-known CFF; everything located under the curve is perceived as flickering, everything above it is perceived as fused (de Lange 1957).

Investigations of the de Lange curve proved to be very fruitful in emphasizing the close connections which exist between flicker and the activity of the visual system.

In this paper, we will refer to recent researches which deal either with CFF alone or with the entire de Lange curve; we will consider different parameters which may play a role in depressing flicker fusion thresholds without mental fatigue being involved in these experiments.

2.   Results and discussion

2.1. *Parameters of the intermittent stimulation which influence flicker fusion thresholds*

It is well known that CFF is influenced by the mean intensity of the flickering light, its light-to-dark ratio, its size, its colour, the retinal locus of stimulation, the illumination of the surrounding field (especially in that this determines the level of adaptation) (Brown 1965, Piéron 1961). The de Lange curve, also, is affected by the mean intensity of the intermittent light (Kelly 1964, Babel *et al.* 1969), the size of the test field (Brown *op. cit.*, Ratliff 1965), and its colour (de Lange 1958). Spatial contrast between test and surrounding field acts on the low frequency range of the curve while high frequencies are not or only slightly affected (Levinson 1964, Babel *et al.* 1969) (Figure 1).

Ergonomists who are working on flicker are well aware of the fact that all these parameters must be controlled and kept constant, throughout the experiment, by appropriate experimental control.

2.2. *Effects of the method of determining CFF*

We have demonstrated previously that when CFF's were determined by decreasing the frequency of the intermittent light, successive thresholds were rather constant but that the standard deviation around the mean was high, since reaction times varied from one subject to the other. When CFF's were determined by increasing the frequency of the intermittent light, successive thresholds were shifted downward in the form of an exponential curve; this effect could be attributed to the exposure of the eye to intermittent stimuli (see below). Consequently we recommended the use of the method through successive approximations which brought out the most consistent results; in such a method, supra and infraliminar frequencies or modulation amplitude of the light are presented to the subject until two neighbouring values are found, one giving way to flicker, the other to fusion (Rey and Rey 1964).

2.3. *Lowering of CFF by a visual task, in the absence of fatigue*

We showed, in numerous ways, that, if the subject was submitted, between two measurements, to an intermittent visual stimulation of efficient frequency, a drop in CFF could be observed.

■ A 2-minute stimulation with stroboscopic flashes induced a U-shaped fall in CFF, according to the frequency of the light; high and low frequencies being less effective than middle range frequencies (Rey and Rey 1965); we failed to demonstrate like other authors (Brown *op. cit.*), that the stimulation of one eye depressed CFF in the other eye (Rey and Stoll 1965). Alpern (1961) was able to lower CFF, with infraliminar frequencies, and to raise it, with supraliminar frequencies, but we never succeeded in reproducing this last experiment (Rey and Rey 1965).

■ A 2-minute stimulation, with sinusoidally modulated light induced a fall in fusion thresholds, all along the de Lange curve, with infraliminar as well as with supraliminar frequencies (Rey *et al.* to be published).

■ A prolonged stimulation with stroboscopic flashes provoked an exponential decline in CFF; no further fall could be registered after 10 to 15 minutes; recovery exhibited also an exponential time course (Rey and Rey 1965).

■ The performance of a visual task, in which intermittent stimulation could occur (for instance, a Bourdon or a reading test) was followed by a decrease in CFF; this depression depended upon the speed at which letters were moving in front of the subject's eyes and not upon mental load; in our experiments, which lasted for 20 minutes, auditory tasks of the same content did not produce any significant drop of CFF (Rey and Rey 1964).

2.4. *The primary relationship between flicker fusion thresholds and the activity of the visual system*

2.4.1. *Retina and optic nerve*

Breukink (1962) was the first author who recorded systematically the de Lange curves of a great number of ophthalmological patients; distortions of the curve, according to this author, were assumed to allow a differential diagnosis between diseases of the visual system. Examining 75 patients with diseases of the retina or of the optic nerve, we found a linear correlation between visual acuity and CFF (Figure 2); using flicker perimetry for high and low frequencies of the de Lange curve, we were able to evaluate, quantitatively, visual field losses with great precision; we showed that fusion thresholds, at least in the low and the high frequency ranges, were influenced by the extent and severity of lesions rather than by their location in such or such layer of the retina or in the optic nerve (Babel *et al.* 1969, Rey *et al. unpublished data*).

Figure 2. Regression line and confidence limits: correlation between visual acuity and CFF decrease in 72 patients.

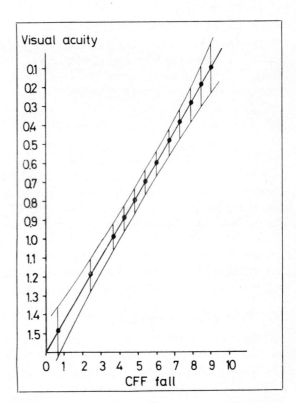

We verified that these distortions of the perceptive de Lange curve in patients with retinal diseases originated from the retina; a curve of desynchronization can indeed be built up, in several ways, from ERG responses to flickering light; these curves are very similar to the perceptive curves and show that the retina attenuates more high frequencies than low frequencies. Moreover, these curves can be shifted in the same manner as perceptive curves when the stimulating conditions are changed (Rey *et al.* 1968, Babel *et al.* 1969). Comparing curves in normal and treated rats, we observed that desynchronization thresholds were depressed when either receptors or bipolar cells were destroyed; in such a case, the averaged ERG was disturbed too (Rey *et al.* 1969) (Figure 3).

117

Figure 3. Desynchronization curves in normal rats (upper curves) and in rats with bipolar destructions (lower curves); the transmittance of neutral filters at threshold is plotted versus frequency.

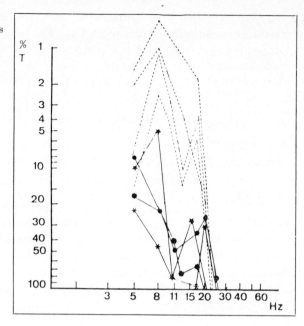

2.4.2. *Higher structures in the visual system*

Battersby (1951), like other authors comparing patients to control groups, found out that lasting depression of CFF appeared in the case of involvement of the higher visual pathways, from the retrochiasmal region to cortex. It is reasonable to assume that the de Lange curve would be similarly affected by damages in higher regions of the visual system.

2.5. *Effects on flicker of brain injuries outside the visual system*

Even though reductions of CFF have been observed in the case of frontal lobe lesions (Halstead 1947, Young 1949), Battersby failed to discover any significant drop, in frontal lesions occurring 4 to 6 years before testing; he concludes that frontal lobe injuries induce only transient CFF depressions in comparison to visual injuries which may be followed by permanent CFF decrease (Battersby *op. cit.*). According to Chandler (1966), fusion thresholds, as they are determined with flicker perimetry, vary as a function of CNS lesion characteristics such as focal versus diffuse, progressive versus static, lobe localization severity of disorder. In 42 cranial traumas, we detected pathological de Lange curves, which were distinguishable, in central fixation, from the curves obtained in ophthalmological patients; however, CFF's were not depressed in proportion to the severity of the cases (Thorens and Meyer *unpublished data*).

2.6. *Drugs as another variable*

In our own experiments on normal subjects, we measured a significant depression of the de Lange curve, with substances such as benzydamine, oxyphenbutazone, indométhacine, méfénamic acid and acétyl-salicylic acid. Placebo did not provoke any decrease in the reference curve. Even though two subjects only were tested in this series of experiments, results were quite consistent; flicker thresholds depression was more pronounced with higher dosage (Meyer and Rey *unpublished data*). In Chandler's (*op. cit.*) trials, however, drugs commonly administered to brain-damaged people did not greatly affect CFF.

Some recent work devoted to the effect of d-amphetamine or barbiturate, on the EEG occipital response to flickering light, reinforces previous findings on CFF reported by Simonson and Brozek (1952). EEG responses to photic stimulation appeared to be sensitive to a frequency selective d-amphetamine effect which stimulated responses to all frequencies above 13 Hz and which increased over the 2-hour period of recording; at lower frequencies, however, no significant difference could be detected between d-amphetamine and placebo (Montagu 1968). It is interesting to compare these results with the ones obtained, by the same author, with barbiturate : at 11 and 13 Hz, he observed a marked depression of the EEG responses ; at lower frequencies, barbiturates stimulated EEG responses ; at higher frequencies, the depressive effect was masked by the stimulating effect of barbiturates on spontaneous activity. The author concludes that these drug opposite effects may well be attributed to the respectively inhibiting and activating action on the cells of the CNS mediated by the reticular formation ; he does not exclude, however, some indirect influence of d-amphetamine through circulatory changes (Montagu 1967).

3.   Conclusion

In short, the following variables may affect CFF and the de Lange curve without fatigue being involved.
- Parameters of the intermittent stimulation.
- The method of determining thresholds.
- Characteristics of the task apart from fatiguing aspects.
- Dysfunctions of the visual system and other brain structures.
- Drug and other effects.

To claim that a decrease in CFF or other fusion thresholds is due to mental fatigue implies that all other factors are controlled or eliminated. To take care of the first two parameters is relatively easy, since it depends upon the experimental set-up and technique of measurement.
- Light source : sinusoidally modulated light or rectangular stimuli are preferable to strobo-scopic flashes since the light to dark ratio can be maintained, in the former methods, at 50 per cent for all frequencies.
- Photopic state of adaptation to light : this must be maintained constant throughout the experiment by the steady illumination of the background field. Absence of contrast between test and surrounding fields is recommended.
- A fixation point should be provided in the middle of the test field.
- Binocular testing should be avoided since the response of the individual eyes may be different.
- The size of the test field should not exceed one to two degrees, to facilitate central fixation.
- The most certain method for determining fusion thresholds is the method of successive approximations. Quick testing is necessary to avoid visual fatigue.

To take care of point 3 is fairly easy in the laboratory. Tiring tasks should be selected which do not involve any exposure of the eye to intermittent stimuli. In industry, however, workers might be exposed to intermittent visual stimuli of different kinds such as moving details, running machines, fluorescent lighting and so on ; as a consequence, the ergonomist may miss these effects which cannot be attributed to the task itself.

The usual practice in measuring the influence of one factor on some physiological function is to compare a tested to a control group. To take care of point 4, it is necessary to submit all subjects, before starting an experiment on flicker and fatigue, to a careful ophthalmological examination and to ask them about brain injuries or other troubles of the CNS. However careful the ophthalmological examination, one has to keep in mind that flicker fusion thresholds were demonstrated to be highly sensitive to tiny defects of the visual field, while other clinical tests were negative (Babel *et al.* 1969, Babel and Stangos 1969, Rey *et al. unpublished data*).

Still more important than these injuries, which usually are responsible for long-lasting CFF depressions, are those variables which provoke transient changes in CFF. We mentioned that substances, used for their anti-inflammatory properties, depress CFF and other fusion thresholds even at therapeutic dosage. Since this effect grows with concentration, the time at which, after absorption, CFF is measured, may affect results. Among transient effects are the activating ones of central temperature and exercise on CFF ; according to the author's hypo-thesis, this activating action could be related either to a direct influence of temperature on nerve cell activity or to an indirect influence through circulatory changes (Grivel 1969). It is obvious that CFF and other fusion thresholds are sensitive to so many variables that some of them may be completely overlooked in a before-after-work type of experiment.

From the above-mentioned experiments, it appears that in most cases of flicker disturbance, the visual system is concerned. But, if all precautions are taken in order to eliminate the influence of any other factor, a decline in CFF following a tiring mental task can reasonably be attributed to mental fatigue. Such a relation, however, takes for granted that the concept of fatigue, as postulated by Schmidtke or Grandjean, is not distinguishable from the concept of lack of alertness. Such an approach has at least the advantage of placing the problem of mental fatigue in a field of investigation, to which prominent physiologists have made important contributions.

References

ALPERN, M., and SUGIYAMA, S., 1961, Photic driving of the critical fusion frequency. *Journal of the Optical Society of America*, **51**, 1379–1385.

BABEL, J., et STANGOS, N., 1970, Lésions oculaires iatrogènes. L'action d'un nouveau médicament contre l'angor pectoris. *Arch. Opht.* Paris (in press).

BABEL, J., REY, P., STANGOS, N., MEYER, J.J., and GUGGENHEIM, P., 1970, The functional examination of the macular and perimacular region with the aid of flicker fusion thresholds. *Docum. Ophthal.* (in press).

BATTERSBY, W.S., BENDER, M.B., and TEUBER, H.L., 1951, Effects of total light flux on critical flicker frequency after frontal lobe lesion. *Journal of Experimental Psychology*, **42**, 135–142.

BATTERSBY, W.S., 1951, The regional gradient of critical flicker frequency after frontal or occipital lobe injury. *Journal of Experimental Psychology*, **42**, 59–68.

BREUKINK, E.W., 1962, De Frequentiekarakteristiek van het Menselijk oog onder Normale en Pathologische Omstandigheten. (*Thesis*, Utrecht.)

BROWN, J.L., 1965, Flicker and intermittent stimulation. In *Vision* and *Visual Perception* (Edited by C.H. GRAHAM) (New York, London, Sydney: John Wiley and Sons, Inc.).

CHANDLER, P.J., PARSONS, O.A., and MAJUMDER, R.K., 1966, Flicker discrimination in relation to nature and severity of CNS dysfunction. *Acta neurol. Scandinav.*, **42**, 558–566.

GRANDJEAN, E., 1967, *Précis d'Ergonomie*. (Bruxelles: Presses académiques européennes; Paris: Dunod.)

GRIVEL, F., 1969, *Personal communication.*

HALSTEAD, W.C., 1947, *Brain and Intelligence* (Chicago: University of Chicago Press).

KELLY, D.H., 1964, Sine waves and flicker fusion. In *Flicker* (Edited by H.E. HENKÈS and L.H. VAN DER TWEEL) (The Hague: Dr. W. Junk).

LEVINSON, J., 1964, Nonlinear and spatial effects in the perception of flicker. In *Flicker* (Edited by H.E. HENKÈS and L.H. VAN DER TWEEL) (The Hague: Dr. W. Junk).

DE LANGE, H., 1958, Research into the dynamic nature of the human fovea-cortex systems with intermittent and modulated light. *Journal of the Optical Society of America*, **48**, 777–789.

MEYER, J.J., et REY, P., 1969, Effets de quelques antiinflammatoires sur la courbe de de Lange. (*Unpublished data.*)

MONTAGU, J.D., 1968, The effect of d-amphetamine on the EEG response to flicker in man. *European Journal of Pharmacy*, **2**, 295–300.

PIERON, H., 1961, *La Vision en Lumière Intermittente*. Monographies Françaises de Psychologie, No. 8, CNRS, Paris.

RATLIFF, F., KNIGHT, B.W., TOYODA, J., and HARTLINE, H.K., 1967, Enhancement of flicker by lateral inhibition. *Science*, **158**, 392–393.

REY, P., et REY, J.P., 1964, La fréquence de fusion optique subjective: comparaison de trois méthodes de mesure, avant et après le travail. *Le Travail Humain*, **26**, 135–145.

REY, P., et REY, J.P., 1964, Pourquoi certaines tâches visuelles abaissent-elles la fréquence de fusion optique subjective ? *Le Travail Humain*, **26**, 293–304.

REY, P., and REY, J.P., 1965, Effect of an intermittent light stimulation on the critical fusion frequency. *Ergonomics*, **8**, 173–180.

REY, P., et STOLL, F., 1965, Action de la stimulation lumineuse intermittente sur la fréquence critique de fusion optique subjective. *Ergonomics*, **8**, 263–265.

REY, P., MEYER, J.J., et STANGOS, N., 1968, Les caractéristiques d'atténuation de l'oeil, mises en évidence par l'ERG, chez le palin. *Helv. Physiol. Acta*, **26**, CR 234–236.

REY, P., MEYER, J.J., THORENS, B., and STANGOS, N., 1969, The attenuation characteristics of the rat's retina under normal and pathological conditions. *Proceedings of the ISCERG Symposium on ERG*, Istambul, Sept., 1969 (in press).

SCHMIDTKE, H., 1965, *Die Ermüdung* (Edited by H. HUBER) (Bern und Stuttgart).

SIMONSON, E., and BROZEK, J., 1952, Flicker fusion frequency: background and applications. *Physiological Review*, **32**, 349–374.

SIMONSON, E., and ENZER, N., 1941, Measurement of fusion frequency of flicker as a test for fatigue of the central nervous system. *Journal of Industrial Hygiene and Toxicology*, **23**, 83.

THORENS, B., et MEYER, J.J., *Unpublished data.*

YOUNG, K.M., 1949, Critical flicker frequency. In *Selective Partial Ablation of the Frontal Cortex* (Edited by F.A. METTLER) (New York: Hoeber).

Dr. Borg is a psychologist. His experience has included
clinical psychology and military applications, and he
has worked in the U.S.A. He is now Director of the
Institute of Applied Psychology, University of
Stockholm, Sweden.

Psychological and physiological studies of physical work

G. Borg

1. Introduction

It is sometimes stated that fatigue arising from heavy muscular work concerns problems mainly of physiological interest and that the methods which depend upon performance are of a psychological nature. Both of these statements are debatable. Problems of fatigue are equally of psychological and physiological interest. Further, some methods for studying performance might well be classified as both physiological and psychological.

One reason for classifying problems of fatigue and strain due to heavy muscular work as psychological is that they belong to the same 'descriptive level' and 'explanatory level' in a macro-micro-continuum from sociology to biophysics as most psychological ones. Also, with regard to 'theoretical level' and 'level of measurement', these problems might be classified as psychological. To define fatigue merely in physiological terms seems as impossible as to do so with such concepts as colour, emotion or motivation. Fatigue is a subjective state of a person with both physiological and psychological aspects. This argument is also true for the study of muscular performances. The fact that there are physiological causes 'behind' the achievements is of course true for any kind of human performance. When the somatic 'causes' are fairly easy to measure, as is the case with some peripheral morphological and physiological variables, and when most of the variance between individuals seems to depend upon these easy measured variables, the performances have attracted the attention of physiologists rather than psychologists.

In 1958 we started to study perceptive and psychophysical problems of muscular work. The studies have expanded to other areas, such as various performance problems, work curves and adaptation functions, methods of determining physical working capacity, motivation for physical work, psychopharmacological, psychophysiological and clinical problems, and also problems concerning the importance of physical fitness for some mental functions.

The field of muscular work provides good examples of the need for both psychological and physiological methods. A physically stressing situation, to which a subject tries to adapt, may be studied with regard to the perceptual, the performance and the physiological (and biochemical etc) responses. These three different kinds of stress indicators, or effort continua, complement each other. Figure 1 depicts this simple situation and the rivalry between psychology and work physiology.

Singleton stresses, in his paper in this book, the need to obtain some measurements of the effort of the subject and to be able in some way to separate performance and effort. This is a very important question. Too often it is taken for granted that the effort for different tasks or by different individuals has been the same. This may occasionally be true, but it is very difficult to determine. In most cases it is impossible or extremely difficult to estimate the degree of achievement motivation or the biological costs for a certain activity. However, when we are studying physical work as in an ordinary ergometer test, the situation is fairly simple, the peripheral physiological endowments can be measured and the effort of the subject to some extent estimated.

When we want to identify an industrial task that might be too strenuous for the individual, the perceived 'difficulty' of the work can easily be obtained and used as a stress indicator. Measurements of apparent force and perceived difficulty can also be used for a quantitative evaluation of the degree of physical stress and thus complement the physiological measures.

Figure 1. The rivalry between psychology and physiology (mainly work physiology). Of the three indicators of physical stress (the perceptive, the performance and the physiological) in an observer (O) in a certain stimulus situation (S) the performance indicator is sometimes classified as both psychological and physiological.

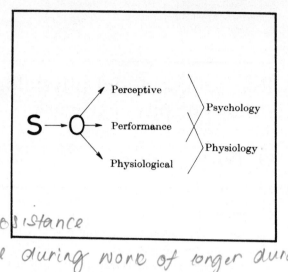

psychological indicator such as perceived stress or strain is studied with psychophysical ratio scaling method.
- able to determine perceived resistance
- perceived exertion of fatigue during work of longer dura

2. Perceptual indicators of physical stress

As a complement to physiological indicators of somatic stress, psychological indicators such as perceived stress or strain were studied with the aid of psychophysical ratio scaling methods. These methods functioned very well and enabled us to describe the subjective force (perceived pedal resistance) during short work periods (less than one minute) on a bicycle ergometer, and also the perceived exertion or fatigue during work of longer duration (several minutes), with power functions with an exponent of 1·6 (Borg and Dahlström 1959, 1960; Borg 1962).

For adapting the work intensity to a subject, this non-linearity must be kept in mind. Heart rate rises linearly with work load, but the perceived exertion, similar to the lactic acid concentration, rises according to a positively accelerating function. If we want to avoid this acceleration and obtain subjectively equal increases of load, the work intensity ought to increase by smaller and smaller steps.

For practical use and differential studies of perceived exertion a simple rating scale has been worked out. The latest scale in use consists of fifteen scale values from 6 to 20. The odd values from 7 to 19 are anchored with the aid of verbal expressions such as 'very, very light' (7), 'somewhat hard' (13) and 'very hard' (17). This scale was constructed to obtain a linear relation between ratings and work load and, since pulse rate rises linearly with work load, also between rating and pulse rate. This objective was achieved and in several Swedish and American normal groups, correlations between pulse rate and rating of ·80 and ·90 have been found (Borg 1962, Skinner *et al.* in press ; Bar-Or *et al.* to be published).

The relation between ratings of perceived exertion and heart rate changes with age and also with pathological conditions. In Figure 2 the pulse rate–rating relation is shown for four different age groups. The relationship is linear and the difference between the age groups quite large. The older subjects rate the degree of exertion to be higher in relation to heart rate than do the younger subjects. This change with age seems mainly to depend upon the decrease of the maximal heart rate with age (Borg and Linderholm 1967).

In a study of patients with vaso-regulatory asthenia it was found that they gave lower ratings in relation to heart rate than a healthy control group. In patients with coronary heart disease there was the opposite relation. The increase of ratings of perceived exertion was greater in relation to the increase of heart rate in all patient groups studied than in the control group (Borg and Linderholm 1969).

Perceived difficulty has also been studied for intellectual work (Borg and Forsling 1964, 1965). Measurements of perceived difficulty were obtained for test items in an intelligence test, and found to be highly correlated with the ordinary z-scores of difficulty obtained from solution frequencies. It was thus demonstrated that perception measurements can also be used in psychometrics.

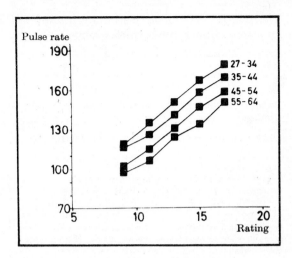

Figure 2. Lumber workers' mean pulse rate in relation to ratings of perceived exertion, in four age groups. Reproduced, with permission, from *Acta Medica Scandinavica* 1967.

2.1. *A note on quantitative semantics in connection with psychophysics*

The interpretation and the precision of verbal expressions is often difficult to determine and varies between cultural groups, etc. A new way to study these semantic problems is to apply modern psychophysical methods to estimate the intensities of the perceptions behind the use of different words (Borg 1962, 1964). In a study of physical exertion it was found that a subject generally uses the word 'laborious' when the perceived work is three times as hard as when he uses the word 'light'. The dispersion of the perceptive values for 'rather laborious' was greater than for 'very laborious', showing that 'very' denotes a higher precision than 'rather'.

The high correlation between values of perceived exertion achieved with the above rating scales supports the use of verbal reports to quantify (and maybe even set norms for) physical stress in work evaluation.

2.2. *Relative response and stimulus scales*

The psychophysical ratio methods do not make possible direct inter-individual comparisons. The functions obtained only give relations between perceptions, but not 'absolute' levels. To allow inter-individual comparisons a solution was proposed some years ago (Borg 1961, 1962), in which the range was used as a frame of reference and equated for all subjects. In the general biological equation $R = a + c(S - b)^n$ (Borg 1962), the constant c may be fixed arbitrarily. To get individual measurements of c is the main problem for inter-individual comparisons. According to the model, c is solved thus:

$$c = \frac{R_t - a}{(S_t - b)^n},$$

where R_t and S_t stand for terminal($t =$ maximal) values for the response (R) and for the stimulus (S). The constants a and b, showing the starting point of the curve, might in some cases be zero. Since a is 'no response' depending on S and since the range $R_t - a$ is the frame of reference and used as a unit for interprocess comparisons, the relative intensity of a response (RR_x) becomes, after inserting the expression for c in the general equation and multiplying by 100:

$$RR_x = \left(\frac{S_x - b}{S_t - b}\right)^n \times 100.$$

If we want to calculate a relative stimulus intensity, as e.g. in the prediction of an effective therapy for an individual from some physiological measurements, we can start from response measurements. We can use the human being as a measuring instrument—as in psychometrical determinations of the difficulty of test items—and determine the intensity of the stimulus.

This relative or subjective stimulus intensity (RS), may then be calculated in the same way as above, but from response measurements:

$$RS_x = \left(\frac{R_x - a}{R_t - a}\right)^{\frac{1}{n}} \times 100.$$

Figure 3 shows how a response measurement is transformed to a relative stimulus score. Two different functions, A and B (e.g. two different individuals) are compared, one function linear (A) and the other negatively accelerating (B).

Figure 3. A model showing how response (R) measurements may be transformed to relative (subjective) stimulus (RS) scores. In the figure two different functions, A and B (e.g., two individuals or two modalities) are compared, one function linear (A), and the other negatively accelerating (B). The a denotes the starting point (e.g., a rest value) of the curve and R_t the terminal value (e.g., the highest measurable value).

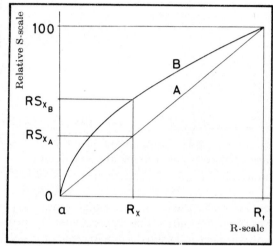

This way of getting relative response or stimulus measurements might be useful in many kinds of interprocess comparisons as, for instance, when we want to compare different kinds of indicators of somatic stress (heart rate, lactic acid concentration, blood pressure, rating of perceived exertion, etc) from different work situations and with different individuals involved.

The concept of 'arousal' may also be treated in this way. Various physiological measurements are often used as indicators of arousal. Raw scores may be somewhat misleading for inter-individual comparisons, because the intensity should be evaluated in a relative way, depending upon the position of the score in each individual's possible range. The form of the stimulus (S)–response (R) function may also vary much between individuals, thus further decreasing the validity of the scores. In most psychological situations it is, however, impossible, or at least very difficult, to assess any S–R function and to determine the possible range in question. We may then have to rely on studies in other modalities, for example on physical work, where these assessments are more easily done. If, for example, heart rate is used as an indicator of arousal or stress in an emotionally stressing psychological test situation, individual data about the range of heart rates may be 'borrowed' from a work test to get information on some relevant limits of the system, together with the heart rate–stimulus functions. This would enable the calculation of relative response or stimulus scores according to the principle above.

3. Working capacity, perceived exertion and maximal performance

Measurements of physical working capacity may be obtained from ratings of perceived exertion in much the same way as those obtained from pulse rate during physical work. A certain reference level in the rating scale is then used, for instance $17 =$ 'very hard', and a work load is determined which gives this rating after a certain duration of work.

In several studies in Sweden it has been shown that this measurement of physical working capacity provides good predictions of such field criteria of working capacity as results from skiing competitions, cross-country running or heavy lumber work. The correlation between this measurement and that obtained from heart rate is moderate, and the two measurements thus complement each other and give together the best possibility for field predictions.

Behaviour measurements of physical working capacity are easy to get for short work periods, but it is more difficult for work of longer duration, where the result not only depends upon muscular strength and motivation, but also upon circulatory and respiratory adaptation. A drawback in most of the performance tests of endurance capacity is that the total time for which the subjects are tested may vary markedly. In extreme cases one subject may work for just a few minutes, and another subject for half an hour. The test will then measure different things physiologically as well as psychologically. To correct this drawback, the following procedure may be performed: instead of using a physically equal test situation, as when the same work loads are used, or the same time–work load schedule, the loads in the suggested test are adjusted according to the responses of the subject to obtain a subjectively equal test situation. Initial work loads are chosen according to various physiological and anatomical data, and the increase in work load is then adapted to the stress responses of the subject. A special programme is worked out and the subject is guided from both heart rate and ratings of perceived exertion as 'feed-back variables' (Borg 1969). In this way approximately the same total test time may be obtained and the test will measure corresponding physiological and psychological factors for all individuals.

A new performance test of physical working capacity is the CSET, the Cycling Strength and Endurance Test (Borg 1962, 1969). The test is constructed to make work conditions approximately equal. An important purpose of the test is to obtain measurements both of muscular strength and circulatory adaptation and also motivation for physical work. During the test a series of intermittent maximal strength thresholds are determined for brief spells on the bicycle ergometer. Each sequence lasts about 45 seconds, during which time the work load increases continuously until the subject cannot pedal any more. This is followed by 15 seconds' rest. A work curve is thus obtained, in which the initial level indicates muscular strength; the regression of the curve and the final level endurance capacity; and some specially calculated measurements, such as the residual variation, motivation for physical work. Figure 4 gives mean values from CSE-tests on a group of conscripts and a group of lumber workers together with individual data from a world champion in cross-country skiing. Very high reliability and validity coefficients have been obtained in several different studies, for instance reliability coefficients between ·96 and ·99, and validity coefficients about ·50 and ·60, against such criteria as wages per day and in lumber work measurements of fitness of military personnel.

3.1. *An approach to work motivation*

For the total variance in a simple ergometer task, the endowments of the subjects and their motivation are the two most important factors. Since the physiological and morphological factors which influence performance can be measured and 'subtracted' from the total variance, the rest should be a motivation variance. Figure 5 shows a typical regression line of performance on physical endowment. The residuals from this regression may be looked upon as indicators of individual work motivation. Some studies have been carried out, and others are in progress to validate this approach (Borg 1964, 1969, Borg *et al.* 1968).

3.2. *Self-paced intensity levels*

Physiological measurements of maximal working capacity and of the consequent biological costs are useful in some situations, but in others the operator has to adapt to a submaximal intensity level. Here he often changes the work load or the rate until it becomes just right according to how he perceives it. Such self-paced preference levels should be studied more frequently. As an example, in an ergometer study on working capacity (Borg 1962), a preferred work load was determined by having the subjects (lumber workers) set an intensity level that was felt to be just right in comparison to 1 hour's lumber work. A high reliability coefficient was found ($r_u = \cdot 90$) and a significant correlation with the field criterion (wages per day in lumber work) of ·35.

Figure 4. CSET-curves for a group of conscripts ($n = 173$) and a group of lumber workers ($n = 57$). The figure also gives individual scores of a world champion in cross-country skiing, whose work curve lies more than three standard deviations (S.D.) above the group means.

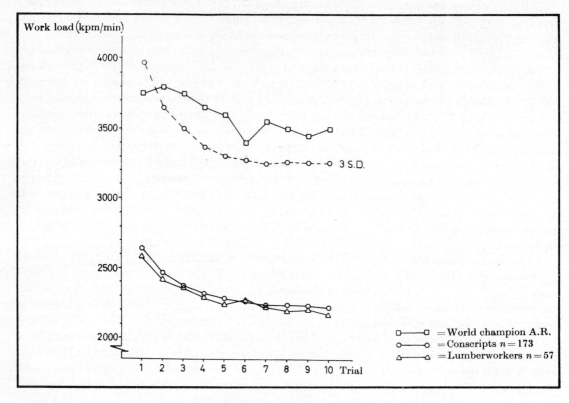

Figure 5. A theoretical model showing how measurements of work motivation may be obtained as residuals from the regression line, assessed for the predictions of performances from somatic endowments. Two subjects are shown, one (1) with a high work motivation (mo_1) and one (2) with a low work motivation (mo_2).

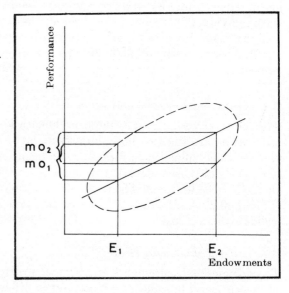

4. Some psychomotor performance indicators of physical stress

Performance measurements such as muscular thresholds or highest work loads for very short work periods relate to maximal working capacity. Another kind of performance measurement covers scores from such psychomotor performances as reaction time and hand steadiness. Some of these measurements are sensitive also to physical stress or 'arousal'. In some of our studies of physical working capacity referred to above, the subjects had also to take part in a test of visual reaction time. It was then found that the reaction time tended to be shorter immediately after physical work than before. This problem has been studied more thoroughly by Sjöberg (1968), who used work on a bicycle ergometer to vary 'arousal level'. The hypothesis was that the inverted U-relationship between level of arousal and performance level should hold true. The performance variable used was a visual choice-reaction task. The results of the experiment were in agreement with the hypothesis, 'performance being most efficient at medium arousal level and successively less efficient with both increasing and decreasing arousal'.

In studies of hand–arm steadiness (Borg, to be published), the inverted U-relationship was not found, but instead a continuously less efficient performance with physical stress. The hand–arm steadiness was very sensitive to this kind of 'arousal' and might therefore be used directly as a stress indicator and thus indirectly connected to the physical working capacity.

5.   Some comparisons between psychological and physiological indicators

One advantage of physiological and performance measures over many psychological ones is that they use dimensions and units which are similar to, and often as stable as, those used in physics and the other natural sciences. This should contribute towards their validity.

In terms of predictive power, performance measures are often superior to physiological measures. For instance, this is the case with performance measurements of physical working capacity, where physical endowments and motivational factors play an important role. With physiological measures we may have either a specialized variable—such as heart rate—or a number of physiological measures, with consequent uncertainties about the best mode of integrating them into a good overall parameter. (In a sense, the performance measurement itself is a very useful, complex, 'gestalt' of a large number of variables.) In addition, the validity of physiological measurements may be reduced by the difficulties of interprocess comparisons, as discussed earlier.

There is one area—taste—in which good psychophysiological comparisons can be made. Borg *et al.* (1967) made direct psychophysiological comparisons by combining psychophysical results on taste with neural responses obtained directly from the taste nerve, the chorda tympani, related to operations in the middle ear. High correlation between the two measures was obtained, contributing towards a validation of the psychophysical methods. Such opportunities are, however, rare.

When it comes to the reliability of the various perceptual, performance and physiological measurements, there do not seem to be any big differences. A high reliability is often shown in psychological performance measurements especially, but it is also found in the others. Guilford (1959) claims that physiological measurements related to personality often have a low reliability. His examples are, however, not especially good, and even if there is sometimes low reliability, this is often the result of too small samples or too few trials having been used.

6.   Conclusion

There is a need for much more study of performance problems of both a psychological and a physiological kind. In Figure 1 the three different kinds of indicators of physical stress, or the three effort continua, namely the perceptual, the performance and the physiological, were described. The first two are often said to belong to psychology and are studied by psychological methods. However, performance studies also belong to physiology, in particular work physiology, and it is necessary to stress the importance of this somewhat neglected field, the study of work performances.

References

BAR-OR O., BORG G., BUSKIRK E.R., and SKINNER J.S., 1970, Physiological and perceptual indicators of physical stress in 40–59 year old men. (*In press*).

BORG G., 1961, Interindividual scaling and perception of muscular force. *Kgl. Fysiograf. Sällsk. i Lund Förh.*, **12**, 117–125.

BORG G., 1962, Physical performance and perceived exertion (Lund: Gleerups).

BORG G., 1964, Bestämning av motivationens inverkan på fysisk prestation. *Nord. Psykiatr. Tidskr.*, **18**, 591–696,

BORG G., 1968, The three effort continua in physical work. *Proceedings of the XVIth International Congress of Applied Psychology, Amsterdam*, 394–397 (Edited by SWETZ and ZEITLINGER).

BORG G., 1970, Hand–arm-steadiness as a function of physical stress. (*In press*).

BORG G., and DAHLSTRÖM, H., 1959, Psykofysisk undersökning av arbete på cykelergometer. *Nord. Med.*, **62**, 1383–1386.

BORG G., and DAHLSTRÖM H., 1960, The perception of muscular work. *Umeå Vetenskapl. Bibl. Skriftserie*, **5**, 1–26.

BORG G., DIAMANT H., ZOTTERMAN Y., and STRÖM L., 1967, The relation between neural and perceptual intensity. A comparative study on the neural and psychophysical response to taste stimuli. *Journal of Physiology*, **192**, 13–20.

BORG G., EDSTRÖM C-G., and MARKLUND G., 1967, Arbetsmotivation. Differensen mellan observerade och förväntade fysiska prestationer. *Psykol. Und., Klin. Psyk. Lab., Umeå Univ.*, 6.

BORG G., EDSTRÖM C-G., and MARKLUND G., 1967, En cykelergometer för fysiologiska och beteendemätningar. *Psykol. Und., Klin. Psyk. Lab., Umeå Univ.*, 10.

BORG G., and FORSLING S., 1964, A psychophysical study on perceived difficulty. *Education and Psychology Research Bulletin, Umeå Univ.*, 1.

BORG G., and FORSLING S., 1965, Bestämning av svårighetsgraden hos testuppgifter utifrån ett upplevelsekriterium. *Rapp. Ped.-Psykol. Inst., Umeå Univ.*, 3.

BORG G., and LINDERHOLM H., 1967, Perceived exertion and pulse rate during graded exercise in various age groups. *Acta Medica Scandinavica*, **472**, 194–206.

BORG G., and LINDERHOLM H., 1969, Exercise performance and perceived exertion in patients with coronary insufficiency, arterial hypertension and vaso-regulatory asthenia. *Acta Medica Scandinavica*, **474**, 187, 17–26.

GUILFORD J.P., 1959, *Personality* (New York: McGraw-Hill).

SJÖBERG H., 1968, Relation between different arousal levels enduced by graded physical work and physiological efficiency. *Report from Psychology Laboratory, University of Stockholm*, 251.

SKINNER J.S., BORG G., and BUSKIRK E.R., 1970, Physiological and perceptual characteristics of young men differing in activity and body composition. In *Fitness 1969*. (Athletic Institute–1970, Chicago, Illinois). Symposium presented in Champaign-Urbana, Illinois in 1969.

Dr. Edwards holds degrees in psychology. Since 1960 he has taught and pursued research in the Department of Ergonomics and Cybernetics at the University of Technology, Loughborough, England. He has extensive experience as an applied psychology and ergonomics consultant to industrial companies.

Techniques for the evaluation of human performance

Elwyn Edwards

1. Introduction

Prior to a review of performance evaluation techniques, a few comments are offered in an attempt to clarify certain points which arise in the consideration of performance criteria.

1.1. *Psychology and physiology*

It is a long time since the dualistic distinction between Mind and Body has been regarded as a particularly useful basis for distinguishing between psychological and physiological methods. A more recent conceptual distinction between a black box approach and an analytic one is perhaps more in line with current practice, although not entirely satisfactory. For example, the determination of the relative efficiencies of different rates of physical working might be carried out wholly by means of black box concepts, but would not usually be classified as psychological.

Since these anomalies arise purely from the retention of classification labels despite the abandonment of the original classification system, a certain amount of untidiness might be expected.

Consequently it is concluded that while the traditional distinction between psychological and physiological studies may be of historical interest, it is of no value as a basis of taxonomy of assessment techniques in line with current thinking relevant to the problems encountered in ergonomics.

1.2. *The interpretation of multi-dimensional studies*

It is extremely convenient for the investigators when every part of a multi-dimensional study points in the same direction, as sometimes happens. However, it is always possible that such might not be the case. Equipments ranked, for example, in respect of users' preferences may be differently ranked in respect of the energies expended during use. Although such a result might cause inconvenience, it would be a mistake to conclude that any contradiction was involved, least of all any contradiction between the results of psychological and physiological assessments. There is no *prima facie* reason why the application of different criteria should not produce different results ; we are merely obtaining different answers to different questions. The practical solution to the resulting incompatibilities in design recommendations must obviously be based upon an ordering of the criteria themselves in order to achieve a compromise which is optimal in the context of any given set of prevailing circumstances.

1.3. *The objectives of performance evaluation*

Human factors specialists may be called upon to answer at least three different types of question regarding human performance.

To evaluate a total system in relation to a given specification.

e.g.   *Are we yet ready to attempt a manned landing on Mars?*

To evaluate a particular individual in relation to his peers, or to some definable standard.

e.g.   *Should Mr. X be awarded a commercial pilot's licence?*

To evaluate equipments or procedures in relation to one another.

e.g.   *Which of these displays better facilitates accurate navigation?*

The objective of any given performance evaluation to some extent constrains the selection of appropriate techniques. It therefore follows that the ergonomist must apprise himself of the nature of the existing objective before embarking upon a programme of assessment.

2.   Classification of assessment techniques

Three broad categories are used to classify assessment techniques. These are defined as follows.
■ Direct Achievement Measurement
   Definition : Evaluation derived by direct measurement of the total achieved performance level of a man–machine system or of some aspect of that level.
■ Operator Loading Measurement
   Definition : Evaluation derived by the measurement of the 'cost' to the operator of the achievement of a given performance level.
■ Correlated Function Measurement
   Definition : Evaluation derived by the measurement of a function not itself a criterion of performance, but known to be highly correlated with such a criterion.
A list of assessment techniques in each of these three categories is set out in Table 1.

Table 1. A taxonomy of assessment techniques.

1.	**Direct achievement measurement**
1.1	Absolute scores
1.2	Speed scores
1.3	Precision scores
1.4	Error scores
1.5	Information scores
2.	**Operator loading measurement**
2.1	Extracted outputs
2.2	Secondary tasks
2.3	Alternative tasks
2.4	Time function changes
2.5	Operator reports
3.	**Correlated function measurement**
3.1	Variability
3.2	Operator reports
3.3	Observer reports

3.   Direct achievement measurement

3.1. *Absolute scores*

Certain types of performance (e.g. athletic field events ; the game of golf) may be evaluated in terms of simple absolute scores. No further information is required to evaluate performance at such activities.

3.2. *Speed scores*

In a few cases (e.g. athletic track events) the speed score serves as an absolute score and requires no further amplification. More usually, speed represents only one aspect of the total assessment (as in typewriting) and requires to be considered in conjunction with other measurements.

3.3. *Precision scores*

It is sometimes possible to express the divergence between an achieved performance level and a pre-determined criterion as a continuous function. Such a method is employed in certain forms of scoring tracking performance (e.g. when scores of the form $\int |e|\, dt$ are employed) and in such games as bowls (British style).

3.4. *Error scores*

Discrete, rather than continuous, functions may be used as an alternative method of expressing divergence from a criterion. The use of discrete categories may be an experimental contrivance (e.g. in certain types of tracking scores using time-on-target) or may be a necessary feature of the system under study (e.g. in the study of keyboard performance).

Error scores are of especial importance in the evaluation of signal detection performance either in the context of signal *vs.* no-signal (e.g. in radar watch-keeping) or in the context of signal differentiation (e.g. inspection and sortation). Here the relation between errors of commission and those of omission are of particular interest.

Accident studies provide a source of information regarding errors, although, for numerous reasons, including the lack of homogeneity in the events termed 'accidents', it is doubtful whether such studies could ever be particularly fruitful as a method of assessment.

3.5. *Information scores*

The mathematical theory of communication provides a useful technique for measuring the rate at which an operator handles information. Such a score provides a single index in place of separate speed, precision and error scores. There are obvious advantages in having a single index, but it is necessary to demonstrate its relevance to any specific practical problem.

3.6. *Multiple scores*

In a large number of situations, several Direct Achievement Measurements may be employed to assess some partial aspect of an integrated performance. In such.cases it is necessary to avoid the allocation of too much emphasis upon that which is most easily specified and measured. The significance of any partial measure needs careful investigation.

4. Operator loading measurement

4.1. *Extracted outputs*

Certain outputs from the human body may be examined in either of two ways (cf. Fogel 1963). They may be conceived as evidence about the *modus operandi* of the organism (usually regarded as physiological study) or they may be conceived as black box outputs shown empirically to be correlated with certain inputs or with other types of output (usually regarded as psychological study). Extracted outputs include ECG, EEG, EMG, GSR, blood pressure, O_2 uptake, skin temperature, pupil diameter, etc. Although many of these measures have been shown to be related to the physical or mental load placed upon an operator, a good deal of further research is required to develop techniques of interpretation. Many aspects of extracted outputs are discussed elsewhere in this volume.

4.2. *Secondary tasks*

It has been argued that the extent to which a task loads an operator may be measured by the introduction of a quantifiable secondary task which is applied until the operator is at full capacity, i.e. performance decrements occur. There are difficulties with this method in its simplest form particularly in respect of the interference which tends to occur during the performance of two simultaneous activities. Nevertheless, the technique provides a valuable form of assessment in many cases.

4.3. *Alternative tasks*

Instead of applying a simultaneous secondary task it is sometimes advantageous to employ an alternative task which is performed after and in place of the primary activity as an index of the latter's cost. Such an alternative task would be one known to show a substantial decrement as a result of continued loading (e.g. flicker fusion frequency).

4.4. *Time function changes*

Some elements of skilled performance exhibit progressive deterioration consequent upon continued repetition and thus serve as indices of performance cost. Such deterioration is usually described as 'fatigue'.

4.5. *Operator reports*

Verbal reports from experienced performers may be used as measures of task difficulty. Such reports are best obtained by the use of systematic interviews or questionnaires.

5. Correlated function measurement

5.1. *Variability*

Improvement in the level of skilled performance is invariably accompanied by an increase in the consistency of the methods by which the performance is achieved. In view of this, the amount of variability serves as a method of assessment.

5.2. *Operator reports*

To a limited extent, the verbal accounts from operators themselves serve as a means of performance assessment. The most exploited technique is that of Critical Incidents, whereby accounts are gathered of events which caused particular difficulty or were thought to be 'near misses'.

5.3. *Observer reports*

Assessments provided by experienced observers are widely used for such complex skills as flying or driving. The assessment is likely to be improved by training the observer in the use of systematic methods of observation and recording.

6. Relative merits of performance assessment techniques

6.1. *Direct achievement measurement*

There are obvious advantages in the utilization of direct objective measurements, thereby avoiding any source of error resulting from the introduction of intervening arguments. However, in all but the simplest cases of performance assessment, direct achievement measures for total skills are unavailable. Undue emphasis upon the measurement of a single aspect of the total skill (e.g. the simple reaction time of motor car drivers) may lead to extremely unreliable conclusions.

The total assessment of certain skills is closely approached by the determination and combination of a number of separate scores. Thus, for example, typewriting performance can be almost completely assessed by the evaluation of Speed and Error Scores. There remains the combinational problem. Whether a 10 per cent increase in speed associated with a 10 per cent increase in error rate results in a net gain can only be decided in the context of a specific problem area. Again, although certain amounts of such 'trade-off' can normally be achieved, there are limits beyond which performance cannot be varied.

Detection performance serves as a further example of a skill requiring two primary assessment indices, in this case Error Scores of omission and those of commission (i.e. misses and false positives). Again, the trade-off problem arises. In some instances speed of response also effects the primary indices, in that performance is optimized when operators are constrained to respond within a period which is neither too short nor too long.

Tracking performance may be assessed by means either of Error Scores or of Precision Scores usually associated in some way with time. It may be shown that the assessment of tracking performance is a function of the type of scoring technique employed. Information Scores provide, to a limited extent, a solution to the combination of speed, precision and error, and such scores have been utilized in the assessment of a wide variety of tasks including simple and disjunctive reactions, sorting, tracking and verbal recollection. In any real-life situation, the propriety of the Information Score requires empirical justification.

6.2. *Operator loading measurement*

Assessment techniques in this category have two distinct uses. Firstly, they are of direct use in establishing whether the load upon an operator is appropriate in respect of his comfort, well-being and safety, with due regard being paid to the expected duration of performance.

Secondly, they provide a means of indirect performance assessment which might be employed when direct measurement is inconvenient or impossible. Thus, for example, secondary tasks have been used to assess driving performance under different road and vehicle conditions since no direct index of such a complex skill as motor vehicle driving is available.

In certain cases, assessment by loading measurement may, consequent upon its sensitivity, be superior to direct measurement. Highly skilled operators may, for example, perform equally well on two equipments, the relative merits of which may then be established only in terms of loading. (See, e.g., Benson *et al.* 1965.)

6.3. *Correlated function measurement*

These methods also serve when direct measurement is inconvenient or impossible. Again, there is some evidence to suggest that their sensitivity may, in certain circumstances, make them superior to the available direct achievement techniques. (See, e.g., Simmonds 1960.)

7. The objectives of performance assessment

Three types of objective have been distinguished (Section 1.3. above). Assessment techniques must be selected in relation to the particular objective.

7.1. *System evaluation*

The assessment of a total system demands the comparison of its performance with a set of pre-determined criteria. Thus, only the direct achievement measurements of speed, precision, etc. are relevant here to establish adequate system performance. Should the system prove itself inadequate, then it may well be that other types of assessment might be employed in the course of further detailed research.

7.2. *Operator rating*

Whilst all types of technique are technically suitable for this purpose, it is doubtful whether any technique other than Direct Achievement Measurement or Observer Reports would be otherwise acceptable for formal purposes. The latter are extensively used in the assessment of, for example, aircraft pilots and motor vehicle drivers. The basic problems in this area are, of course, associated with the acceptability in law and by the population to be assessed, of any given set of techniques.

7.3. *Comparative evaluation*

All three groups of assessment techniques have been widely used for this purpose. Provided that Direct Achievement Measurements are viable, then these provide perhaps the most con-vincing proof of differences between equipments or procedures. The importance of selecting the appropriate parameters must be emphasized. Should, however, these measurements fail to indicate differences, further studies should be pursued to investigate possible differences in Operator Loading.

References

BENSON A.J., HUDDLESTON H.F., and ROLFE J.M., 1965, A psychophysiological study of compensatory tracking on a digital display. *Human Factors*, **7**, 457–472.

FOGEL L.J., 1963, *Biotechnology: Concepts and Applications* (N.J.: Prentice-Hall).

SIMMONDS D.C.V., 1960, An investigation of pilot skill in an instrument flying task. *Ergonomics*, **3**, 249–54.

Dr. Rolfe has psychology degrees, and since 1959 has pursued research in the Royal Air Force Institute of Aviation Medicine, Farnborough, England. His major interest is in the effects on aircrew of changes in flight techniques and equipment.

The secondary task as a measure of mental load

J. M. Rolfe

1.   Theoretical background

The human operator is currently seen as a single channel data processing system, having its limitations in the central decision mechanism which must be allowed a finite time to process one stimulus-response before a second can be accepted. Evidence for this view has come from the study of the phenomenon known as the psychological refractory period. The extensive study of this phenomenon in the last twenty years has stemmed mainly from the work of Craik (1948). Other workers have extended the investigation of this aspect of behaviour; in particular Hick (1949) who, by employing different sensory modalities for the presentation of successive stimuli, was able to demonstrate the central locus of the refractoriness, and Welford (1952) who set out the theoretical implications of the refractory period. Davis (1956, 1957, 1959 and 1962) argued on the basis of extensive experimental work, that the single channel concept could only be held to be feasible when applied to man's ability to deal with information which required attention. Adams (1964) took a similar view, and basing it on his own researches (Adams and Xhignesse 1960, Adams and Chambers 1962) put forward the suggestion that the one channel decision mechanism could be a valid concept providing its role was defined as the resolution of event uncertainty but not temporal uncertainty.

Bringing together the more theoretical studies of the human operator and the applied areas of psychological research, Broadbent (1957) elaborated a mechanical model of the single channel concept which conformed to the data obtained from his own and other workers' research into multiple task performance using the auditory channel. Knowles (1963) acknowledged a similarity with Broadbent's ideas when he compared the operator to a multiplex communication system using a single channel to transmit messages from several sources to several destinations. So long as the channel was connected to a given source and a given destination, messages from other sources to the same or other destinations could not be transmitted.

Garvey and Henson (1958) also drew upon the concept of limited capacity in relation to multiple task performance. In their view, 'It is reasonable to suppose that, with all other conditions equal, the greater the error the subject perceives the more effort he expends in attempting to reduce it and the less of his capacity remains available for simultaneously contending with secondary tasks or other circumstances which demand his attention'.

Figure 1. Brown's model for using a second task to occupy reserve capacity.

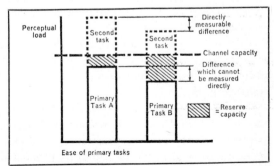

Brown (1964) used the diagrammatic model shown in Figure 1 to illustrate the function of the secondary task. In his words :

'On the model, the man's total channel capacity is drawn at some arbitrary point on the ordinate representing amount of information (in practice of course, it may vary with his state of arousal and other aspects of behaviour). The difference between his capacity and the perceptual load imposed by some imaginary task is his reserve capacity. This clearly depends upon the imposed load, but the important point is that if it exists at all, reserve capacity, and also perceptual load, cannot be measured directly from performance errors on the task in question since, by definition, the man will make no errors.

'The diagram also demonstrates the most convenient method of occupying reserve capacity so that perceptual load becomes measurable. The man is required to perform a second task concurrently with the primary task which is being evaluated. On the one-channel hypothesis the load imposed by the dual task can be arranged to exceed his limited capacity so that errors occur.'

The above brief review has attempted to provide the framework within which the secondary task technique has been developed and employed. It is essentially an applied research tool but one which has its origins in basic experimental psychology. The theoretical basis has been explored above but recognition must also be given to experimental work which has sought to examine the areas of divided attention and the effects of distraction (Woodworth and Schlosberg 1954, Kreezer *et al.* 1954). These spheres of investigation reach back to the nineteenth century. A particularly relevant example of this is the work of Lehman (1901) who studied the distracting effect of mental arithmetic on the performance of a physically demanding task. This work must be considered a forerunner to the more often cited investigations of Bornemann (1942) who examined the degree of impairment accompanying the simultaneous performance of two tasks.

1.1. *Definitions*

It is essential at this point to make some attempt at defining the terms used in this area of research. The basic experimental task under investigation and to which additional tasks are added is usually referred to as the primary task. The additional tasks are referred to as secondary tasks. One other convention is in relation to the use of the terms 'loading' and 'subsidiary' task. If the subject is instructed to aim for error-free performance on the secondary task at the expense of the primary task the second task is called a loading task (a point of some confusion here is that the primary task in this condition is only primary to the experimenter. To the subject it can appear as the secondary task). If the subject is instructed to avoid making errors on the primary task the secondary task is called a subsidiary task.

In relation to the role of the secondary task in the loading task and subsidiary task strategies Knowles (1963) made the following points.

'The first use of the secondary task (the loading task case) is to compensate for any deficiency in the loading of the primary task and to simulate aspects of the total job that may be missing. Ordinarily there is little interest in the secondary task performance *per se*. The secondary task is used to bring pressure on the primary task with the idea that as the operator becomes more heavily stressed his performance on difficult tasks will deteriorate more than will his performance on easy tasks. In this first application of second task then, the emphasis is upon stressing the primary task. . . .

'In the second application of secondary tasks as subsidiary tasks, the task is used not so much with the intention of stressing the primary task as with the intention of finding out how much additional work the operator can undertake whilst still performing the primary task to meet the system criteria.'

It must be noted at this point that not all experimenters have found it necessary to employ one or other of the above rationales. There are experiments in which subjects have been instructed to do as well as possible on both tasks. Equally, not all experimenters have found it necessary to record what instructions were actually given to their subjects.

2. The application of secondary tasks

2.1. *The evaluation of equipment and procedures*

In applied research the experimenter often wishes to measure the relative merits of a number of differing methods of performing the same task. Another problem is that of attempting to discover the differences in work load various phases of an operation impose upon an operator. In such situations the inclusion of a secondary task has been shown to be capable of indicating differences between the conditions under investigation, which were not apparent when the task variants were performed alone.

Garvey and Knowles (1954) used a mental arithmetic task to load subjects' performance in a study which compared six different methods of presenting one hundred discrete visual stimuli and their associated response buttons. Garvey and Henson (1958) studied the effect of variations in display sensitivity on the performance of a compensatory tracking task, including in their investigations trials in which the subject simultaneously performed a second visual task or a mental arithmetic task. Garvey and Taylor (1959) used similar secondary tasks in a comparison of subjects' performance on a compensatory tracking task when the subjects had control via an acceleration-aided, acceleration or position control system. In all the above experiments the loading tasks were successful in as much as they increased the performance differential between the alternative task arrangements, leaving the more efficient variant relatively unaffected and causing a deterioration in performance on the least efficient tasks.

Walker *et al.* (1963) made use of an auditory shadowing task to increase the performance load on a subject tracking with an acceleration control system in which the response lag could be varied. The presence of the loading task increased the effect of the lag, bringing about a marked deterioration in performance.

Rolfe (1963) used a second task to load subjects in an experiment which compared four visual displays for the speed and accuracy with which they could be read. The second task, a continuous compensatory tracking task, proved effective in maintaining the load on subjects between readings forcing them to keep their reading times as short as possible.

Dougherty *et al.* (1964) used a loading task in the comparison of two aircraft instrument display systems: a conventional instrument panel and a pictorial display generated on a cathode-ray tube. As they performed a standard flight profile in a simulator, subjects were required to read out digits which appeared on a separate display. The results obtained indicated no significant differences in performance on the two display configurations when the digit reading task was absent or at a low rate of presentation. However, as the digit presentation rate was increased performance on the pictorial display remained stable, while errors on the conventional panel increased.

The experiments described above sought to induce, by means of secondary tasks, changes in the level of primary task performance indicative of the effect of increasing the operator's work load. However as was indicated earlier what may be required is some indication of the amount of extra work the operator can undertake whilst maintaining a proficient level of performance on the primary task.

Day (1953b) employed a two-choice light cancelling subsidiary task to examine the effect of changes in the size of the target area on the performance of a tracking test. The experiment was unable to demonstrate any significant difference in secondary task response times which could be related to changes in primary task difficulty. This study suffered from the major drawback that the author failed to present any examination of the effect of the secondary task on the tracking task. Had he done so he may well have shown that, despite instructions, it was the primary task which suffered a deterioration in performance due to the presence of the secondary task. This view is supported by an earlier experiment by Day (1953a) which used the same apparatus and which showed that if the difficulty of the primary task was kept constant and the difficulty of the secondary task was varied in terms of signal frequency, secondary task performance was unaffected by these changes. However, primary task performance decreased significantly in a linear manner related to the increased frequency of signals on the secondary task.

137

Poulton (1958) employed a subsidiary task in an experiment to measure the order of difficulty of two dial watching situations in which the subject was required to respond each time the pointer reached certain points. One task involved watching two dials and the other six dials. The total rate of response required was the same in both cases. The second audio task involved the use of short term memory in a checking test. The results showed that when dial watching was combined with the auditory task, watching six dials gave significantly more errors on the auditory task than when watching two dials.

Benson *et al.* (1965) used a subsidiary light acknowledging task to compare the effectiveness of two display configurations in a compensatory tracking task. The displays were one which was purely numerical and one in which the numerical display was supplemented by a scale and pointer indication. Whilst compensatory tracking performance was unaffected by the display used, secondary task performance was significantly affected. Subjects missed significantly more secondary task signals when using the purely numerical display than when using the numeric plus scale and pointer display. Primary task performance was degraded by the presence of the secondary task. Physiological measures taken during the experiment indicated that the presence of the secondary task increased the level of the physiological responses thus implying changes in the level of the subjects' state of arousal.

Rolfe (1966) used the same experimental situation as that described above to examine performance in a task requiring the subject to change the indicated value on the display to another demanded by the experimenter. The presence of the secondary task increased setting times and the number of errors made on the displays. The increased times and errors were not significantly different between the displays but the type of error made was. On the purely numerical display overshoot errors predominated; whilst on the numerical display plus scale and pointer, levelling early was the most frequent mistake.

Allnutt *et al.* (1966) compared compensatory tracking performance with either a scale and pointer display or a numerical display when the subject was provided with an acceleration order control. A subsidiary task was present in the form of a peripheral light detection task. The results obtained indicated that the performance on the numerical display was more susceptible to disturbance than that obtained on the scale and pointer display.

In a later experiment Rolfe (1969) examined the pattern of secondary task response occurring during individual setting operations. The results showed that increases in response time on the secondary task could be related to covert decision making as well as motor response in relation to performance of the setting task. Again physiological measures were recorded and these indicated similar patterns of response variation to that obtained with the secondary task. Level of response again showed an increase when the secondary task was added to the task situation.

Kalsbeek (1970) described two evaluations employing secondary tasks. In one, two forms of aircraft engine instrument, circular or vertical scale and pointer, were compared. In the other pilots flying a four-engined commercial aircraft simulator were studied whilst making instrument landing approaches with varying degrees of difficulty introduced into the flying task.

Not all experiments have utilized the loading or subsidiary task rationales. In some cases no priority was given to either task, the subject being instructed to achieve the best possible performance on both tasks.

Olson (1963) used a second visual task to increase the realism of an experimental investigation of the effect of different instrument panel layouts and information presentation rates upon performance. As Figure 2 shows not only did the different arrangements affect performance on the head-down instrument monitoring task, they also affected performance on an independent head-up pursuit tracking task, simulating vehicle steering by external cues. Whilst some part of the result could be attributed to changes in the relative position of the two tasks in the subjects' visual field, brought about by the different instrument panel arrangements, the author noted that his results indicated that the more data sources using the same sensory channel, the less information the operator can be expected to handle.

The above experiments demonstrate the value of the secondary task as a means of comparing alternative items of equipment. In the context of comparing various task conditions the secondary task has also been of utility.

Brown and Poulton (1961) attempted to measure the spare capacity of car drivers by giving an auditory subsidiary task to the driver to perform whenever he could. The experimenters chose two sections of road in the City of Cambridge where there was a difference in traffic density; residential areas with light traffic and shopping areas with heavy traffic. It was hypothesized that the mental load on the drivers would be greater in heavy traffic, so spare capacity to deal with the auditory task would be smaller there than in the light traffic and in consequence performance on the secondary auditory task would show a decrement. As predicted, spare capacity was significantly smaller in the condition of higher mental load; i.e. when driving in shopping areas. The authors pointed out that the validity of the technique depended upon there being no effect of the subsidiary task on driving. The only check on this in the first experiment was to score the average speeds in the two conditions. There was no significant change in this score.

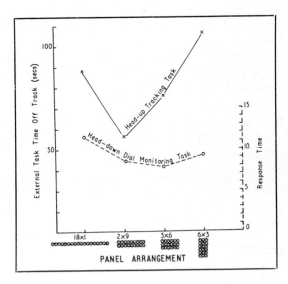

Figure 2. Olson's experiment demonstrating the effect of panel arrangement on the performance of two simultaneous tasks.

To test the value of a secondary task as an indicant of fatigue, Brown (1962a) carried out an experiment using as subjects police drivers from the City of Cambridge. The procedure was the same as in the previous experiment and the drivers were tested at the beginning and end of a spell of duty. All subjects were tested in the afternoon when one group of drivers was coming off duty and when the second group of drivers was about to commence duty. Performance on the auditory task whilst driving was worse for the subjects who were about to commence their duty than it was for the drivers who had just finished. The author pointed out that there were several explanations for this pattern of results. One of these was related to the number of hours before the test the subjects had been awake. A measure of this factor revealed that the men from both groups got up about the same time of day irrespective of whether they were to begin work at 8 a.m. or 4 p.m. Those subjects who were starting duty appeared to fill the day with activities such as gardening and decorating. The author suggested that it was possible that the general fatigue from these off duty activities transferred to driving with much greater effect than that produced by the driving itself.

Brown (1965) when comparing two alternative forms of subsidiary task for their effectiveness as indicants of driver fatigue found significant increases in the time to complete the test driving circuit when the additional tasks were present. Brown (1966) used a subsidiary task to study the performance of trainee public service vehicle drivers. It was found that the 'reserve capacity' of the ultimately successful trainees measured, in terms of secondary task performance, was significantly higher early in training than that of the candidates who later failed the course. However, there was a consistent tendency for concurrent performance on the subsidiary task to produce increases in the subjects' mean driving time and steering wheel activity.

Brown *et al.* (1967) employed an interval presentation task (Michon 1966) and found that its presence significantly increased steering wheel activity.

In a later experiment Brown *et al.* (1969) applied the knowledge of secondary task techniques to an investigation of the interfering effects of concurrently using a radiotelephone whilst driving. Using gaps of varying width as tests of driving skill it was concluded that the presence of the second task produced a relaxation of criteria and an impairment of perception indicating that central rather than peripheral mechanisms were being interfered with.

In a different applied situation Ekstrom (1962) sought to measure pilot work load in the X.15 rocket aircraft simulator by having the operators perform an additional self-paced push-button light cancelling task. The push-button scores were converted into an operator loading index to demonstrate differences in the primary task difficulty. The result of the evaluation showed that the pilot could complete all the various phases of the flight profile, but that a more automatic system demanded less of the pilot's attention, particularly during the highly critical release from the parent aircraft, after engine burn-out and re-entry into the atmosphere.

Knowles and Rose (1962) evaluated the ability of a two man crew to effect simulated lunar landings. A secondary light cancelling task was employed and the scores on this task were converted into a measure of percentage workload. It was found that, with practice, acceptable landings were achieved fairly consistently, but that the general level of operator load for the two men was severe. Moreover, as Figure 3 shows, secondary task performance revealed differences in work load between members of the two man crew and showed that the particular control mode under consideration was unsatisfactory because of the extreme build-up of operator load during the last few seconds of the landing.

Figure 3. Knowles and Rose estimation of operator load in a simulated lunar landing task.

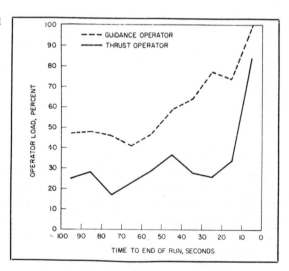

2.2. *The investigation of operator learning*

It is often argued that with practice the perceptual and attentional demands of the task are reduced and in consequence less of the operator's capacity is taken up by the task. Unfortunately, such a belief cannot be supported from the conventional learning experiment, for all that is demonstrated in this situation is that performance of the task to be learned improves. However, if a secondary task is introduced into the experiment, it should be possible to demonstrate the reduction in the primary task's demand by showing an accompanying increase in the level of performance on the secondary task.

Bahrick *et al.* (1954) conducted an experiment to investigate the value of extra-task performance as a measure of learning a primary task. It was reasoned that proprioceptive and tactile stimulation associated with one motor response might come to elicit the next response when repetitive sequences were learned. This kind of motor sequence learning would permit subjects to respond more effectively to other concurrent stimulation and would, therefore, be revealed in improved extra-task performance. To test this hypothesis subjects were trained either on a repetitive or a random version of a motor task. Concurrent performance on an extra task in the form of an auditory arithmetical subtraction task was required either early or late in practice on both versions of the motor tasks. The results showed arithmetic performance was comparable for the random and repetitive groups if the secondary task occurred early in practice. However, arithmetic scores were superior for the repetitive group when the second task was added late in practice. The authors concluded that the results indicated extra-task performance on a mental arithmetic task could be used to measure an aspect of learning on a visual motor task which was not reflected by the scores of the motor task itself, and that this aspect of learning was dependent upon the repetitiveness of the task.

The change in behaviour which takes place during the learning of repetitive tasks has been labelled automatization. Broadbent (1958) cited experiments which demonstrated that some tasks with practice become less dependent on the intake of information from the outside world. His interpretation was that 'in a sequence which is often repeated in the same order, the first events convey as much information as the entire sequence, and the other input events need not occur for the output to remain satisfactory. The other events become, in the language of information theory, redundant. So if a stimulus side of a task contains redundancy the worker will after practice still perform the task while taking in fewer events. Because of this phenomenon, the capacity required for efficient performance will be reduced, and neural mechanism will be set free for the performance of other tasks'.

Bahrick and Shelly (1958) examined the value of a secondary task to provide an index of automatization in a time sharing situation. During prolonged practice of a repetitive task they attempted to relate the degree of redundancy of a task to the degree of automatization that could be reached during extended practice. In the experiment concurrent performance of a visual and an auditory serial reaction task was measured at three stages of practice on the visual task. There were four versions of the visual task differing in degree of redundancy of the stimulus sequence. The degree of redundancy of the visual task had no effect upon performance when the task was practised alone, all versions of the task being performed at an equally high level of proficiency. On those occasions when the secondary auditory task was introduced the greatest decrement in performance took place on the random version of the task and progressively reduced through the lower redundant version and higher redundant version to the lowest decrement taking place on the repetitive version of the visual task. Performance on the auditory second task with all four conditions of the visual primary task was not significantly different. The authors concluded that the results supported the hypothesis that the redundancy of stimulus sequences permitted a change from extroceptive control of responses to proprioceptive control and that performance in time sharing situations provided a useful index of this process of automatization.

Can a point be reached where, through practice, the primary task has become so completely automatized that a second task can be performed without interference or deterioration taking place on either task? Baker et al. (1951) studied the interference of one task on another after varying degrees of practice on the first. The primary task was a complex coordination test which involved the learning of a motor skill. The interfering secondary task consisted of appropriate forward and backward movements of a lever in response to the appearance of a light. The experiment involved six matched groups of male students, and was divided into six stages. One group had the interfering task from Stage 1 onward; another from Stage 2 and so on. The interfering task produced an increase in the time necessary to make settings on the primary task. At each stage of learning there was an initial sharp increase in time score when the interfering task was introduced for the first time. Although the performance decrement on the primary task was reduced as practice with both tasks present continued, it never reached the level of the performance of those subjects who had not been given the interfering task.

Approaching the topic of learning and secondary task performance from a different basis, Garvey (1960) took a previously developed analogue computer model describing human operator behaviour in the learning situation (Garvey and Mitnick 1957) and attempted to extend its application to a situation in which the operator was forced to perform secondary tasks. The primary task took the form of compensatory tracking and three alternative secondary tasks were employed; namely, mental arithmetic, a visual detection task, or a five choice response task. No differentiation between the effects of the various secondary tasks was reported but all three brought about a deterioration in the quality of tracking task performance. Describing the differential effects of learning and secondary task performance Garvey stated: 'In general the results of the study indicate that the type of mechanism which may be substituted to provide performance analogous to the human operator differs as a function of the amount of practice the subject has had with the system. At the beginning of training the subject performs analogously to a one-integrator system. As training progresses the subject picks up an integrator, so to speak, and performs as a two-integrator system. When secondary tasks are added, however, the second integrator drops out leaving the performance characteristic of the man analogous to a one-integrator system'.

2.3. *The investigation of secondary task techniques*

Perhaps because it is an essentially applied technique there have been few recorded attempts to assess the value of different types of secondary tasks or different procedures. In the main, experimenters appear to have been satisfied if their particular method was successful. However a number of experiments do deserve noting in this context.

Schouten *et al.* (1962) did attempt an examination of the value of different secondary tasks in an experiment to investigate some of the problems involved in measuring perceptual and mental load. A standard task was used, to which was added a number of alternative second tasks. In the standard task, two tones were presented in random order to the subject who responded by pressing the appropriate foot pedal. This arrangement was chosen in order to leave the subject's eyes and hands free to carry out a second task. The procedure was to obtain from each subject a measure of his maximum performance on the first and second tasks when carried out separately. These maximum performance values were then used to establish the subject's 100 per cent level. Subjects were instructed to carry out two tasks simultaneously and measurements were made to determine the degree of mutual impairment. All the second tasks (assembling nuts and bolts, inserting pins in holes, maze tracking, mental arithmetic and spontaneous writing) showed a general decrement in performance as a function of increasing performance on the first task. However, the degree of impairment was not uniform for all tasks. The experimenters commented in their discussion that even with maximum performance in the first task, a considerable performance could be obtained on a second task, and that it might be some indication of the operator's faculty for interweaving intermittent quasi-simultaneous tasks.

Kalsbeek (1964) examined the effect on a primary task of a secondary task the component parts of which, movement, positioning and choice, could be separated. When the secondary task required only movement, or movement and positioning, the effect on the primary task was mainly one of slower performance. However, when choice was also part of the secondary task errors tended to occur on the primary task.

Kalsbeek sees the role of the secondary task as a stress which acts to distract the subject in the performance of the primary task. A second feature of Kalsbeek's work is his use of physiological measures (Kalsbeek and Ettema 1963, 1964 and Kalsbeek 1968), and the finding that heart rate variability (sinus arrhythmia) may be employed as an index of mental load.

Huddleston and Wilson (1969) assessed the value of four different subsidiary tasks as indicants of the effect of a lag introduced into the control dynamics of a simulated aircraft. All four secondary tasks involved the manipulation of visually presented numerical information. The results obtained indicated differences in the effectiveness of the four subsidiary tasks. Two of the tasks, checking if the sum of two numbers was odd or even or detecting the occurrence of the same digit separated by one other digit, e.g. 5,8,5, were able to indicate differences between the lagged and unlagged control condition, not apparent when the secondary task was absent. The two most effective secondary tasks in the above experiment demanded the use of short-term memory.

Other experimenters have also found that tasks including memory are affected by the presence of a second task. Murdock (1965) examined the effect of a secondary card-sorting task on the performance of a short term memory task involving the recall of lists of common English words. As the subsidiary task became more demanding so the number of words recalled decreased.

Michon (1966) examined the perceptual load imposed by a range of primary tasks, employing as his measure of load the regularity with which subjects could tap. He found that secondary task performance discriminated between primary tasks and indicated an order of difficulty which agreed with the subjects' rating of task difficulty.

Hilgendorf (1965) sought to discover if secondary task performance could be related to known variations in primary task information content. In four experiments using a fairly simple but quantifiable primary task it was found that a secondary 'attention' task was lawfully related to information input on the primary task. However, the presence of the secondary task interfered with primary task performance.

Glucksberg (1963) used a two task situation in which the secondary task could be presented via the same sensory channel as the primary task or via one of two different sensory channels. The primary task took the form of visual pursuit tracking. The secondary task could be varied for its content, simple or choice response, and for its mode of presentation, visual, auditory or cutaneous. The results obtained from a series of experiments showed the subject performance on the primary task was only impaired, in comparison with the performance of a control group which performed the primary task alone, when the secondary task was presented via the same sensory channel as the primary task. Whilst Glucksberg omitted any analysis of the secondary task performance, it would appear from examination of the data provided that performance on the secondary task was also influenced by sense mode in the two task conditions. In the two task condition all three modes of presentation showed increased response times in comparison with control groups with the increase in visual response times higher than the other sense modalities.

Another area in which some investigations have been undertaken is in relation to determining if the interference that accompanies a second task occurs as a function of stimulus or response. O'Hanlon and Schmidt (1964) compared the performance of two groups of subjects performing a visual vigilance task with a secondary vigilance task present. One group was instructed to respond to the second task and the other group was instructed to ignore it. Performance on the primary visual vigilance task was impaired by the presence of the second task, regardless of whether the second task was to be responded to or ignored.

Trumbo *et al.* (1967) attempted to determine the source of the interference between a primary visual tracking task and two forms of a secondary number anticipation task. The presence of the secondary task resulted in a marked interference of the tracking task particularly in relation to the timing of responses on the task. In addition the results indicated that one locus of the interference was in the response requirement of the secondary task. Thus, subjects required to anticipate stimulus items and subjects required to make the same responses, but in a free response-choice condition, showed the same loss of tracking proficiency.

In two further experiments Noble *et al.* (1967) sought to discover if the presence of a secondary task impaired the learning of the primary task. The results indicated that whilst a decrement in performance did occur when the secondary task was present this did not prevent an improvement in performance taking place. There was no support for an interpretation of the interference solely in terms of competition between peripheral responses. This was demonstrated by the fact that although one of the groups of subjects was required to make free responses at a rate in keeping with the anticipatory responses of the group performing the secondary task only the latter group's primary task performance deteriorated by a significant amount in the two task condition. The experimenters interpreted the above results as indicating that the source of the interference between the two tasks lay in the selection and implementation of responses.

The above interpretation favouring the response component of the tasks, is contradictory to that held by Triesman and Gaffen (1967). Their investigations led them to argue that the interference between primary and secondary task occurred at the perceptual level rather than the response. The conflict between interpretations may be one which arises because of differences between the tasks employed in the experiments and particularly in relation to the complexity of the response. When the response is of a discrete nature following a complex discrimination task the perceptual argument may be the case, but when the response is continuous, requiring the operator's attention in order to achieve the required end point, the demands of the response component may also make themselves apparent. The above arguments are also in keeping with those that suggest that the single channel mechanism is not only concerned with the making of decisions but also with monitoring responses (Welford 1967).

3. Discussion

If the above analysis of the results obtained from secondary task experiments is correct it would appear that the technique is a sensitive indicant of primary task demand when perceptual-motor activity makes up the demand. However, all is not well with the secondary task technique. It and its exponents are to be criticized for the following shortcomings.

- The extent to which the value of secondary task experiments have been degraded by poor experimental procedures.
- The near absence of any attempt to validate the technique and its findings outside the laboratory.
- The repeated finding that despite instructions the presence of a secondary task depresses performance on the primary task.

3.1. *Experimental shortcomings*

Whilst admitting that this is a highly applied field of investigation it must be noted that at times the experimental procedures employed have left much to be desired. For example, in the loading task situation subjects are instructed to attempt to perform the secondary task at the expense of the primary task. Consequently it might be expected that performance on the primary task will suffer in the two task situation, while the secondary task should be performed at the level of proficiency similar to that obtained when the task was performed on its own. This rationale was used by Garvey and Knowles (1954), yet they apparently did not measure secondary task performance in isolation in order to determine if, when performed in company with the primary task, its performance level altered. The experiment of Garvey and Henson (1958) lacked any analysis of the secondary task used, whilst that of Garvey and Taylor (1959) made only a brief and insufficient reference to the secondary tasks.

In the subsidiary task situation, it will be recalled that the subject is instructed to attend to the secondary task only when the primary task allows. Thus it would be expected that if the instructions were obeyed the level of performance on the primary task would be no different when the secondary task was added, but that performance on the secondary task would show a decrement in the two-task condition. The requirement for the successful use of the subsidiary task technique is a measure of both primary and secondary task performance level when performed separately. Yet Knowles and Rose (1962) and Ekstrom (1962) appear to have used this technique and made no attempt to discover if the presence of a secondary task did affect primary task performance, by using control runs in which the secondary task was absent. Similarly Michon (1966) makes no reference to any attempt to determine if performance of the interval production task influenced primary task performance.

In the present context reference must be made to Poulton (1958) as this experiment is often quoted as evidence that the presence of a secondary task need not degrade primary task response. The experiment showed no significant difference between the performance of an experimental group which performed the primary task with the secondary task and that of a control group who performed only the primary task. However, experimental and control groups were given entirely different amounts of practice before the scored runs. The experimental group's measured performance was preceded by two practices of about one hour each. The control group before each experimental period practised for up to five minutes on the relevant task condition. As both groups did produce comparable performance on the dial watching task it may be argued the control group reached a level of proficiency after five minutes' practice which was only reached by the experimental group, having the secondary task, after two hours' practice. Thus the presence of the secondary task did have an effect because it retarded learning.

3.2. *Lack of validation of secondary task results*

This attempt at a comprehensive review of secondary task studies contains only one programme of research in which the technique was employed in the 'real world', i.e. Brown's experiments at Cambridge. In consequence there is an absence of any indication that results obtained in the laboratory (the term includes the use of simulators) actually relate to performance in the actual environment of utilization. It may of course be argued, and in some cases correctly, that the secondary task is essentially a laboratory tool which acts as a substitute for the additional load encountered or anticipated in the actual situation. This being so there still remains the need to conduct follow-up trials to check that the laboratory results are correct. Again there appears to be an absence of any such trials being widely reported.

These omissions in methodology may mean that the secondary task has reliability as a measuring technique but no validity as a measure of the efficiency of equipment design or the perceptual motor load a potential system may be expected to impose on the operator.

3.3. *Interference with performance by the secondary task*

The two above criticisms are faults created by the users of the technique rather than the secondary task itself. This last criticism relates directly to the technique and its philosophy. The problem that besets the technique, particularly the subsidiary task method, is that, to be effective as a measure of the broader concept of work load, primary task performance should not alter as a result of the addition of the secondary task. The technique expects that a performance decrement will take place when the two tasks are performed concurrently but argues that by employing a particular strategy and by giving specific instructions it is intended that the degradation in performance should be confined to the secondary task. But the deterioration is not confined to the secondary task. Early studies in series of experiments have often claimed to be successful in this respect, but as the measuring techniques have been improved so an impairment in primary task performance has become apparent. An excellent example of this is the work of Brown (e.g. Brown and Poulton 1961 and Brown 1966).

A possible explanation for the decrement in primary task performance in contradiction to instructions may lie in the argument that as the secondary task increases the perceptual motor load imposed upon the subject it may be considered as acting as a task-induced stress. If this is the case then its effect on performance should conform to other situations, in which a stress is known to be present. Miller (1960, 1961) has listed seven particular mechanisms which typify an operator's reaction to a stress situation, including the condition of sensory overload. The mechanisms are, omission, error, queueing, filtering, approximation, synchronizing multiple channels and escape. A number of the above categories of response may be seen to be present in the experiments reviewed here when the secondary task was present. Their occurrence could be taken to indicate that the two-task situation was stressful as the subjects' ability to handle information was being overloaded.

Regarding the additional task as a stress is not a new attitude, for Conrad (1965) has argued in terms of speed and load stress, while Kalsbeek (1964), Sterky and Eysenck (1965) and Eysenck and Thompson (1966) have made use of the idea of the secondary task as a distraction stress. However, if the actual title to be given the stressor is ignored the acceptance that the second task stresses the operator does provide a possible explanation for the deterioration in primary task response. In several studies of subjects' performance under fatigue (or prolonged work stress) a broadening of the criteria for acceptable performance was noted (Drew 1940, Bartlett 1948, Broadbent 1958). Buckner and McGrath (1961) also showed that increased work load engendered an increase in the limen for detectable error. It can therefore be suggested that the stress imposed by the presence of the secondary task changed the subjects' criteria of what constituted acceptable primary task performance. If that change was not appreciated, as has been argued in other situations, then the subjects would continue to be confident that they were actually doing as well in the two-task situation as they were when performing the primary task by itself. In passing it is worth noting that the above suggestions are a reminder that a statistically significant difference in performance need not be so perceptually.

Brown (1968) has also adopted a critical attitude towards the subsidiary task. He sees the technique as having shortcomings both as a measure of the influence on performance of changes in the subjects' level of arousal and as a measure of the difficulty of alternative tasks and items of equipment. In the former condition he argues that attempts to measure the result of changes in arousal, e.g. due to fatigue, will be confounded since the addition of a subsidiary task also affects arousal. In the comparison of different tasks he points out that the difficulty of a task can be due to a number of factors, e.g. information load, information speed, rate of responding, stimulus compatibility and memory load. An assumption that changes in the difficulty of the primary task due to any of these factors will have a comparable interaction with the subsidiary task are in Brown's opinion not justified.

Can the above shortcomings be overcome if the responsibility for attention sharing is in the subject's control and if the mechanisms of control are influenced by the presence of the secondary task? One possibility is to employ a technique where access to the secondary task is determined by the quality of primary task performance. In this scheme the secondary task is only presented to the subject when he is performing within given acceptable limits on the primary task, but the subject is rewarded for the quality of his response on the secondary task.

An alternative approach is the concept of the cross-adaptive task proposed by Kelly and Wargo (1966).

'A major problem of the secondary task technique is how to weight measurements of the two separate tasks, since improvement in one frequently goes with decrement in the other. The self-adjusting simulator technique makes it possible to standardize performance on the primary task by adjusting secondary task difficulty as a consequence of primary task performance. When primary task performance is above standard secondary task difficulty is increased; when the primary task performance is below standard secondary task difficulty is decreased. The desired performance level on the principal task is thus forced, while the adjusted difficulty of the secondary task in forcing standard performance indicates the workload of the man, or the bandwidth available for doing other tasks.'

The idea of the secondary task is not one to be dispensed with. However, the shortcomings of the technique should be recognized and if possible allowed for. Having shown that the secondary task brings about a deterioration in primary task performance does not necessarily mean that the technique is of no value. For example, when comparing two alternative primary tasks it may be the case that the same degree of deterioration takes place on both and thus parity of performance is maintained in the two-task condition. However, if in this situation the secondary task can differentiate between the two variants, then it can be argued that the technique is informative.

4.   Conclusion

The final word however must be that the secondary task is no substitute for competent and comprehensive measurement of primary task performance. The technique should always be looked upon as a means of gathering additional information rather than an easy way of obtaining primary information.

References

ADAMS J.A., 1964, Motor skills. *Annual Review of Psychology*, **15**, 243–285.
ADAMS J.A., and CHAMBERS R.W., 1962, Response to simultaneous stimulation of two sense modalities. *Journal of Experimental Psychology*, **63**, 198–206.
ADAMS J.A., and XHIGNESSE L.V., 1960, Some determinants of two-dimensional tracking behaviour. *Journal of Experimental Psychology*, **60**, 391–403.

ALLNUTT M.F., CLIFFORD A.C., and ROLFE J.M., 1966, Dynamic digital displays: a study of compensatory tracking with an acceleration order control. *Royal Air Force Institute of Aviation Medicine Report No.* 374.

BAHRICK H.P., NOBLE M., and FITTS P.M., 1954, Extra-task performance as a measure of learning a primary task. *Journal of Experimental Psychology*, **48**, 298–302.

BAHRICK H.P., and SHELLY C., 1958, Time sharing as an index of automization. *Journal of Experimental Psychology*, **56**, 288–293.

BAKER K.E., WYLIE R.C., and GAGNE R.M., 1951, The effects of an interfering task on the learning of a complex motor skill. *Journal of Experimental Psychology*, **41**, 1–9.

BARTLETT F.C., 1948, The measurement of human skill. *Occupational Psychology*, **22**, 31–38.

BENSON A.J., HUDDLESTON H.F., and ROLFE J.M., 1965, A psycho-physiological study of compensatory tracking on a digital display. *Human Factors*, **7**, 457–472.

BORNEMANN E., 1942, Untersuchungen über den Grad der geistigen Beanspruchung. *Arbeitsphysiologie*, **12**, 142.

BROADBENT D.E., 1957, A mechanical model for human attention and immediate memory. *Psychological Review*, **64**, 205–215.

BROADBENT D.E., 1958, *Perception and Communication* (London: Pergamon Press).

BROWN I.D., 1962a, Measuring the spare 'mental capacity' of car drivers by a subsidiary auditory task. *Ergonomics*, **5**, 247–250.

BROWN I.D., 1962b, Studies of component movements, consistency and spare capacity of car drivers. *Annals of Occupational Hygiene*, **5**, 131–143.

BROWN I.D., 1964, The measurement of perceptual load and reserve capacity. *Transactions of the Association of Industrial Medical Officers*, **14**, 44–49.

BROWN I.D., 1965, A comparison of two subsidiary tasks used to measure fatigue in car drivers. *Ergonomics*, **8**, 467–473.

BROWN I.D., 1966, Subjective and objective comparisons of successful and unsuccessful trainee drivers. *Ergonomics*, **9**, 49–56.

BROWN I.D., 1968, Criticisms of time-sharing techniques for the measurement of perceptual-motor difficulty. *Paper to International Congress of Applied Psychology, Amsterdam.*

BROWN I.D., and POULTON E.C., 1961, Measuring the spare 'mental capacity' of car drivers by a subsidiary task. *Ergonomics*, **4**, 35–40.

BROWN I.D., SIMMONDS D.C.V., and TICKNER A.H., 1967, Measurement of control skills, vigilance, and performance of a subsidiary task during 12 hours of car driving. *Ergonomics*, **10**, 665–673.

BROWN I.D., TICKNER A.H., and SIMMONDS D.C.V., 1969, Interference between concurrent tasks of driving and telephoning. *Journal of Applied Psychology*, **53**, 419–424.

BUCKNER D.N., and McGRATH J.J., 1961, A comparison of performances on single and dual sensory mode vigilance tasks. Human factors problems in air sea warfare. *Tech. Report No. 8. Human Factors Research Inc.: Los Angeles.*

CONRAD R., 1955, Some effects on performance of changes in perceptual load. *Journal of Experimental Psychology*, **49**, 313–322.

CRAIK K.H.W., 1948, Theory of the human operator in control systems. *British Journal of Psychology*, **38**, 142.

DAVIS R., 1956, The limit of the 'psychological refractory period'. *Quarterly Journal of Experimental Psychology*, **8**, 13–38.

DAVIS R., 1957, The human operator as a single channel information system. *Quarterly Journal of Experimental Psychology*, **9**, 119–129.

DAVIS R., 1959, The role of 'attention' in the psychological refractory period. *Quarterly Journal of Experimental Psychology*, **11**, 211–220.

DAVIS R., 1962, Choice reaction times and the theory of intermittency in human performance. *Quarterly Journal of Experimental Psychology*, **14**, 157–166.

DAY R.H., 1953a, Reaction times during a difficult tracking task. *Flying Personnel Research Committee Report No.* 845.

DAY R.H., 1953b, The effect on a difficult co-ordination task of the frequency of signals. *Journal of Experimental Psychology*, **5**, 159–165.

DREW G.C., 1940, An experimental study of mental fatigue. *Flying Personnel Research Committee Report No.* 227.

DOUGHERTY D.J., EMERY J.H., and CURTIN J.G., 1964, Comparison of perceptual workload in flying standard instrumentation and the contact analog vertical display. *Janair Tech. Report D.228-421-019: Bell Helicopter Co., Forth Worth, Texas.*

EKSTROM P.J., 1962, Analysis of pilot work loads in flight control systems with different degrees of automation. *Paper presented at the I.R.E. International Congress on Human Factors Engineering in Electronics, Long Beach, California.*

EYSENCK H.J., and THOMPSON W., 1966, The effects of distraction on pursuit rotor learning, performance and reminiscence. *British Journal of Psychology*, **57**, 99–106.

GARVEY W.D., 1960, A comparison of the effects of training and secondary tasks on tracking behaviour. *Journal of Applied Psychology*, **44**, 370–375.

GARVEY W.D., and HENSON J.B., 1958, Interactions between display gain and task-induced stress in manual tracking systems. *U.S. Naval Research Laboratory, Washington, N.R.L. Report* 5204.

GARVEY W.D., and KNOWLES W.B., 1954, Response time patterns associated with various display-control relationships. *Journal of Experimental Psychology*, **47**, 315–322.

GARVEY W.D., and MITNICK L.L., 1957, An analysis of tracking behaviour in terms of lead lag errors. *Journal of Experimental Psychology*, **53**, 372–378.

GARVEY W.D., and TAYLOR F.V., 1959, Interactions among operator variables, system dynamics, and task-induced stress. *Journal of Applied Psychology*, **43**, 79–85.

GLUCKSBERG S., 1963, Rotary pursuit tracking with divided attention to cutaneous visual and auditory signals. *Journal of Engineering Psychology*, **2**, 119–125.

147

HICK W.E., 1949, Reaction time for the amendment of a response. *Quarterly Journal of Experimental Psychology*, **1**, 175.

HILGENDORF E., 1965, The indirect measurement of task difficulty. *Australian Department of Supply Report* ARL/HE4.

HUDDLESTON H.F., and WILSON R.V., 1969, An evaluation of the usefulness of four secondary tasks in assessing the effect of a lag in simulated aircraft dynamics. *Royal Air Force Institute of Aviation Medicine Report No.* 479.

KALSBEEK J.W.H., 1964, On the measurement of deterioration in performance caused by distraction stress. *Ergonomics*, **7**, 187–195.

KALSBEEK J.W.H., 1968, Measurement of mental work load and of acceptable load; possible applications to industry. *International Journal of Production Research*, **7**, 33–45.

KALSBEEK J.W.H., 1970, Objective measurement of mental workload: possible applications to the flight task. In *Problems of the Cockpit Environment*. Advisory Group for Aerospace Research and Development AGARD CP No 55.

KALSBEEK J.W.H., and ETTEMA J.H., 1963, Continuous recording of heart rate and the measurement of mental load. *Ergonomics*, **6**, 306.

KALSBEEK J.W.H., and ETTEMA J.H., 1964, Physiological and psychological evaluation of distraction stress. *Communication at the 2nd International Ergonomics Conference, Dortmund.*

KELLY C.R., and WARGO M.J., 1966, Cross-adaptive operator loading tasks. *Human Factors*, **9**, 395–404.

KNOWLES W.B., 1963, Operator loading tasks. *Human Factors*, **5**, 151–161.

KNOWLES W.B., and ROSE D.J., 1962, Manned lunar landing simulations. *Hughes Aircraft Company, California. Tech. Memo.* T.M. 728.

KREEZER G.L., HILL J.O., and MANNING W., 1954, Attention. *WADC. Technical Report*, 54–155.

LEHMAN A., 1901, *Die physichen Äquivalente der Bewusstseinserscheinungen* (Leipzig: O.R. Reisland).

MICHON J.A., 1966, Tapping regularity as a measure of perceptual motor load. *Ergonomics*, **9**, 401–412.

MILLER J.G., 1960, Input overload and psycho-pathology. *American Journal of Psychiatry*, **116**, 695–706.

MILLER J.G., 1961, Sensory overloading. In *Psychophysiological Aspects of Space Flight* (Edited by B.E. FLAHERTY) (Columbia University Press).

MURDOCK B.B., 1965, Effects on a subsidiary task on short-term memory. *Journal of Applied Psychology*, **56**, 413–419.

NOBLE M., TRUMBO D., and FOWLER F., 1967, Further evidence on secondary task interference in tracking. *Journal of Experimental Psychology*, **73**, 146–149.

O'HANLON J., and SCHMIDT E.A., 1964, The effect of the level of vigilance of an adjacent secondary vigilance task. *Tech. Report No.* 750–4. *Human Factors Research Inc., Los Angeles.*

OLSON P.L., 1963, Variables influencing operator information processing. *Human Factors*, **5**, 109–116.

POULTON E.C., 1958, Measuring the order of difficulty of visual-motor tasks. *Ergonomics*, **2**, 234–239.

POULTON E.C., 1965, On increasing the sensitivity of measures of performance. *Ergonomics*, **8**, 69–76.

ROLFE J.M., 1963, The evaluation of a counter-pointer altimeter display for the United Kingdom altimeter committee. *Royal Air Force Institute of Aviation Medicine Report No.* 253.

ROLFE J.M., 1966, A study of setting performance on a digital display. *Royal Air Force Institute of Aviation Medicine Report No.* 365.

ROLFE J.M., 1969, An evaluation of the effectiveness of a secondary task using psychological and physiological measures. *Royal Air Force Institute of Aviation Medicine Report No.* 473.

SCHOUTEN J.F., KALSBEEK J.W.H., and LEOPOLD F.F., 1962, On the evaluation of perceptual and mental load. *Ergonomics*, **5**, 251–260.

SENDERS J.W., 1966, The estimation of pilot workload. In *The Human Operator and Aircraft and Missile Control*, AGARD.

STERKY K., and EYSENCK H.J., 1965, Pursuit rotor performance as a function of different degress of distraction. *Life Science*, **4**, 889–897.

TRIESMAN ANNE, and GAFFEN GINA, 1967, Selective attention: perception or response. *Quarterly Journal of Experimental Psychology*, **19**, 1–17.

TRUMBO D., NOBLE M., and SWINK J., 1967, Secondary task interference in the performance of tracking tasks. *Journal of Experimental Psychology*, **73**, 232–240.

WALKER N.K., DeSOCIO E., MOWBRAY H., and DURR L., 1963, The effect of a particular stress on one man's performance of various tracking tasks. *USA Meds Contract No.* DA-49-193-Md-2369, *Interim Report.*

WELFORD A.T., 1952, The 'psychological refractory period' and the timing of high speed performance—a review and a theory. *British Journal of Psychology*, **43**, 2–19.

WELFORD A.T., 1967, Single-channel operation in the brain. In *Attention and Performance* (Edited by A.F. SANDERS) (Amsterdam: North-Holland Pub. Co.).

WOODWORTH R.S., and SCHLOSBERG H., 1954, *Experimental Psychology* (London: Methuen).

Dr. Rabideau holds psychology degrees, and has varied academic and industrial experience. He has a long association with human engineering aspects of military and space systems, and now is a Senior Member of Technical Staff, Defense Systems Division, Bunker-Ramo Corporation, Fairborn, Ohio, U.S.A.

Observational techniques in systems research and development

G. F. Rabideau

1. Introduction

In the design and development of military systems, the human factors engineer uses observation as a source of design data and for the evaluation of system performance at various stages in its development. It is the purpose of this exposition to demonstrate how and why.

1.1. *Definition*

Observational techniques are defined in this discussion as approaches and procedures for abstracting and measuring (to the various extents possible in given situations) criterion aspects of human behaviour that are observable and/or inferable phenomena in 'real world' systems. In this context, then, observation and several other associated techniques that will be described all follow a non-interference principle which 'tells it like it is'. By this it is meant that independent variables, or conditions are not employed in the sense that an experimenter would so do in order to systematically measure their effect on selected dependent variables that measure in some ways specified human and/or system outputs, or behaviours. Thus, the observations that are made of human and/or system behaviour are based on a more-or-less 'in vivo' situation and it can be presumed that the contingencies that occur and the problems that are experienced are samples that are reflective of those that the system will encounter in replications of the same functional sequence.

One may wish to inquire at this point why the 'hands-off' rule. There are at least two reasons, one practical and the other somewhat theoretical. In engineering practice, the human factors specialist often is denied the opportunity to make certain experimental or test checks simply because of practical, or mandatory, constraints on budget and time, or the required facilities and equipment may be unavailable. On the theoretical side, there are circumstances when so little is known about the interaction between the input and output variables of a complex system that observation of pertinent human performance is required to define questions open to experiment. From the foregoing assertions, the reader could be led to conclude that observational techniques deal with extremely unstructured situations. On the contrary, as will be made clear, employment of observational techniques in system design and development typically involves clearly specified, unambiguous objectives and procedures. It is the system under development that may be somewhat tentative.

Another facet of observation also requires definition. This is the 'what' aspect of the techniques. At one extreme, the observer might not know what behavioural phenomena he could expect to see, as was the case in the contract studies of stressful situations done for the U.S. Army by Human Resources Research Office (1963); or, he might have at hand a complete description of the expected or required behaviour, perhaps as operator sequence diagrams (Kurke 1961).

It is appropriate, also, to define the engineering purposes of human behaviour observation. It may be stated that the objectives may be threefold: to observe the behaviour of man as a man–machine operator/maintainer and hence as a system component in a dynamic assemblage of people, equipment, environmental control and/or modification sub-systems in order to provide data useful for (1) establishing the nature of human performance characteristics in a given system context; (2) determining which system design alternatives show greatest promise; or (3) to verify that the developed system will be able to satisfy its mission performance requirements.

1.2. *Description*

Observation techniques are divisible into two major categories on the basis of their data sources. These may be designated 'direct' and 'indirect' observation, respectively. Since this differentiation is important to the validity of obtained data each category is described below.

Direct observation consists of all procedures for observing human behavioural phenomena that are directly collectable by the observer himself or by the observer from objective recordings of the subject performance. This form of observation may be characterized as requiring no interference with the ongoing observed processes, generally also involving the observer as the abstractor of the observed phenomena, i.e., he determines relevance of phenomena, recording detail to desired or prescribed levels, and providing his interpretations and inferences when the task calls for them. Direct observation may be performed in either real-time or non-real-time, that is, as the phenomena occur, or from video or audio recordings. The latter case enables time compression or expansion, as in industrial engineering memomotion or micromotion techniques (e.g. Mundel 1955). Real-time observation itself can be discontinuous, in the use of activity sampling or in recording behaviour only at important stages.

Indirect observation may or may not interfere with an ongoing behavioural sequence. It generally involves the solicitation of data concerning the observed behavioural phenomena from the observed operator, from his supervisor, or from other individuals who are present during the period of observation. All of these sources inevitably yield information that has been abstracted and subject to interpretations and inference that are not the observer's own. This could affect the validity of the information, unless procedures for its original acquisition are carefully delineated. Nevertheless, indirect observation, no matter what its source (reports, evaluations or critical comments from active system participants) can valuably augment direct observation in system evaluational processes. For example, if an observer notes a delay in the completion of a task he may not be able to deduce the reason for that delay. By questioning the operator, he might discover that the operator experienced some difficulty in deciding what a correct procedure should be, and the operator might attribute the difficulty to an unclear written procedure, or instruction.

Data obtainable from indirect observational sources generally may be described as discontinuous in character, but, as in the above example, this may be all the observer needs. A drawback to this approach prevails when the reporting of indirect observations is on a non-real-time basis, since an operator may forget, or at least err in his recall, with respect to details of a set of procedures that he had completed only an hour or two previously. Thus, the observer has to choose, in advance, between such data loss and the interruption to the operator caused by the alternative of real-time indirect observation.

Some operator behaviour in systems is of a discontinuous nature. Corrective maintenance activities are common examples. Consequently one can only naturally observe maintenance personnel performing troubleshooting and repair tasks when appropriate items of equipment exhibit failure symptoms. Evaluation activities of human factors engineers, however, do include contrived or simulated equipment malfunctions, but one would have to philosophically view these situations as experiments or tests. (Observation, both direct and indirect, might still be employed to obtain some of the test data.)

In some short-term, intermittent system operator tasks, near-real-time indirect observations are quite practical because one may arrange for the operator to call an observer/monitor or, alternatively to enter written observations on specified aspects of his own task performance in some kind of log.

What has been said thus may cause the reader to conclude that observation is a laborious and time-consuming technique. And this certainly has been true over the early years of its extensive utilization as a human factors tool in large military system development. Repeated observation of virtually every job or personnel system required large commitments of time, as did the reduction and analysis of the voluminous data obtained. There was, at that time, considerable room for improvement in techniques, some of which has since been accomplished.

Sampling of system behaviours is an obvious course to pursue in order to reduce the magnitude of observational task burdens. Sampling can be used to greatest advantage primarily when one's objective is to sample how an operator is allocating his time, when tasks are of short duration, or when one only needs to observe human behaviour involved in certain system mission segments which are determined to be most critically related to mission success.

2. Employment of observation in system development

2.1. *Relationship between stage of system development and purpose served by observation*

Shapero and Erickson (1961) divided system testing and evaluation into exploratory, resolution and verification.

The first of these, exploratory evaluation, is characteristic of much of the applied sciences experimental work that is devoted to establishing functional relationships between various independent and dependent variables associated with human performance capabilities and limitations, the feasibility of certain advance, or tentative system design concepts, etc.

Resolution, on the other hand, is much more limited in terms of variables of concern to the evaluator. This kind of evaluation simply seeks to predict which of two or more potential system design configurations will (if selected) result in a more effective system when measured against some selected mission criterion variables, against cost, development time, etc.

Verification is determining that the system, as finally designed, fabricated, assembled and exercised by means of simulated operational system missions, fulfills those functional performance requirements imposed by specified mission objectives. Because this is largely an 'either-or' process, all test results which reveal certain potential degradations in long-run system effectiveness have to be critically analysed to determine the gravity of the problems, costs of remedial action, and necessary corrective measures.

Meister and Rabideau (1965) developed a matrix (Table 1) to compare the important characteristics of these three 'test types'.

Table 1. Major differences among test types

| Characteristics | Types of tests | | |
	Exploratory	Resolution	Verification
Typically performed in	Predesign and early development	Early development	Throughout development
Control of independent variables	High	Moderate	Low
Number of measures recorded	Few	Few to many	Many
Repeatability of test conditions	High	Intermediate	Low
Number of conditions compared	Any reasonable number (e.g. factorial design)	Few (gross configuration differences)	One (comparison with performance standard)
Control over test environment	High	Moderate	Low
Number of dependent variables	Few	Few to many	Many
Factors initiating test	Ambiguity of system inputs/outputs	Need for design decision	Need to verify system adequacy
Part/system testing	Part	Part to sub-system	Sub-system to system
Resemblance to operational conditions	Low	Moderate to high	High

It can be seen from the table that control over independent variables goes from high to low as system development proceeds, while the number of measures recorded pursues exactly the opposite trend. Also, the resemblance of successive test situations to the eventual (or expected) operational conditions becomes greater as the total system is developed.

The utilization of observational techniques seems to vary over the three stages of system evaluation. Observation is probably most useful in system verification testing and least valuable in exploratory types of studies: mainly because of the poor control over independent variables and the large numbers of required system performance measurements that typify later system evaluation. This is not to exclude observation from exploration but merely to suggest that each experimenter considers himself an expert observer, following his own unwritten ground rules for the observations that he makes that (sometimes) lead to worthwhile, testable hypotheses. It may even be claimed that the experimenter employs *indirect* observation, to the extent that he accepts the observations, deductive conclusions, and/or inferences contained in reports prepared by other experimenters.

It is the large middle ground covered by resolution evaluation that is less clear-cut, for the continuing utilization of observational techniques. Two incontrovertible facts influence this situation. First, engineering resolution of design tradeoffs often allows only brief time between the identification of the alternatives and the choosing of one solution. Second, advanced computer technology has made it possible to devise dynamic simulations of subsystem design problems, and although human factors engineers have been able to perform ever-increasing amounts of man-in-the-loop simulations where appropriate, it has been difficult to accurately estimate and simulate potentially heavy task loads.

We should note also an important supporting role that observation can and should play in human factors experimental work during resolution. Some pertinent information does not lend itself well to easy detection, scoring or measurement in such simulations. Such behaviour may include discrete procedural errors; they will almost always involve subjective data such as experiences of difficulty and feelings of confidence. These items of information can and should be obtained by formal, indirect observation techniques.

2.2. *Dimensions of the technique*

Continuous detailed description of all behavioural phenomena may be required sometimes, but in any well-organized and effectively designed system, it should suffice to observe only operator errors, that is his deviations from prescribed procedures. (Such an approach would not, of course, be applicable to continuous tracking tasks, but only to behaviour which is 'discrete' or procedural in nature.)

With a check-list of the sequence of operator tasks and subtasks, required performance time characteristics (where pertinent) and a means of keeping track of elapsed time, the observer could be expected to detect and record at least the following: (1) performance of procedural steps out of the prescribed 'correct' sequence; (2) the making of an incorrect response; (3) the omission of a prescribed response; or (4) performance of responses out of coordination with a team. The observer would also note performance times and observable difficulties that the operator appeared to have as a function of configuration of displays and controls, workplace layout, and specific environmental factors. These latter aspects can be conveniently observed only when the observer is recording 'live' performance as opposed to abstracting his observations from some recording medium, e.g. motion picture film, of course.

It may be necessary to monitor auditory media which the system operators must use in the performance of their tasks, especially when important mission functions of the system are concerned with information acquisition, processing, and transmission; as would be the case with a 'manual' (as opposed to an automated) air defence system. Frequency and duration of communications circuit usage would be recorded for cross-telling, forward-telling and feedback of information, and message content analyses performed from either live transmissions or from tape recordings.

The dimensions of indirect observation are highly dependent upon the depth of analysis required. During post-test debriefings of operators the observer would seek to verify his observations concerning symptoms of difficulty, and, to determine whether the operator has any additional information to provide to supplement the objective observations. For example, the operator might make specific comments about the clarity and completeness of the written procedures he had used, and he could indicate whether the design characteristics of the equipment's information displays were satisfactory from the standpoint of ease and reliability of interpretation.

2.3. *Corollary techniques and aids*

A standard technique is the checklist of required operator procedures. The AC Spark Plug Division (1962) personnel subsystem test plan for the Titan II missile guidance system and the Space Technology Laboratories (1960) personnel subsystem test plan for the Atlas missile system both employed observer checklists, each checklist covering some system operator's task performance requirements for specified mission segments, e.g. receiving inspection, assembly, checkout, periodic inspection/maintenance, countdown and launch, etc.

Debriefing questionnaires are sometimes employed by the observer as a means of minimizing the time required for post-test interviews of operators. Checklist data and certain of the operators' subjective responses to questionnaires can direct interviews towards those tasks that were determined to cause the most critical problems.

Rating scales also are useful in at least two ways. They may be used to provide subjective estimates of the skills, motivational levels and cooperativeness of the personnel who are used as system operators in system verification tests. Such scales are often completed by the supervisors of the operators who are being observed. Marks (1961) describes these aids in a technical report. The other use of rating scales is in estimating the performance characteristics of the total system which is under development. These kinds of data are of potential significance to the assessment of cause of operator performance deficiencies that are detected during the individual test operations. This use of rating scales is discussed more fully by Meister and Rabideau (1965). It should be noted, however, that not all operators' errors influence system performance; many possible errors are counteracted by automatic (or human) checking devices.

2.4. *Observation as a function of system development stage*

Stages of system development in which observational techniques may be systematically and beneficially employed consist of: design; integration of system components and subsystems; and the total system verification of validation.

During system design, applicable observational techniques and procedures may include the verification of link analysis data that are developed by human engineers to check the relative effectiveness of alternative control-display configurations, panel layouts, and physical workspace arrangements and accesses. These simple tests often are performed with only crude, readily fabricated and assembled mockups of the proposed design features and parameters; more elegant materials and components are, of course, possible. Observations typically concern body and limb positions, reaches and accesses to particular (spaces for) components. The sequential nature of the 'links' between man–machine elements in such a work area simulation can be checked by having the test subject follow the task analysis descriptions of procedural sequences that are defined during the design process.

The observed phenomena may be manually recorded, or still photos or motion pictures may be used.

A principal objective of this application of observational procedures to data gathering is that of providing data that will support analyses of design problems and, thus, the eventual improvement, or even optimization, of the resultant design.

Some system design processes may permit the utilization of simulators to test the effectiveness of design characteristics of complex control-display subsystem dynamics. This was previously mentioned in this discussion. Observation in such cases is relegated to more subordinate roles such as providing direct observations which test conductors may make about the fidelity of the simulation and the probable validity of the obtained operator performance data. Also, the test subjects may provide additional indirect observational data concerning their subjective estimates of task difficulty, dynamic aspects of information feedbacks concerned, for instance, with performance of tracking tasks.

Two factors influence the significance of data obtained via such evaluational techniques during system design. First, the extent to which any of the subsystem 'models' represented in these tests will resemble the corresponding subsystems of the ultimate, operational system, including both procedures and hardware characteristics, is virtually impossible to predict. Second, the test subjects often are engineers, human factors personnel, or technicians. One expects that operational system personnel will differ in terms of specific aptitudes, educational backgrounds and training from these test subjects. Furthermore, during system operations, the operational crews will have the advantage of complete documentation of operating and maintenance procedures.

It is important to view the testing that is characteristic of the design stage as an iterative process, just as design itself often is. This means that evaluation, including the utilization of data gathered by means of observational techniques, should be repeated in cycles that correspond to the rates with which system design is refined and 'milestones' are reached.

There is a scheduling problem here (not to mention a budget problem for human factors work support on some programmes). The human factors evaluations must be so designed and scheduled that their results, design criteria, and recommendations are available to designers in time for required tradeoff decisions. This then is the strongest single argument that favours simple straightforward evaluation that makes a high degree of use of direct and indirect observation.

Integration may be defined as the assembly of an initial operational system model. During the integration stage the human factors engineer can make widespread use of observational techniques to ascertain if human engineering standards have been met. An associated human factors evaluation activity will consist of observing system interface tests (assembly or connection of major system equipment items) and the proof tests that demonstrate the performance capabilities of the system. In some tests, opportunity for direct observation may be drastically curtailed or completely barred by the nature of the system. For example, in the case of an aircraft system, test flights will only permit limited recording of inflight data germane to human factors evaluation, these generally being available on a post-flight basis. In such cases there is a substantial need for indirect observational data obtainable from the aircrew personnel who perform the flight tests. Here, then, the human factors specialist can design checklists and questionnaires for use in post-flight debriefing situations.

For these checklist activities, the observations are not of behavioural processes. Rather, the observer must identify and describe cases where the system design conflicts with the relevant section of the military human engineering standard. For instance, a maintenance access opening on an aircraft may be found to be smaller than standard dimensions. Even here, provided access were not too severely limited, the design might still be acceptable to the military customer. Such analyses would typically make use of rating scales for assessing severity of problem. Another useful technique could involve the employment of human performance reliability estimates on the basis of the design features constituting the problem situation. Central data stores on operator reliability are in existence. A description of these is provided by Altman (1965).

In the final, validation operations, the human factors engineer determines that none of the system design features, including equipment, procedures and people, will serve to significantly deter or degrade required task performance by the system's human components. It is difficult at best to relate minor deviations in operator task performance to system output variables in any valid fashion. More often than not, however, the directly and indirectly observed behavioural phenomena, which demonstrate errors and delays in the performance of tasks by system operators, create situations that make it difficult for operators to perform their assigned tasks. The observer-analyst then must make an assessment of the criticality of the probable long-run effects of such phenomena on other criterion variables than mission performance, adequacy and safety, for example cost-effectiveness of system operations, subsystem down time versus system availability, and system manpower demands upon pools of skilled personnel. In brief, the analysis of observed human performance errors and deficiencies in terms of relating these to meaningful criterion measures is a deductive process that resembles a detective story.

2.5. *Current state-of-the-techniques and problems*

The 1960's have seen a vast and rapid growth in the evaluational activities associated with the development of large systems, particularly in conjunction with military systems development in the U.S.A. When suddenly confronted with test requirements characteristic of the large strategic missile programmes mentioned earlier in this paper, the existing human factors groups were about as poorly-prepared as would have been the 'brass-instrument psychologist' of the early 1900's. Nevertheless, these pioneers, armed with checklists, stopwatches, and courage, valiantly stepped forward to meet the challenge. Keenan *et al.* (1965) document the state-of-the-art in this area for the first half of the 1960's.

This was followed by both individual development of advances in techniques, supporting equipment and aids, and group examination of common problems inherent in such evaluational activities. A recent example of the latter is the October, 1968 Human Factors Testing Conference sponsored by the Air Force Human Resources Laboratory at Wright-Patterson Air Force Base (see Kincaid and Protempa 1969). At that meeting, representatives of the Air Force, Army, Navy, NASA and private industry heard twenty presentations on different aspects of human factors test concepts, models, techniques, existing and recently conducted programmes, as well as problems involved in the management and conduct of tests.

Along the lines of individual developments several items are of note because they reflect the state-of-the-art (as opposed to technique). They concern observational data collection, recording of data, and analytic techniques. Each of these will be viewed briefly below.

Human factors specialists have long recognized the importance of sampling, or some form of discontinuous data collection. Sampling techniques themselves are not new; one could point to the pioneering human engineering work of Christensen (1950) in the late 1940's and to the development of the industrial engineering ratio-delay technique for measuring work activity levels (see Barnes 1958). The need in large-scale evaluation is for a means whereby a single observer can monitor the task progress of several system operators in real-time or near-real-time. Askren *et al.* (1968) described the development and initial evaluation of PAARS (Personnel Activity Analysis Radio System), a technique for data collection that is supported by voice-radio communication where the observer monitors the task performance of several system operators from a single, central console. Each operator has a remote, light-weight portable radio transceiver which he uses to report information concerning his task activities, problems, etc. This is, of course, an application of an indirect observation technique.

Tape recorders have seen extensive use in recording of field survey data concerning human factors aspects of tactical air operations problems. Rabideau and Ritchie (1968) report a critical incident survey involving interviews with 321 pilots and other U.S. Air Force personnel. However, one of the greatest disadvantages of real-time, continuous recording of behavioural phenomena whether by video or audio means, is the time required for abstraction of significant information. During the final interviews of the survey reported by Rabideau and Ritchie (1968) a tryout was given to intermittent use of the recorder, e.g. only recording question and response when a problem was reported by the interviewee. This technique has not been subjected to a validation as yet, however.

Berridge reported in Kincaid and Protempa (1969) the development of DACOLS, or Data Collection System at the Air Force Armament Development and Test Center. This device can permit recording of the elapsed time involved in multiple numbers of events such as operator tasks. Activities are coded and recorded on magnetic tape, and a radio time base can supply synchronization between test director and remote data collectors. Because the time and activity data are digitally-recorded they may be readily transferred to standard computer input tape. Berridge described a data reduction programme which provides line by line printout of data frames, including durations of activities as a function of the DACOL's switch button inputs.

Data analysis has not yet made full use of computer-assisted (or performed) statistical analyses. Some of the difficulties have been mentioned above: complex relationships between human performance errors and the large number of system performance criteria; changes in system personnel and procedures; and changes in hardware.

Perhaps the human factors man should relinquish, for the present, this goal of predicting the contribution of the human engineering to the overall performance of the system. He could then concentrate his efforts on verifying the adequacy of manning, training, procedures and equipment design for certain critical mission segments. Cotterman and Wood (1967), for example, describe a statistical approach concerned with pilot reliability for a single mission segment, based upon performance data obtained in a lunar landing simulator. Again, space does not allow a detailed description of the technique but it appears to offer much potential in situations where the most critical tasks in a given mission segment are of a continuous, or tracking nature. However, we see nothing in prospect at present for a similar statistical analysis of the relationship of specific discrete tasks, or procedures, to mission success or operator performance reliability prediction, at least not in the context of system performance verifications—even by constraining operations to be specified mission segments.

Table 2. Advantages and disadvantages of instrumentation and manual recording methods

	Advantages	Disadvantages
Instrumentation Recorder	1. In many applications has greater sensitivity and accuracy.	1. Often highly specialized.
	2. Can measure things human cannot; not limited to data received by senses.	2. May be costly, bulky and difficult to transport and install.
	3. Avoids human error and data bias.	3. Subject to catastrophic failure.
	4. Permits repetitive playback of data under same or varied conditions.	4. Requires continuing maintenance.
	5. Provides rapid, on-line data processing and recording.	5. Cannot readily react to changing evaluation conditions.
Human Recorder	1. Can perform certain evaluations (e.g. ratings) the machine cannot.	1. Subject to error and bias (as a result of data interpretation).
	2. Can interpret as he records.	2. Often less sensitive than machine.
	3. Often less expensive than equipment.	3. Must be trained to perform recording task.
	4. Needs little or no maintenance.	4. Generally slower at data recording than instrumentation devices; easily overloads.
	5. Comparatively compact.	5. Handles only peripheral (sensed) data.
	6. Adapts flexibly to changing evaluation conditions.	6. Possibility of 'Hawthorne effect'.
	7. Does not fail catastrophically.	
	8. Comparatively mobile.	
	9. Little or no interference with on-going system operations.	

3. Summary

An attempt has been made to identify and describe possible objectives which observational techniques may satisfy, in evaluating man–machine systems. The techniques have been discussed in the context of military system design and development, since it is there that they have had their most extensive employment, but they can be applied usefully to human performance in any 'system', government-sponsored, commercial or others.

What the human factors specialist is confronted with in an evaluational situation such as that referred to in the above paragraph is a series of tradeoffs he must make that will serve to define his data gathering and recording techniques. A comparison of the advantages and disadvantages of instrumentation and manual data acquisition methods, borrowed from Meister and Rabideau (1965) is presented in Table 2.

In a paper contained in the testing conference documented by Kincaid and Protempa (1969), Cyrus Crites listed and described criteria that are appropriate as a conclusion to this paper. We would suggest that the development of improved observational techniques be assessed by how far they will achieve the following aims.

- Objectively measure human performance; this should result in the expression of phenomena in terms of time, distance and task difficulty.
- Provide data that are useful to a system design effort. Obtained and reported information should be germane to specific system test objectives and in a format that is meaningful to engineers who compile and/or interpret test reports.
- Be capable of utilization during test activities. Data collection procedures must be such as to permit data collection at the work sites of individual system operators without delaying scheduled test operations.
- Produce minimum interference with test activities thereby assuring valid information about human task performance.
- Possess the potential for widespread use: techniques should be easily modifiable for use in a variety of physical and functional settings.
- Possess characteristics required of a test 'tool'. Orientation of the technique should be toward the definition of operational problems rather than toward research and development problems.
- Be usable even by personnel who lack extensive training: training for use must be simple and clearly identified.
- Generate data that can be utilized by System Programme Offices, contractors, Air Training Command, and Air Force Operational Commands. The information requirements of each of these agencies and organizations should be capable of being satisfied by data gained by the use of the technique.
- Be adaptable to fast reaction situations. Preliminary steps necessary to starting data collection should be brief, e.g. so as to maximize the collection of unscheduled maintenance task performance data, following unexpected system breakdown.
- Provide results that have operational significance. Techniques should be able to support the determination of time required for launch, for vehicle turnaround, to reach operational ready status, for loading of weapons, etc.

References

AC SPARK PLUG DIVISION, 1962, Personnel subsystem test and evaluation plan, Titan II inertial guidance system. *AFBM Exhibit* 60–20*A*.

ALTMAN J.E., 1965, *Human Error Analysis*. Paper presented at the Quality Control Conference, Pittsburgh, October.

ASKREN W.B., BOWER S.M., SCHMID M.D., and SCHWARTZ N.F., 1968, A voice-radio method for collecting human factors data. *AFHRL Technical Report* 68–10, Wright-Patterson Air Force Base, Ohio.

BARNES R.M., 1958, *Motion and Time Study*, 4th Edition (New York: Wiley).

CHRISTENSEN J.M., 1950, A sampling technique for use in activity analysis. *Personnel Psychology*, **3**, 361–368.

COTTERMAN T.E., and WOOD M.E., 1967, Retention of simulated lunar landing mission skills: a test of pilot reliability. *AMRL Technical Report* 66–222, Wright-Patterson Air Force Base, Ohio.

HUMAN RESOURCES RESEARCH OFFICE, 1963, HumRRO interim bibliography of reports. 1 July to 31 December 1962. *HumRRO, George Washington University*, Jan.

KEENAN J.J., PARKER T.C., and LENZYCKI H.P., 1965, Concepts and practices in the assessment of human performance in Air Force systems. *AMRL Technical Report* 65–168, Wright-Patterson Air Force Base, Ohio.

KINCAID J.P., and PROTEMPA K.W., 1969, Human factors testing conference, 1–2 October 1968 (Chairman: M.T. SNYDER). *ASD Technical Report* 69–   (in publication), Wright-Patterson Air Force Base, Ohio.

KURKE M.I., 1961, Operational sequence diagrams in system design. *Human Factors*, **3**, 66–73.

MARKS M.R., 1961, Development of human proficiency and performance measurements for weapons system testing. *ASD Technical Report* 61–733, Wright-Patterson Air Force Base, Ohio.

MEISTER D., and RABIDEAU G.F., 1965, *Human Factors Evaluation in System Development* (New York: Wiley).

MUNDEL M.E., 1955, *Motion and Time Study*, 2nd ed. (Englewood Cliffs, N.J.: Prentice-Hall).

RABIDEAU G.F., and RITCHIE M.L., 1968, Human engineering critical incident survey of tactical air operations in Viet Nam. *AMRL Technical Report* 68–127, Wright-Patterson Air Force Base, Ohio.

SHAPERO A., and ERICKSON C.J., 1961, Human factors testing in weapon system development. Paper presented at the *Missile and Space Vehicle Testing Conference, American Rocket Society*, Los Angeles, 13–16 March.

SPACE TECHNOLOGY LABORATORIES, 1960, Integrated test plan for WS 107A–1, operational system test facility. OSTF No. 1 and No. 2. Supplement: personnel subsystem test plan annex, *STL Report GM* 6300. 5–1060.

Dr. Chiles has a psychology degree. Since 1952 he has
carried out research into performance measurement and
environmental stress, first with the U.S. Air Force
Aerospace Medical Research Laboratories, and now with
the Civil Aeromedical Institute, Oklahoma City.

Complex performance: the development of research criteria applicable in the real world

W. Dean Chiles

1. The problem of performance assessment

The number of articles in the psychological and human factors literature whose titles contain references to human performance gives a misleading picture as to the true status of research in this area. When the topic of interest is the real world of work as represented by systems such as those found in aviation, the extrapolative chain that must be constructed to get from the typical experimental situation to the task confronting the pilot or air traffic controller is truly formidable . . . and probably seldom justified.

Two of the reasons for this state of affairs stand out because, in my opinion, something can be done about them. First, as has been pointed out in a compelling article by Chapanis (1967), researchers on human factors problems have not made the break from the traditional methodologies and theories developed in the academic environment. Thus, in their attempts to achieve the cleanliness of the ideal doctoral dissertation, numerous variables are either excluded or held constant. In many cases, the factors that have been neutralized for purposes of neatness of experimental design are potentially of greater relevance to performance in real world jobs than the variables being manipulated in the experiment.

The second reason that much of the performance literature is of limited value grows out of the reluctance on the part of the researcher to deal with experimental tasks that approximate the level of complexity of the demands placed on the human operator as a part of the man–machine system. Such complexity does not readily lend itself to neatly conceptualized explanations or theorizing. Therefore, the response of most performance researchers has been to retreat from this problem and to deal with simple, single tasks in the vain hope that individual measures can be developed, each of which will provide an adequate assessment of one or more of the many psychological functions exercised by the man in performing his job. The inherent fallacy in this approach is the fact that, by definition, it does not provide measures of the ability of the operator to time-share tasks that involve psychologically disparate functions. And, at least in systems such as those found in aviation and space flight, time-sharing (and decision making) are the primary abilities of the human operator that make him such a valuable component of man–machine systems. Because the man can apply his information-processing and decision-making skills in a time-shared manner to its various sub-systems, the *man*–machine system works whereas, at a given level of technological development, a *machine* system would not.

In our defence, I think it is fair to say that this unfortunate situation is not entirely the fault of the human factors scientist. In all too many instances, research that could have been relevant to the development of methodology was prevented from achieving that goal because of the pressures of the moment. The information being sought as the primary purpose of the experiment was required immediately to answer some question critical to the development of a specific system judged to be essential to national defence, public safety, or some other eminently worthwhile effort. A second point to be made in our defence is the fact that the facilities and resources required to conduct research on realistically complex performance have not been, nor are they now, available to more than a very few researchers. Substantial capital outlays are required for the very complex and sophisticated measurement systems necessary to achieve the requisite levels of precision and control.

This symposium bears witness to the factor that lies at the heart of the problem; namely, performance assessment criteria that are applicable to the real world have not as yet been identified. We do not know how to answer the question, 'What dependent variables yield measures that permit the operationally meaningful, quantitative description of the performance of the man–machine system?' However, as scientists, we should not lose sight of the fact that even if we had such measures handed to us on a stone tablet, we would still be a long way from being able to specify or understand the laws that govern the behaviour of the human operator as a component of the man–machine system (see Wallace 1965 and Weitz 1961).

2.   A new methodology: experimental control and behaviour taxonomy

I conceive of the problem as being basically one of developing a new methodology synthesized of two ingredients. The first ingredient consists of those elements of the traditional laboratory approach that are necessary to assure the degree of experimental control and precision that will, at a minimum, provide for the repeatability of the experimental situation. Thus, we must be able to specify and, hence, control the time-history of the sequence of events to which the subject responds. The second ingredient is supplied through the identification of the behaviour, or classes of behaviour, that the operator exercises in the system or systems to which we wish to generalize.

In selecting the specific laboratory customs to be observed, we must proceed with great care; the slavish application of mandates originating in the academic laboratory may 'throw the baby out with the bath water'. I will not dwell on this point because it has been very well covered by Chapanis (1967).

2.1. *The operator's tasks*

In selecting the various components of the second ingredient, I know of no guidelines that will, if followed, guarantee that the resultant task-complex can be rigorously defended against accusations that the functions performed by the operator in the real-world system are properly represented. Ultimately, this procedure is a matter of the judicious application of expert opinion or professional judgment.

Keeping in mind the fact that they are supplied without any written guarantee, what guidelines are available? If we take the direct approach, we could borrow a copy of the designer's time-line analysis which enumerates all of the functions performed by the man in the system. If we then implement this list by building the appropriate task apparatus, we will have built a simulator. Clearly this is what happens if the realism of the task complex is pushed to the extreme. However, when realism is tempered with the cautions of good experimental design dictated by the laboratory approach, we find that in addition to the fact that its initial and operating costs are quite high, the off-the-shelf simulator is not suitable. If the simulator does the job it was designed to do, that is, if it can be substituted for the real system as regards proficiency development and maintenance, then, in terms of the complexity of the situation confronting the operator (and the researcher), it is the same as the system. The primary difference, which makes the simulator so valuable as a training tool in both aviation and space operations, is that it generally does not harm anyone if the operator makes an error. The important point is that, if we knew how to measure performance in the simulator, the problem posed by this symposium would have already been solved. We would know how to measure performance in the system! (For further discussions of this point see Grodsky (1967), Chiles (1967a, b) and Chiles *et al.* (1968).)

Returning to the designer's time-line analysis, how do we decide how many and which of these functions should be measured? One source of relevant information is the catalogue of environmental variables that are likely to impinge upon the system and perhaps affect operator performance. For example, if the operating profile of an aircraft or other vehicle contains periods of substantial levels of vibration, we might want to select tasks from the designer's list that are dependent upon or involve visual acuity and manual control. Thus, any direct physical interaction between the system and the environment should be closely examined in relation to the possibility that effects may be transmitted to the human operator. It is in this area that we are most likely to find performance effects that are peculiar to particular sensory or motor functions.

2.2. *The relevance of tasks in laboratory studies*

When we move out of the realm in which possible direct mechanical effects of the environment may permit us to select certain tasks over others, we quickly find out, in bold-faced type, that **LABORATORY RESEARCH HAS LET US DOWN AGAIN.** All too much of that research, it turns out, has dealt with operationally trivial independent variables. Very seldom is information available about the way in which the performance of a task varies as a function of the realistic manipulation of operator workload. We can find little or nothing in the literature about the experimental interactions among simultaneously performed (i.e. time-shared) tasks. And, because of the way in which most environmental and psychological stress studies have been conducted, we find little that is useful as regards the sensitivity of tasks to operationally meaningful variables. (See Chiles 1966 and Chiles *et al.* 1968.) Clearly, we would not want to measure psychological functions that are so resistant to the impact of environmental and/or procedural variables that the system would certainly fail before they were compromised. Conversely, we would not want to measure functions that are so sensitive that changes in their performance are produced by factors that patently pose no threat to the integrity or the safe operation of the system. When stated in this rather crude form, it is obvious that psychological functions of the sort referred to in this latter case should be readily identifiable as not being critical to system performance. But, a statement made by Chapanis in his discussion of the selection of dependent variables in laboratory research suggests to me that the behavioural scientist may have difficulty on this point; he may find it very hard to resist the temptation to include a task '. . . on the basis of what is most likely to yield a significant result, and so a publishable paper' (Chapanis 1967).

Another aspect of the problem that demands our careful attention is the fact that the operator tends to assign priorities to the different tasks for which he is responsible. If he is well trained, the operator will certainly do this in the man–machine system, and, within the limits of his capabilities, he will exert all of the effort that is necessary to protect the performance of those tasks he knows to be critical to the safe or satisfactory operation of the system. The subject probably behaves in an analogous manner in the laboratory *if* we have successfully instilled in him the proper motivation. There are two implications of this hypothetical characteristic of the operator. The first implication is that he is unlikely to permit serious levels of impairment or decrement to be exhibited in his performance of what he considers to be critical tasks. The second implication is that, to the extent that the increased effort on such tasks uses up his available 'channel capacity', the probability of decrements in concurrently performed tasks will increase.

2.3. *The importance of time-sharing*

This leads me to the very firmly held conviction that time-sharing *must* be included as a basic element in our final assemblage of tasks. It is required for two reasons. First, as I noted earlier, time-sharing is almost a defining characteristic of the human operator's job in the man–machine system. And, second, measures that relate to the ability of the operator to divide his attention among tasks involving the exercise of psychologically disparate functions will, I am convinced, ultimately prove to be the most useful indices of operator performance in the man–machine systems context (see Chiles *et al.* 1968; and Chiles and Jennings 1969). By *psychologically disparate*, I mean that the performance of the different tasks must be based on different 'mental or perceptual sets'; the operator must, in effect, be required to shift his 'mental gears' when he moves from attending to one task to focusing on another. I believe that I am prepared to go so far as to say that the selection of the tasks is not extremely critical so long as time-sharing is required of the operator.

3. Measuring the operator's contribution to the system

Let us assume that we have identified a set of behaviours that we, as experts, consider to be adequate to describe the man's contribution to the performance of a particular system. Our next step is to devise or, where possible, adapt tasks that will afford quantitative measures of the quality of those behaviours. Unfortunately, we again find the performance literature to be deficient. The past pre-occupation of the behavioural scientist with single tasks, one after another, as a way of achieving measurement universals has meant that seldom are important methodological characteristics of the tasks known, e.g. their reliabilities and correlations with other tasks. Thus, such information must be developed as a part of the on-going programme. Although the tasks developed by the factor analytically-oriented researchers can be thoroughly described on these counts (see Fleishman 1967 and Parker 1967), specific implementations of such tasks are typically highly artificial from the subject's point of view. And, whatever the source of the tasks to be used, we must at all times remain cognizant of the fact that we want the subject to react to the situation as being in the realm of work; not in the realm of laboratory experimentation or play. This is especially critical if we plan to use trained operational personnel as subjects. Other than paying appropriate attention to the realism of the tasks as this factor relates to motivation, the criteria to apply in developing the specific mechanizations of the tasks are essentially those we would use in the laboratory. The resultant devices should yield reliable measures; the stimulus sequences should be manipulable, specifiable and, of course, repeatable.

When the manufacturer delivers our task ensemble as a finished product, we must then construct an 'operating profile' that defines specific task combinations and the durations of the sub-periods for which those task combinations are to be performed. If we turn to the man–machine system for assistance in carrying out this procedure, we will typically find that the system has been reasonably well human engineered, and, therefore, at no point under normal operating conditions will the limits of the capabilities of the man be exceeded or even encroached upon. Only under some emergency conditions will the demands placed upon the operator approach the overload point. This is perhaps just another way of saying that the human operator is extremely resourceful and resilient, but, probably the determination of the limits of his resourcefulness is where the need for precise human factors information is most critical. Thus, although there may be exceptions, we shall want to devise task combinations that vary the performance demands over the range from relatively light to near-overload conditions. Obviously, the achievement of the high demand combinations must be done in a manner that is meaningful both as regards the man–machine system and in relation to the effect of such conditions on the motivation of the subject. We must not present the subject with a situation that he has reason to view as being contrived or consisting of pure 'monkey' business on the part of the experimenter.

And now we put our measurement system to work and the data begin to roll in. What will these data look like? They will look like reaction time measures, per cent correct measures, error measures: in short they will look like the kinds of measures that have been collected in the laboratory from the beginning of laboratory time. The only difference will be in their antecedents; they will have been collected under conditions that have in the past been seldom used in the laboratory: conditions that more closely resemble those to be found in the operational situation. It is my opinion that, at least for the future I can foresee, the basic data of the performance researcher will continue to be measures of response time, response accuracy and, where appropriate, response amplitude. From these measures and their various combinations, numerous derived metrics can be developed, but I am not optimistic that asking a computer to go through all of their possible permutations and variations will bear particularly edible or otherwise useful fruit.

4. Predicting the effects on system performance

How, then, do we translate these measures into quantities so that their possible implications for system design and operation can be determined? I am not so bold that I will attempt to give any sort of complete or even direct answers to this question. But, I will try to describe what I consider to be the general nature of those answers.

Ultimately, we want to make quantitative statements about the efficiency and the reliability of the system. In the realm of efficiency, we want to be able to answer questions about the impact of human factors variables on things like the initial cost of the system; the cost of operation; the cost of maintenance; and someone must be concerned about the costs of personnel selection and training programmes. In the realm of reliability, we must be able to answer questions relating to both the mean time to failure and the mean time between failures of the system with all of the implications of these factors for both safety and convenience. At the present time, the behavioural sciences are, at best, in a position to deal with such questions only on an ordinal scale. If we are asked which of two system designs will be the safer, we have a fighting chance of giving a defensible answer. On the other hand, if we are asked *how much* safer design A will be than design B, we are in a very weak position. Unless there were something about one of the two designs that indicated very clearly that under some conditions it placed impossible demands on the operator, the best we could hope to give the designer is a probability estimate. Hopefully, we would eventually be able to give the designer the following kind of answer. If, in the operation of the system, this particular configuration of emergency conditions should arise, then with system design A the probability is, say, ·02, and with design B it is ·10, that the operator will lose control of sub-system Z. If the designer knows the probability that this emergency configuration will arise in the alternative designs and can specify the impact of loss of control of sub-system Z on system operation, we have given him what he needs to know to make his design decision.

I believe that it is fair to say that the behavioural and human factors scientists are not the only ones whose houses are in something less than perfect order. It is only recently that reliability engineering has been given the attention that it deserves by the designer. Thus, our inability to provide the designer with precise human factors data has stemmed from two causes. One of these has been the primary burden of this paper: the failure of the human factors researcher to deal with system-relevant boundary conditions in his experiments. The other lies with the designer; in important respects, he has been building systems better than he knows how. He has succeeded through over-design and by adding a liberal portion of the extreme adaptability of the human operator.

In large measure, significant progress in this very difficult area is dependent upon contributions from the design engineer with, perhaps, an assist from the human factors scientist. For example, although it is clearly the final criterion, defining a *safe* man–machine system as one that is free from accidents or incidents is not, without further refinement, very helpful in trying to quantify safety as a system performance metric. The computer might be of considerable value in attacking this specific problem. For example, with an appropriate programme to define the operation of the system, sufficient iterations of operational profiles could be run to develop a probability distribution that could then be used to define a metric for safety. If the human factors scientist can then provide appropriate descriptors of operator performance to be added to the computer programme, an interim solution to the problem might very well result. Obviously, this sort of thing has been and is being done, at least in principle, on an isolated basis. What has not been done is the execution of a concerted attack with a view towards generality.

But the *if* that preceded the contribution of the human factors scientist to this enterprise is a very large one. We have a long way to go before we can be content to sit back and hurl epithets at the engineers because all is not going well.

References

CHAPANIS A., 1967, The relevance of laboratory studies to practical situations. *Ergonomics*, **10**, 557–577.
CHILES W.D., 1966, Assessment of the performance effects of the stresses of space flight. *AMRL-TR-66-192, Wright-Patterson AFB, Ohio.*
CHILES W.D., 1967a, Methodology in the assessment of complex performance: Introduction. *Human Factors*, **9**, 325–327.
CHILES W.D., 1967b, Methodology in the assessment of complex performance: Discussion and conclusions. *Human Factors*, **9**, 385–392.

CHILES W.D., ALLUISI E.A., and ADAMS O.S., 1968, Work schedules and performance during confinement. *Human Factors*, **10**, 143–196.

CHILES W.D., and JENNINGS A.E., 1969, Effects of alcohol on complex performance. *FAA, Office of Aviation Medicine Reports*, Washington,D.C. (in press).

FLEISHMAN E., 1967, Performance assessment based on an empirically derived task taxonomy. *Human Factors*, **9**, 349–366.

GRODSKY M.A., 1967, The use of full scale mission simulation for the assessment of complex operator performance. *Human Factors*, **9**, 341–348.

PARKER J.R., Jr., 1967, The identification of performance dimensions through factor analysis. *Human Factors*, **9**, 367–374.

WALLACE S.R., 1965, Criteria for what? *American Psychologist*, **20**, 411–417.

WEITZ, J., 1961, Criteria for criteria. *American Psychologist*, **16**, 228–231.

Section 2 -Techniques

Épilogue

1. The physiology/psychology distinction

The emphasis throughout this book is on 'mental' tasks, rather than those involving physical energy expenditure, and we do not need to reiterate the arguments for increasing concern with mental workload in modern man–machine systems. Nevertheless, in terms of our original question on the relative contributions of physiology and psychology, it must be noted that several of the techniques which seem to offer most promise for an attack on these problems do not emanate from a purely psychological background. The electrophysiological techniques, and in particular the various methods of observing cortical activity, are increasing in importance. Sinus arrhythmia is another example of a technique which emphasizes the disappearance, discussed in several other parts of this book, of the traditional academic barriers between the two basic disciplines. The need to develop such areas common to physiology and psychology is one of the major conclusions to be drawn from this whole set of papers. Thus, in a slightly different context, Borg's attempt to scale the perception of heavy physical workload is another step towards the integration of psychological and physiological aspects of work situations.

In the prologue to this section, we mentioned that the physiology/psychology distinction in techniques is seen often as identical with the well-being/performance distinction in criteria. There is some degree of correlation, but it is certainly not universally true. The electrophysiological techniques can provide measures of efficiency, and secondary tasks are measures of operator loading. Once again, we have to assert the importance of a joint approach to assessment: we have the examples of psychophysiological techniques referred to above, and in any case, in the long term, both well-being and performance criteria must be equally germane to system effectiveness.

There remains still the problem of combining a number of measures into a useful description of human activity, for it is most unlikely that any single technique will ever be adequate on its own. In Borg's paper, it was suggested that a performance measure has a special utility by virtue of being a complex gestalt of a number of underlying variables. This is certainly true, in that performance is determined by an array of separate factors, but in turn each separate performance measure represents only one facet of the operator's effectiveness, and, as Edwards points out, the optimum combination of separate achievement measures is typically elusive. Chiles' emphasis on the complexity of operator performance in advanced systems is relevant to this argument: it is notable that he singles out *time-sharing* as the salient feature of most tasks, and even suggests that *any* task might be used in a simulation study, as long as it exhibits that characteristic. At the very least, this justifies much of the work on dual task techniques. Though, as both Rolfe and Kalsbeek demonstrate, these latter techniques are capable of more sophisticated results, with careful design of the experiment. Further, Kalsbeek's studies of 'distraction stress' provide another technique for dissecting the structure of complex skilled performance.

2. Arousal

Our concern with mental tasks leads naturally to this widely-discussed, though as yet little-understood, plane of human functioning. While it is now clear that a single scale of arousal is an over-simplification, it would be most useful to extend research on the various physiological measures as arousal indicators. Wisner points out that not only is there divergence of performance measures and physiological parameters, but that the latter are often not consistent within themselves. Obviously, progress is essential, both at the empirical and conceptual levels. It seems quite possible that certain measures are more useful in specific environments or tasks, and, ultimately, we must progress beyond correlational studies to a more complete understanding of the underlying physiological mechanisms.

Certainly, more empirical evidence of the occurrence of the U-shaped performance/arousal curve is needed. At present, its acceptance seems similar to the idea of plateaux on learning curves in psychomotor tasks: there are few cases of clear observation of the phenomenon, but it is a plausible explanation for several other situations. Thus, Wisner's emphasis on under-load and over-load is important, and it relates to the assertion that the ergonomist's aim is not to eliminate stress completely, but rather to reduce it to an acceptable, and indeed stimulating, level. Another important feature emphasized by Wisner is the tendency for physiological measures to reflect under-load or over-load in advance of performance changes. While Chapanis has rightly warned us of the dangers of over-sensitivity, this is a potential asset for critical situations.

3.   Comments on specific techniques

Perhaps the most impressive of the techniques more recently introduced into ergonomics is the study of evoked cortical potentials. Some of the papers have shown methods of determining the direction of attention, and, as Groll-Knapp suggests, there is great promise for investigating practical situations where signal detection is a primary feature of the operator's task. Equally, there are many industrial inspection tasks where it is unlikely that ECP's are of use: for example, where the inspected objects move on a conveyer line, and it is impossible to measure the time at which a 'signal' appears. It is also clear that the technique will be limited to the laboratory situation for some time yet, and that the repeated presentation of signals, which is essential to the averaging process, may in itself modify the nature of the subject's task.

Several other electrophysiological techniques have been reviewed, and it is essential that they be further tried and tested in a range of ergonomics investigations. The specific purpose of a measure may well change from one study to another: thus, EMG may be used as an arousal indicator, to detect specific muscle group activity, or to evaluate posture. It is only by pursuing such varied applications that we shall be able to build up a reliable body of knowledge to support practice. The same exhortation can be applied to the specialized techniques of critical fusion frequency, sinus arrhythmia and secondary tasks, where the papers have set out the extensive precautions which must be taken to ensure reliable results.

While we have emphasized the importance of field-studies in ergonomics, we have tended so far to think only of the application of laboratory measurement techniques outside the laboratory. The importance of Rabideau's paper is that it discusses the observational techniques which make a definite attempt to capture the patterns of characteristic features of real-life performance. In this respect, it may be linked with the papers by de Jong and by Christensen later in this book. The ergonomist must never neglect the valuable information which may be gained by passive observation, rather than active interference and control, in realistic situations. It can be argued that some of the techniques portrayed by Rabideau are suitable only for his particular context of large-scale military systems, but there are still several possibilities, ranging from several of his examples to the recording techniques of the work-study practitioner, which can be exploited in less complex systems. His inclusion of de-briefing interviews and questionnaires, for instance, in the armoury of observational techniques is a useful reminder of their wide importance.

4.   Some general problems

It has been argued already that the validity of a measurement depends in part on the existence of norms. However, there have been several comments in this section that this has not been achieved for many of our techniques. For instance, there are little comprehensive data on the EEG characteristics of normal populations. We should be exploring the possibilities of assembling normative data on a wide range of criteria, as is being attempted in the area of human reliability data (e.g. Swain 1969). In particular, it would be extremely useful to have some norms, however approximate, for the 'under-load' and 'over-load' levels of various electrophysiological measures related to arousal.

We must accept, however, that the construction of norms is seriously hampered by the effects of motivation and morale in specific investigations. During discussion of the above papers, this aspect was raised often. For example, there are many cases of subjects withstanding unusually high stress levels in laboratory experiments because of high motivation, and, conversely, of manpower requirements predicted from laboratory experiments being able to be reduced because of high motivation in military field conditions. Our understanding of these effects is so limited that almost any ergonomics prediction or suggested improvement can be obscured or even reversed in practice. There have been a few attempts (e.g. Berkun *et al.* 1962) to simulate high levels of 'psychological stress', but much more work is needed in this area.

It is always easier to retreat to the development of sophisticated hardware for ergonomics measurements, and there have been several examples of such improvements. The transducing and recording of most parameters, and in most cases even under field conditions, is relatively straightforward. The major gaps appear in the analysis and interpretation of data. All too often, we need to rely on human, rather than automatic, analysis. Prime examples are the large amounts of data from eye-movement recordings and from observational techniques.

Finally, the ergonomist working in practical situations must not neglect the human relations aspects of his investigations. His sponsors and his potential subjects may try to dictate his choice of techniques. In a typical industrial situation, management will tend to emphasize measures of performance, while the employees will favour assessment of well-being. In this final analysis of acceptability, our principles of validity and reliability, and our sophisticated data-collection techniques, are liable to some distortion.

References

BERKUN M.M., BIALEK H.M., KERN R.P., and KAN Y., 1962, Experimental studies of psychological stress in man. *Psychological Monographs*, **62** (15, Whole no. 534).
SWAIN A.D., 1969, Overview and status of human factors reliability analysis. *Report SC-R-69-1248, Sandia Laboratories, Albuquerque, New Mexico.*

Section 3 - Applications

Prologue

A comparison of psychological and physiological criteria in ergonomics research can easily become one more example of ergonomics simply providing a forum for behavioural scientists to explore common interests. By directing the discussion finally towards the implications for applied problems this pitfall has been avoided and the fundamental nature of ergonomics, or human factors, as a technology has been highlighted. Discussions of theories and research of men-at-work are surely sterile unless they lead to significant changes in men's efficiency or well-being at work.

Accepting this challenge, the ergonomist in practice must add new dimensions to what, in the past, has been a rather academic approach. He must temper his contribution to design or production situations with the constraints of criteria from other disciplines: of which not the least is cost-effectiveness. The interesting practice of analysing designs to detect human factors errors only offers the ergonomist intellectual exercise (strictly limited in nature) unless he can demonstrate some economic justification of his labours. In the applied field the ergonomist must have concrete proposals to put forward based on systematic analyses incorporating viewpoints other than his own. Without such proposals his contribution to any project runs the risk of being obscured by the competition of proposals from such as operations researchers, accountants or work study engineers.

Thus, schemes, techniques, procedures and models are necessary to allow the ergonomist in practice systematically to analyse his problem and synthesize a viewpoint which while unashamedly biased towards human factors criteria integrates them with other relevant criteria. In the main, these still await development.

But first there must be data and criteria for the ergonomist's analyses or proposals. At first sight there seems hardly a problem. Such a cursory glance would be deceptive.

Certainly in relation to anthropometrics and basic physiological and biomechanical capacities facts and figures abound. There are, however, fewer established generalizable facts about men's psychological performance in systems and in an ever advancing technology it is these latter data which are the more important.

An equally serious drawback to the endeavours of the human factors practitioner has been the reluctance of his research colleagues to orientate their work towards his need for more sophisticated data and theories if he is to tackle real world problems. The passing of trivial interface problems as the ergonomists stock-in-trade has created a need for new research approaches and theories. Man's efficiency and well-being at work is determined by the interaction of many factors. To be truly effective (and if ergonomics, or human factors, is to be a distinct science with unique theoretical formulations) the research ergonomist must face this problem. How he tackles it will be determined by his own research creativity. But to relieve a pressing need of the practitioner, fundamentally he must: either turn his attention to deriving data in the laboratory, or in the field, which reflects the multifactor determination of man's performance; or look to developing techniques or models which allow the integration of discrete data into more complex formulations.

The growing recognition of the need to question the adequacy of current data from the 'measurement of man' and the success of its application in practice inspired, in part, the previous papers in this collection. In those which follow the points will be taken more specifically and it should be possible to evaluate the seriousness of the deficiencies in current ergonomics research which hinder its practice.

This report on 'the state of the art' of ergonomics practice (for such it must be described) has been drawn from a variety of sources so it includes design and production viewpoints; and portrays the problems of both simple and complex systems.

It is to be hoped that the discussions will motivate a new attack on the basic problems of making ergonomics, or human factors, an identifiable science and a viable technology. They must certainly occasion some degree of admiration for individual workers who, by their creativity and ingenuity, have overcome current deficiencies in data and theory on an *ad hoc* basis, so keeping 'the measurement of man' to the forefront of decisions on man in his working environment.

Dr. de Jong holds engineering degrees. He has wide experience of industrial engineering and ergonomics in engineering and other companies, and is now a Board member of the management consultants B. W. Berenschot Co., Amsterdam. His current research is in co-partnership schemes and shift working.

The applicability of physiological and psychological techniques

J. R. de Jong

individual well being and system effectiveness.

1.   Introduction

Underlying our approach to the choice between psychological and physiological techniques to applied problems are the basic criteria of *individual well-being* and *system effectiveness*. It appears useful first of all to throw some light on the latter concept.

The system effectiveness is determined by the value of the system outputs on the one hand and by the strains resulting from the system's existence and functioning on the other. What occurs can be shown diagrammatically as in Figure 1. The effectiveness with which the system functions appears to be decided by

System costs
with such elements as:

Input costs
(raw materials, energy, etc.)

Costs of physical system components
(tools, machines, buildings etc.)

Personnel costs
(including such facets as skill, motivation, wages, absences and turnover)

Value of system outputs
(main product, by-products, waste products)
with the system variables:

Quality
(including reliability, weight, appeal, etc.)

Quantity

Availability
(delivery dates)

Figure 1. The functioning of man/machine systems.

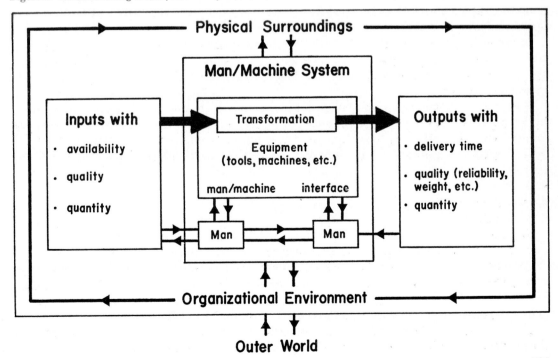

2. The contribution of ergonomics to the system

The aim of the system is to strive for optimization, e.g. by maximizing the output value/cost ratio. This optimization is a problem which has technical, economic and human factors aspects.

In principle, the contribution of ergonomics to optimization may relate to all output variables and all system cost elements, as far as they affect man within the system, and as far as they influence the effect of human action.

From the viewpoint of ergonomics the system elements constitute a range of networks.

- In the first place there are the networks with the elements determining the information that reaches man within the system through the senses.
- Secondly there are the networks determining the other human inputs, for example the inputs resulting from physical conditions at work.
- Finally there are the networks of influences via the human outputs: often mainly through motions and forces which in the first place are aimed at influencing effectively the system inputs, the transformation thereof and the system outputs.

However, one and the same system element, say a relay that is too slow for its purpose, may play a role in several networks.

Individual well-being is closely related to system output value and system costs: via attitudes and motivation it is linked to individual performances and (by absences and turnover) to selection and training costs. Thus system effectiveness and individual well-being, which have both been accepted as basic criteria need not lead to conclusions that are at variance with each other.

Nevertheless in the design and evaluation of man-machine systems conflict may arise through opposing views on work loads and attitudes, and the causes thereof. Such conflicts, however, are merely the result of uncertainty on the part of those involved, and this in turn is due to lack of necessary knowledge.

3. Criteria to be used in considering the application of physiological and psychological techniques

With regard to the people within the system, physiological and psychological techniques may yield data susceptible to limiting the uncertainty referred to. But what factors should determine their application? How is a justified choice between given methods and techniques of investigation arrived at?

In principle the system elements of the networks outlined above are all variables when it comes to considering the methods of investigation that should be applied in respect of a given system, whether existing or to be designed. This is equally true of the allocation of functions to men and to technical means.

Now if we consider man's well-being in the system we are interested in the strains caused by his inputs (relevant information, noise, heat, etc.) and his outputs (motions, forces, oral information, etc.). More specifically how these strains compare with his capabilities. In dealing with this question, we roughly distinguish three areas which in point of fact partly overlap:

- physical load
- information processing
- job satisfaction.

Naturally allowance must be made for the variability of human capacities, i.e. the inter- and intra-individual differences (resulting from such factors as learning and age effects).

Considering system effectiveness, we are interested in the effect of feasible alternatives on system costs and system output value (including such components as reliability and quantity mentioned earlier). There may also be alternatives in respect of the physical system elements (e.g. of the machine interface), the methods used by the operators, the rest periods, etc.

On the strength of these considerations we arrive at the following problem structure as far as the application of physiological and psychological techniques is concerned.

1. *What strain aspects demand our attention?*

This point raises the following questions.

To what extent are the strains involved in the alternatives to be studied, acceptable to the people concerned? How frequent and how long are the rest periods they necessitate?

How far do the strains affect task performance (reliability of observations, correctness of decisions, accuracy of motions, etc.)?

How far do the strains produce a negative attitude to work and the company? How do they affect the motivation to work?

2. *To what extent does available knowledge render the need for a physiological and psychological study superfluous?*

Regarding heavy muscular work (whether or not in combination with heat stress) the question may be asked whether for a given situation to be evaluated, such data as has been collected by Aberg *et al.* (1968), Spitzer and Hettinger (1964), and Rohmert (1966), along with the instructions for its use (Hettinger *et al.* 1968, Rohmert 1960) make it unnecessary to carry out measurements of the oxygen intake, or whether it would be sufficient to determine individual capacities.

The criteria underlying the choice between the use of available literature data and the application of a psychological and/or physiological study (both frequently in conjunction with observational techniques) are :

■ the validity ;
■ the accuracy of results obtained ;
■ the investigation costs ;
■ the damage risks which in terms of individual well-being and system effectiveness are associated with incorrect conclusions.

We shall revert to these factors when dealing with the fourth question. As regards the reliability of general data it is worth noting that Aberg *et al.* (1968), using the formula they developed, arrived at results showing a coefficient of variation of 17 per cent relative to measurements of O_2 consumption. They assumed that a static muscle load component was responsible for at least part of the deviations in question. The results recorded by Hettinger *et al.* (1968) showed less deviation from those obtained by measurements, viz. 4 per cent, 3 per cent, −6 per cent and −2 per cent respectively.

As a general rule the costs attendant on the use of existing knowledge (literature data) will be considerably lower than those ensuing from the application of a physiological and psychological study.

3. *To what extent do the available physiological and psychological techniques cover the entire area of strains?*

Section 2 of this volume provides a review of the techniques that may be employed today. It shows among other things, that for different aspects of strain a choice can be made between physiological and psychological methods, or that a combination of techniques derived from both disciplines is suitable for application. It demonstrates that in the last few years several new possibilities have become available, such as physiological and psychological methods for the assessment of mental strains (Borg and Kalsbeek in this book ; Bartenwerfer 1969).

Also, with instrumentation, there has been marked progress so that, for example, measurements taken in industry have a less disturbing effect (due, among other factors, to the use of transmitters instead of cables).

Nevertheless, there still are gaps especially where studies are to be carried out by a company's own staff.

4.   *What procedure should be adopted in deciding on the techniques to be applied in a given case?*

In the second question formulated above the following factors were indicated as being material in arriving at decisions on investigation methods.

Figure 2. Relationships between individual well-being and system effectiveness

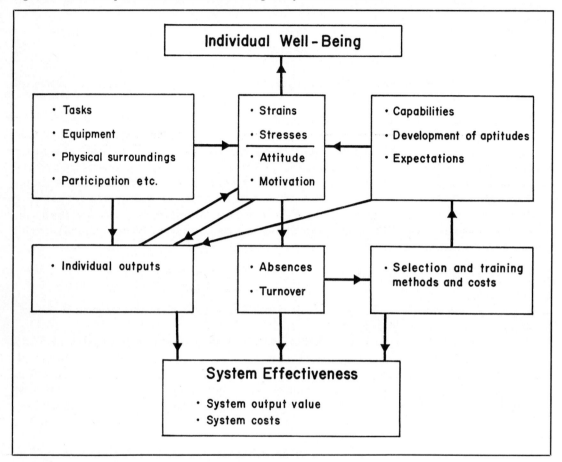

- Validity

    Does the method provide relevant information on the strain component we are concerned with (and possibly on the resultant need for alternation with activities of a different nature or for rest periods)?

- Accuracy

    What is the closeness of estimates to the true values?

- Costs of data gathering and analysis

    What are the costs involved in the use of a given technique? The term costs is understood to include not merely the direct investigation costs but also the costs ensuing from delays caused to others, and the negative (or positive) effect exercised by application of the method in question upon performances *after* the investigation period. Of course the costs depend to a great extent on the equipment required (is this manufactured in quantity?) and on the demands it imposes on the users.

■ Damage risks of incorrect conclusions (for individual well-being and for system efficacy)

To the extent that these risks are more serious, more reliable investigation methods will be called for and higher study costs will be warranted.

Given the problems involved, physiological and psychological techniques are adopted less widely than would be desirable. Thus there is a need for :

research having the object of assessing where gaps continue to exist (e.g. methods permitting effective appraisal in practice of loads on small muscle groups and of combinations of static and dynamic muscle loads) and programming ways of filling those gaps ;

standardization of equipment and methods (both to increase the reliability of results and restrict costs) ; a first step in this direction is the publication issued by the Netherlands Health Organization GO–TNO on the measurement of the physical load in the case of heavy muscle load (Bonjer 1965).

Obviously in dealing with both points international cooperation would offer notable advantages.

4. Studies, system design, and system evaluation and the applicability of techniques

Studies are the 'acquisition of unbiased evidence'. Within the boundary of our subject they may be directed at human strains and the stresses causing them, and at the physical system elements (tools, machine interfaces, physical surroundings). Studies may be conducted because problems present themselves in given man/machine systems—whether in the design stage or operational—or may be based on the need to investigate questions of more general concern.

As regards the use of known physiological and psychological techniques these studies in the majority of cases are hardly subject to any limitations other than those of an economic nature.

Within the framework of system design, however, the limitations imposed are manifold.

Design is 'the creation of new systems'. That is the context in which we place investigations into the effectiveness of developed systems (or parts thereof) with the aid of models, mockups and prototypes (next to the actual designing of hardware and software, i.e. of machines with the related methods, instructions, and training programmes).

In other words, such investigations do not fall under 'studies'.

Now given the task involved, which is that of arriving at well-founded views on the system to be developed, limitations are imposed by :

incomplete simulation of the future system ;

incompletely trained operators, which implies limited skill and limited indications on which to base a prediction of any long term effects.

In such circumstances physiological measurements may not be of much use, while the results of such methods as interviews and questionnaires require cautious interpretation.

Among other things, this applies to the system effectiveness (and particularly the system output) anticipated, and to the satisfactoriness of the tasks that must be performed.

Also, limitations are frequently imposed through a lack of understanding between design engineers and human factors specialists. The following may help to eliminate these limitations : imparting ergonomics information to design engineers ; accepting the rule that designing the hardware must go hand in hand with designing the methods, instructions, and training programmes and training aids to be used ; and, thirdly, directions from clients, such as the 'Human Factors Requirements for the Development of U.S. Army Matériel' (Chaillet 1967).

The limitations mentioned with regard to the development of systems have little or no relevance where the validation of an operating system ('the checking of system or system element performance against an independent criterion') is concerned. Here the application is often mainly limited through disbelief in the usefulness of validation studies on the part of management.

With system design a condition governing the proper application of ergonomics is that the design process should be open to human factors influence from the beginning (i.e. the allocation of functions to hardware subsystems and human tasks) to end (i.e. the evaluation of prototypes, or at least of the final design, if no prototypes are made).

In other words, system design calls for an integrated application of ergonomics.

Now the question arises whether it is admissible to confine oneself to ergonomics aspects in validating a man/machine system.

Is it justifiable to leave technical, economic and social-psychological facets out of consideration? Once the system effectiveness has been accepted as a basic criterion it is obvious that at least the economic aspect should be consciously taken into account. Which leads to the equally obvious deduction that in system validation *all* facets (and their interrelations) should receive attention.

Aside from this, a major argument in favour of an integrated approach is that such an approach is a condition for the correct interpretation of data collected by means of ergonomics techniques. This is true for the outcome of physiological measurements (e.g. of sinus arrhythmia) as well as for opinions—gathered through personal interviews and the like—on the strenuousness of the work, the disturbing effect of the climate, etc.

Hence significant advantages are gained by combining the collection of data on work load, equipment, physical conditions at work, etc. with gathering information on work organization, on the job satisfaction of those concerned, and on the psychological environment. Such data, along with personal details on educational background, age, etc. and company statistics concerning absences and turnover, plus a technical and economic analysis (made by others) are indispensable in obtaining the basis required for a sound evaluation of a man/machine system.

In our view this concept, predicated on the combination of a range of aspects, including ergonomics and social considerations, is in agreement with the necessity of 'humanizing human factors engineering' rightly formulated by Jordan (1963).

References

ABERG U., ELGSTRAND K., MAGNUS P., and LINDHOLM A., 1968, Analysis of components and prediction of energy expenditure in manual tasks. *International Journal of Production Research*, **6**, 189–196.

BARTENWERFER H., 1969, Einige praktische Konsequenzen aus der Aktivierungstheorie. *Z. exper. u. ang. Psy.*, **16**, 2, 195–222.

BONJER F.H. (editor), 1965, *Fysiologische Methoden voor het Vastleggen van Belasting en Belastbaarheid (Physiological Methods for the Assessment of Work Load and Working Capacity)* (Assen: Van Gorcum & Comp.).

CHAILLET R.F., 1967, Human factors requirements for the development of U.S. Army matériel. *Ergonomics*, **10**, 278–286.

HETTINGER Th., PAQUIN K.H., SUCKER G., 1968, *Kalorienverbrauch und Erholungszeitberechnung* (Frechen, Köln: Bartmann-Verlag).

JORDAN N., 1963, Allocation of functions between man and machines in automated systems. *Journal of Applied Psychology*, **47**, 3, 161–165.

ROHMERT W., 1960, *Statische Haltearbeit des Menschen* (Berlin, Köln, Frankfurt a.Rh.: Beuth-Vertrieb).

ROHMERT W., 1966, *Maximalkräfte von Männern im Bewegungsraum der Arme und Beine* (Köln, Opladen: Westdeutscher Verlag).

SPITZER H., HETTINGER Th., 1964, 4th edition, *Tafeln für den Kalorienumsatz bei körperlicher Arbeit* (Berlin, Köln, Frankfurt a.Rh.: Beuth-Verlag).

Dr. Christensen holds psychology degrees. He has long experience of human factors research and development in the U.S. Air Force. Since 1945, he has been a research scientist and administrator in the Human Engineering Division, A.M.R.L., Wright-Patterson A.F.B., Ohio, and is now Director of the Division.

Human factors engineering considerations in systems development

J. M. Christensen

1. Introduction

The purpose of this paper is fourfold. First, the life cycle in the design and development of a typical system will be described. Second, the nature of human factors engineering requirements will be described. Third, these requirements will be related to the systems development cycle and, finally, a brief evaluation will be made of the tools and information available to the human factors engineer.

The partitioning of the development cycle of a system into phases obviously is arbitrary. The scheme developed and used by the United States Air Force is adopted in this paper primarily because of the writer's familiarity with it (Mitchell and Koppel 1965, U.S.A.F. 1963, 1964a and b). The fourphase breakdown that it employs is defined below.

1.1. *Conceptual*

Encompasses the period of time between the determination of the broad objectives of the system until approval is received to go ahead with the Definition Phase.

1.2. *Definition*

Includes a systematic analysis of the system which results in detailed plans regarding system activities, engineering, management, maintenance, deployment and costs.

1.3. *Acquisition*

The period between the award of the developmental contract and acceptance of the system by the customer. Detailed design, test and production are all accomplished during this phase.

1.4. *Operational*

Includes the period from the acceptance of the first unit by the customer until disposition finally is made of the last unit.

2. Requirements

The first significant step in the development of a system occurs during the Conceptual Phase and involves the writing of the 'system requirements'. These are statements establishing the need for, and the functions to be fulfilled by, the system. More specifically, requirements establish and define the goals of the system and the criteria against which the performance of the system will be measured. These enable designers and others associated with the development of the system to judge the relevancy, and increase the precision and specificity, of their actions and recommendations. Finally, requirements serve as a basis for agreements between contractor and customer and between contractor and subcontractors.

Experience has shown that certain problems occasionally arise during this period. Two of the most common problems are associated with the lack of unequivocal and absolutely clear definition of the system and the establishment of appropriate criteria. Extra hours spent initially on these two areas will save significant time and money later. Unfortunately the performance of a system of more than modest complexity is dependent upon so many factors and their interactions that it becomes difficult, if not impossible, to attribute each element of the criterion variance to its proper causative factor(s). It would seem that significantly more resources should be spent on the problem of developing more reliable and valid operational criteria.

Another problem associated with requirements is the fact that their iterative nature sometimes goes unrecognized. The *final* set of requirements that evolves is invariably a compromise between ideal objectives and such realistic constraints as time, money, availability of scientific and technical information, etc.

3. Evaluation of systems approach to development

Even a highly systematic approach to the development of systems will have advantages and disadvantages. The following appear to be advantages of the approach described in this paper. First, the process encourages, in fact demands, delineation of requirements during the very early stages of planning. The mere fact that a systematic procedure is specified forces virtually everyone who has anything to do with the development of the system to begin early and systematically to consider his areas of responsibility appositively with those of others. This includes those responsible for hardware, software, facilities, personnel, etc. The successful systems designer will assure that the working relationships among the members of his team are such that each not only is thoroughly familiar with his own requirements but also is sufficiently familiar with the requirements of the other sub-systems so as to make possible intelligent and effective compromises and trade-offs. It is a truism of the first magnitude that the system that finally evolves invariably is the embodiment of an almost infinite number of compromises.

A systematic approach also allows the derivation of more realistic time estimates and the development of more accurate methods for predicting bottlenecks during the development cycle. More systematic interplay among those responsible for design, production, testing, installation, operation, training, simulation, etc. is also assured.

Management of the entire enterprise becomes more systematic under an approach of this nature and helps assure better integration of the efforts of customer and contractor and contractor and subcontractors. The latter point is of particular importance these days since many large contracts require that a significant proportion of the total effort be subcontracted.

Critics of a highly systematic approach to systems development contend that it fosters 'overstudy' of the problem, sometimes to the extent that nothing is ever built, or, if built, is outdated before it is completed. (Unfortunately, there is no way to determine which of those never-developed systems would have been successful, and which really never should have been developed.)

Other critics claim that a highly systematic approach does not encourage or allow initiative and creativity at the designer's level. However, no objective data have been uncovered that would either support or deny such a contention.

Finally, there is concern in some quarters that the systems approach encourages over-management and undue attention to costs at the expense of technical capability. Again, no data were found to support this position.

In summary, the 'systems approach' is a relatively recent development which arose in answer to a number of needs having to do with technical complexities with costs, time constraints, etc. Time and experience will suggest necessary changes in the procedures that support the systems approach, but it is difficult to see how the modern, ultra-complex systems could have been developed without some model that assures timely and systematic consideration of the myriad of details that constitute a modern man–machine system.

4. Human factors engineering responsibilities during systems development cycle

Major human factors responsibilities during a typical systems development cycle are summarized in Table 1. During initial systems analysis the human factors engineer must contribute along with others to the identification and definition of the general functions that must be fulfilled for the goals of the system to be realized. A corollary of this responsibility includes the preliminary identification, definition and assignment of specific functions to men and/or machines. Hopefully, these assignments can be made in terms of overall criteria of systems effectiveness but at the same time without asking man to perform tasks that are demeaning, unnecessarily dangerous, or unduly boring.

It is probably fair to say that, in the past, considerably more attention has been given to the operators of systems than to those who must maintain and support the systems. At least from the economic point of view this hardly seems wise. Numerous studies have shown that over the lifetime of a complex system from 10 to 100 times the initial cost of the system will be spent maintaining it.

Simultaneously with the analysis of functions, the human factors engineer should conduct an analysis of the flow of information through the system. The purpose is to determine what decisions are made by each individual and what information is required, in what form, to make those decisions. Special care should be taken to see that each individual gets only the information that he needs to do his job. There is a temptation in these days of computers, cathode-ray tubes, etc., virtually to inundate individuals with information. Perhaps this happens because the informational requirements of each position have not been carefully defined or because of the feeling that there is a remote chance that the operator will find a use for certain information. This latter possibility should be the subject of experimental enquiry and not operational 'happenstance'. Finally, too much information or information that is irrelevant to the task quickly ceases to be information and really becomes noise in the system, thus degrading overall performance.

Once the informational requirements have been established, the human factors engineer should work closely in support of the design engineers charged with the responsibility of determining what sensory channel(s) to use, type and amount of redundancy, and other details of the design of the informational displays. Not the least of these considerations is that of the environment in which the job will be performed: will the operator be vibrated or accelerated, will he be hot or cold, etc.?

The previous paragraph emphasized information because that is often the major 'product' of many modern complex systems. One can easily substitute a material product in the above scheme without appreciable adjustments in his thinking. Informational analysis is analogous to the analysis industrial engineers have been performing for decades with their flow process charts.

Experts disagree regarding the proper time during the development cycle at which to begin the analysis of manpower and training requirements. Some would argue that such analyses should begin very early in the conceptual stage. The writer's experience has been that things are in such a state of flux at that time that an attempt to be precise with respect to the number and type of operators and maintenance men results in figures and statements that, at best, will be drastically revised as the details of the system are more clearly defined and, at worst, are misleading to the design engineer. This is not to say that gross deductions should not be made early in the cycle ('the vehicle will have a driver') but detailed specifications probably are best made well into the Definition Phase.

There is a movement underway to make manpower figures available during the conceptual stage. The writer does not take exception to this, but he is concerned with imposing 'qualitative and quantitative personnel requirements' as stringent constraints on the design engineer. What particularly worries the writer is the difficulty of predicting what people can or will do under real operational conditions. For example, the in-commission rates for aircraft in combat zones are often better than in non-combat zones. The maintenance men apparently feel a particular sense of responsibility, involvement and pride under those conditions. At least it is

Table 1. Relationship of human factors engineering activities to systems development cycle.

H.F.E. Activity	Purpose	Primary (P) or Contrib (C) Respons.	Conceptual	Definition	Acquisition	Operational	Methods/Tools Available	Remarks
1. Systems Analysis	Identification and definition of general functions necessary for accomplishment of goals of system	C	●	●			Operations research methods; activity analysis; time line analysis; computer studies. (Also for 1.a. through 1.e.)	Support systems engineer
a. Functions Analysis	Identification, definition and allocation of specific functions to men and/or machines	P	●	●			Functions analysis	Includes maintenance and ground support as well as operators. Coordinated with safety engineers, RAM engineers, and logistics personnel
b. Information Analysis	Analysis of basic flow and processing of information and decisions that will result therefrom	P	●	●			Flow process charts	Special attention to decisions and operations; includes both operations and maintenance.
c. Analysis of Manpower Requirements	Estimate number and performance characteristics of men required to operate and maintain the system; serves as basis for selection and classification standards	P		●	●		Manpower resources reports	Both operations and maintenance; selection and classification coordinated with manpower resources planners.
d. Analysis of Training Requirements	Estimate of nature and amount of training required to bring available personnel up to required standards	P		●	●		Functions analysis; task analysis	Both operations and maintenance. Coordinate with training planners.
e. Analysis of Simulation Requirements	Estimate of simulation requirements for both training and maintenance of proficiency	P		●	●		Mock-ups; dynamic simulation; functions analysis; task analysis	Both operations and maintenance. Coordinate with training planners.
2. H.F.E. Design Recommendations	Develop and supply information regarding human capabilities and limitations relevant to design of equipment and subsystems	C		●	●		H.E. handbooks, specifications, standards, reports; experience with similar systems; mock-ups	Recommendations in terms of relevant systems criteria such as efficiency, safety, training costs, ease of maintenance, reliability, etc. Support design of simulators and ground support equipment, manuals, etc. as well as operational equipment.
3. Development of Procedures	Procedures necessary to operate and maintain system. Identification of critical tasks	P		●	●		Task analysis; job analysis	Operators and ground support personnel
4. Test and Evaluation	Determine how well system meets requirements. Obtain and feedback information from operational use	C		●	●	●	Activity analysis; questionnaires; performance records; critical incidents	Support systems test manager. Evaluation includes operational follow-up.
5. Production Methods	Design of safe, efficient, and accurate methods of fabricating, assembling and testing equipment at manufacturer's plant	C			●	●	Task analysis; activity analysis	Support industrial/production engineers

DEVELOPMENT PHASE

known that, under such conditions, they can and do work effectively and 'happily' 12 to 16 hours a day, seven days a week. The point is that it could be unwise to compromise overall systems performance simply because an analysis of qualitative and quantitative personnel requirement suggests that men may not be able to accomplish the job working a 40 hour week or that men having particular skills, may not be available in sufficient number. Even though it would be possible, with modern technology, to keep virtually 100 per cent of a squadron of P-40 fighters in commission, they would not last long against a modern weapon system even though its in-commission rate were very low. Perhaps, at least for military applications, a designer should be able to expect operators and maintenance men to perform beyond a nominal, routine, 40-hour week, level when the 'chips are down'. The truth is that we know very little about this area and we should exercise caution in imposing constraints on the designer until we do.

Analysis of requirements for simulation must include the requirements for maintenance of proficiency as well as training. This is particularly true now that some systems are becoming so costly that operation of the entire system simply as a training device or as a device for maintaining proficiency is not economically feasible.

The second major human factors engineering activity has to do with detailed design recommendations. (See Table 1.) This traditionally has been called 'knobs and dials' work but in reality is much more than this since it should always include consideration of all man–machine interfaces and the interrelationship of each element of a sub-system with every other element. Similar considerations should be made at the sub-system level. Detailed standards and specifications must be available for this phase of the work since there are simply not enough human engineers to provide one to each design engineer. The large amount of sub-contracting alluded to earlier in this paper suggests a further need for a standard way of doing things so that incompatibles do not occur when all the elements are assembled into a sub-system and these, in turn, are assembled into a system.

Although tremendous strides have been made in the past several years to produce standards, specifications and handbooks containing human engineering design recommendations, the human factors engineer will still find many occasions when he is unable to supply the design engineer with the information that he needs. He should expect, and be capable of, performing 'quick and dirty' studies or surveys to answer such questions and, if time doesn't allow even that, to make an educated guess. His educated guess in his own field of expertise should be better than the decision of a single design engineer who will probably base the decision on how he, one individual, feels about the matter.

Concurrent with detailed planning and design, the human factors engineer must keep an accurate account of the tasks and jobs, both operator and maintenance, that are being generated by each design decision. These data also serve as a valuable source of information for establishing manpower estimates, training and simulation requirements, and manuals of operation and maintenance.

The human factors engineer has several important contributions to make during test and evaluation. Like others involved in the development, this is the time to determine whether or not all the work up until now has really resulted in a system that meets stated requirements. In addition, it seems that the human factors engineer has a special contribution to make in two areas. First, he should help develop reliable and valid criteria against which to judge the performance of the system, calling upon his background in assessment and criterion development. Second, he should assure that the test conditions are as realistic as possible and, particularly, that the operator and maintenance personnel are representative of those who eventually will operate and maintain the system when it finally reaches the field.

The development of more efficient production methods has traditionally been the province of the industrial engineer, industrial psychologist and safety engineer. However, recent work by people like Teel at North American Aviation Inc. has disclosed that the human factors engineer can profitably become involved here too. In particular, he can apply the same principles to the design of equipment, inspection, etc. that he traditionally has applied to operational tasks. He is, as is the industrial psychologist, generally somewhat more aware than is the industrial engineer of the nature and degree of individual differences among people and, hopefully, he is sensitive to the need for each production worker, like each operator and maintenance man, to receive the satisfaction that results from doing useful work well.

5. Tools available to the human factors engineer

Some feel that the value of human factors engineering is severely limited by two factors: information and methods. A moderately careful review of both content and method has convinced the writer that while improvements can be made in both, these are not the primary problem. The primary problem is to get the information that is already available more generally applied and to go out and use the tools already available. This will disclose voids in current content and deficiencies in current methods. It is difficult, for example, to understand why the three-pointer altimeter is still so widely used when Grether disclosed its serious deficiencies two decades ago and offered reasonable, safer alternatives. Similar examples are available from nearly every branch of human engineering.

Let us begin the evaluation of tools by an examination of the area of activity analysis. Activity analysis is here defined simply as the recording and analysis of man–machine activities. Its history can be traced back to at least the time of Moses for the Bible tells us that, while under the Egyptians, if a Jew failed to make his quota of bricks during a day the overseer would take away his straw. Early attempts at precision are vividly illustrated by the fact that in England in 1820 the standard production quota for number 11 pins was 5,540 in 7·6892 hours! Modern time study began under Taylor around 1881 and this was augmented by the work unit (therbligs) efforts of the Gilbreths. These individuals had a profound effect on industrial production methods, incentive plans, and even the design of tools and equipment (for example, recall Taylor's work on shovels). Unfortunately, the notion of human individual differences seemed generally to be ignored as well as considerations of man as a complex social organism that responds to many things other than monetary incentives. Muensterberg of Harvard gave consideration to man as a social being as well as a cog in a huge production complex, but his concern seems to have been fairly well buried by the enormous influx of practitioners, many of questionable capability, into this new lucrative field. It finally culminated in the United States Government forbidding anyone to time the work of a government employee with a stopwatch nor could government contractual funds be used for such studies.

Such studies have become acceptable, however, under the careful eye of the industrial engineer who has brought a distinct air of professionalism to the field. In fact the modern industrial engineer is about as close as one ever comes to being a true systems engineer. Speaking as a psychologist, the writer sincerely hopes that they will give increased attention to men as individuals and to indirect contributors to job success such as morale, satisfaction in a job well done, and so on.

The uses of activity analysis of special interest to human factors engineering are manifold. Requirements for new equipment and systems are frequently suggested after a careful study of the activities of individuals. For example, the writer found as a result of a detailed analysis of the job of the arctic navigator that approximately one-third of the first navigator's time was devoted to making entries in a log and to performing other paper work. Streamlining of procedures, more efficient solution of celestial observations, elimination of non-essential information, etc. have reduced the extent of this onerous burden.

A careful analysis of the activities associated with a tool or a method will usually disclose some way in which it can be improved so as to reduce time requirements or errors or both. The same applies to the layout of workplaces. An analysis of the time devoted to each element in a workplace can serve as a basis for deciding which items should occupy prime space and which should occupy peripheral or secondary space. Recording the time measurements in sequence will allow one to compute 'link' values which can serve as a basis for arranging elements for most efficient sequential use.

The results of an activity analysis can serve also as the basis for accurate task analyses, functions analyses (for both men and equipment) and for job descriptions of increased currency and validity, since one is reporting what *actually* happens on a job.

These same data are useful to selection and classification experts who must define the inherent and acquired (or acquirable) characteristics that they wish their selectees to possess. From an overall manpower resource pool standpoint, it is important that they draw only what is needed in terms of numbers and qualities and not 'overselect' for the described jobs.

Those responsible for the development of programmes in training and maintenance of proficiency can gain increased insight into the true nature of jobs from the results of an activity analysis. Again, the results of an activity analysis disclose what the activities really are as contrasted to what an armchair analyst feels the job is or ought to be.

If a method of analysis is used that provides a detailed description of activities, careful study will disclose sources of error and time delays. The industrial engineer, of course, has been and is pre-eminent in this use of activity analysis.

Finally, someone who makes a detailed analysis of the activities of a man–machine complex almost invariably will gain insights that will suggest improved tools and procedures for getting the job done. There is something about actually *measuring* what goes on in a job that provides insights that are difficult if not impossible to obtain any other way.

The interested investigator can select from many methods of activity analysis (Mundel 1947, Niebel 1955). Many of these were developed by the industrial engineers. No attempt will be made here fully to describe and evaluate all these methods but, rather, points both *pro* and *con* critical about each of them will be recorded, hopefully as an aid to the investigator who is faced with the difficult task of deciding which method of activity analysis to use in a specific set of circumstances.

The first one that we shall discuss briefly is time study. Time study consists simply of timing a task or series of tasks with a stopwatch and deriving standards from the analysis of the resultant data. Allowances based primarily on experience are made for fatigue and delays over which the operator has no control. Time study has been found to be especially useful on repetitive acts such as one finds in a typical production line. Reliability of measures is very high and validity is variable.

To a psychologist it appears that many time studies have ignored two principles that are fundamental to any study of human behaviour. The first of these principles is that of individual differences. There is no skill known to the writer that, upon examination, does not disclose rather impressive individual differences, even when the distribution is artificially truncated. The second of the principles is related to the first and has to do with the problems of sampling. Again, for most skills, there are usually not one or two individuals who can legitimately be considered representative of a population. And taking 25 measures on one individual is quite a different matter than taking one measure of 25 different individuals. Industrial engineers, probably largely from necessity, have tended towards the first of these two alternatives.

Motion study also has its roots in the early history of industrial engineering. One associates the Gilbreths with its development. They developed 17 basic motion patterns which could be used to describe the motion and non-motion ('hold' for example) patterns of virtually any job. While motion study has done much for industry, it, like time study, has some features that bother psychologists (and modern industrial engineers). First, there is probably no validity to the idea of a 'one best way' to do a job. Williams and Mantle were both superb hitters but they went at the job quite differently. Again we see a rather casual treatment of the fact of individual differences. It is true that there is always one way (or, better, one *approximate* way) that can be learned by most operators. However, that is quite different from proving it best for all. At least as serious as this deficiency is the idea that one can put these 'therbligs', as the basic motion patterns are called, together in different patterns to meet different requirements and predict performance even on a job that has never been directly analysed. To do this one must assume that no interaction exists among the elements of motion and as K.U. Smith and others have shown, this simply is not true.

The industrial engineers also originated the 'standard time' notion which is the idea of compiling a bank of times for various task elements on current jobs, using these times to make predictions regarding the time that will be required to perform a job that either did not exist before or at least had not been measured. Its most recent corollary is the 'task bank' idea which is essentially a repository of descriptions of standard tasks and the times required to perform them. The 'standard times' method suffers from the same limitations as the motion study method, as has been realized by the industrial engineer, Niebel, who states '... there is a question as to the validity of adding basic motion times for the purpose of determining elemental times in that therblig times may vary once the sequence is changed' (Niebel 1955).

Process analysis, again according to Niebel, is a '... technique for recording the various steps involved in a process'. When used in the typical manufacturing plant, the analysis results in 'flow process charts' which are really a sort of time history of the manufacture of an item from the raw material stage through final inspection of the finished product. Symbols are appropriately affixed to the chart for the elements of operation, inspection, delay, storage, combined operation and transport. With a little imagination, variations of this basic idea can be used to record and analyse almost any process from flow of information to the development of an entire system. (The 'Pert' scheme was, or at least could have been, based on the flow process chart idea. High speed computers have made these sophisticated variations possible and practical.)

Many avoid the use of interviews and opinion surveys, apparently feeling that they lack both objectivity and validity. Nevertheless, in the hands of a skilled practitioner these can be extremely useful tools, and sometimes are the only tools that an activity analyst will be able to use. The writer has found these techniques very useful in making a broad, frontal attack on a problem, preparatory to focusing on a particular part of the system that needs detailed treatment. (A word of caution: only those who are well-trained in their use should attempt to employ these tools. They appear deceptively easy to construct and use, which is probably the reason that so many worthless surveys have been conducted.) One must be especially careful to assure independence among the respondents since one dominant figure can have an undue effect on the results through his influence on the other respondents.

The critical incident technique was developed by Flanagan (1954) and is a method for eliciting narrative information from individuals regarding accidents or near-accidents that they have experienced. The investigator then analyses these reports with the hope of uncovering causes in terms of such factors as inadequate or inappropriate training, faulty equipment, poorly designed equipment, faulty procedures, etc. The technique has been used with considerable success by many workers (Fitts and Jones 1947a and b, McFarland and Moseley 1954, Vasilas *et al.* 1953). It appears to make one assumption that is rather bothersome and that is that one must assume that critical (unusual?) events are sufficiently typical of a population of events to warrant statistical analysis and the drawing of conclusions regarding that population. It is not unlike taking the I.Q. of an exceptional person and generalizing from his behaviour to the behaviour of all those in the distribution. However, the technique has been used with great success, and merits the investigator's full consideration.

One can gather plain narrative descriptions regarding tasks such as is done in job analysis. If gathered carefully from qualified people, such descriptions can contain much valuable information. Unfortunately, such data are very difficult to reduce and to express quantitatively.

The writer (Christensen 1959) thought in 1947 that he had invented a new method of activity analysis based on sampling theory. It was termed 'activity sampling'. However, it apparently was already in use by industrial engineers under the title of 'ratio-delay' (Heiland and Richardson 1957). Regardless of origin, the technique does seem to be quite useful and is economical in terms of both equipment and observer personnel. Elsewhere (Christensen 1948a and b, 1949) the writer describes a series of studies of the activities of the arctic navigator in which he used activity sampling. The resultant data served as a reliable, objective basis for recommendations regarding needed equipment development, equipment design changes, procedural changes, duty assignments, etc. The results were impressive; the cost was amazingly low.

In summary, then, with respect to activity analysis, there are numerous methods available from which the investigator can choose. None is perfect, but many are adequate. Much more effort should be expended in determining the actual activities of operators and maintenance men under field conditions. The investigator is virtually certain to obtain information that is both surprising and useful. He will generally discover a significant discrepancy between the prediction of what an operator does and what he really does. He will gain detailed insight into the job and almost invariably will gain ideas that will improve both the procedures and equipment used on the job: occasionally an idea leading to an invention will begin this way. It is hard, difficult work, but extremely rewarding for someone who will apply himself.

This seems to be the age of models and computers and no scientist can afford to admit that he does not have access to one or preferably both of these accoutrements. Models and computers seems to have replaced books on experimental design and slide rules and, in some instances, thought. Chapanis, in his lucid and delightful lecture on models while President of The Society of Engineering Psychologists, stated 'models are analogies'. Keeping this statement in mind will prevent one from gross misuse of them.

Actually, models have always been a part of scientific thought or any other type of thinking. Their uses are manifold. Good models can help us to understand complex events, to learn complex skills (trainers, simulators), to serve as frameworks for the planning of experiments, to lead us in our attempts to uncover new relationships, to enable us to make predictions when, for some reason, other means are not feasible, and to serve as actual design tools.

It is important, however, quickly to review some of the limitations of models. A model is never the real thing; otherwise it wouldn't be called a model. This means, as Chapanis suggests, that one is dealing with analogies: and we all learned in Logic I the dangers of this.

In addition, a model occasionally seems to instill an unquestioning and unreasoning loyalty on the part of he who adopts it and the longer the association, the greater the degree to which these characteristics are manifest. They often prevail in the face of overwhelming evidence to the contrary. Such loyalty can eventually lead the investigator into strange and diversionary avenues that are of interest only to him and the temptress who enticed him in the first place.

Models also tempt one to overgeneralize. The model almost invariably is only a partial pattern of the whole and the user should exercise considerable caution in extending it to other seemingly similar situations or even to the system as a whole from the sub-system that it represents.

Finally, models are often accepted *prima facie* and intentions to validate them are neglected. In fact, the extreme in this regard is reached when the investigator rejects empirical fact and restudies the issue until, somehow, the real world is remoulded to fit his model!

A mock-up is another tool available to the human factors engineer. In reality it is a model and as such possesses the advantages and disadvantages recorded previously.

The modern simulator may also be considered a model and is usually distinguished from the static mock-up in that it attempts to reproduce the dynamic qualities of that which it is simulating. The modern computer has become an integral part of many simulators and greatly increases the validity of this particular type of model, enabling one to increase the validity and power of experiments conducted with them. These computer-driven simulators may well revolutionize experimental work in many areas by allowing the investigator to probe the extent of variable interrelationships in greater depth and variety, leaving the experiment as a carefully used and defined tool to test critical points and relationships. Preliminary work will then be done on the computer, not in the laboratory.

Human factors people have been about as faithful as those of other disciplines in the systems development business in getting their findings into texts, specifications, standard and handbooks (Chapanis *et al.* 1949, Department of Defense 1968a and b, Woodson and Conover 1964). Unfortunately the examination of current systems shows undeniably that much of the 'common knowledge' of a decade or two ago is still not being used with any degree of regularity. Meister's recent study (*personal communication*) confirms that engineers often simply are not using human factors information in their design work. The reasons for this unfortunate state of affairs probably can be found in management's attitude toward human factors, the design engineer's background and attitude, the reluctance to try anything new as long as things are going reasonably well, the inappropriateness or irrelevance or unimportance of much human factors data, etc.

Reports of past experience can contain valuable information. The critical incident technique described earlier is an example of a technique that relies on human recall for its data. One is well-advised to keep in mind the fallacy of memory when using reports of past experience and also the extreme reluctance on the part of anyone to indict himself, particularly if in so doing unfortunate things may happen to him.

The human factors engineer must constantly keep in mind that the system on which he is working will have an impact on current and future manpower resources and must ensure that the manpower pool has or will have the necessary operators and support people in sufficient number with sufficient skills when the system is finally placed in the operational inventory.

Examination discloses that there is already much more human factors information available than is being used. Clearly, a breakthrough in applications is needed. It is *not* paradoxical to observe in the very next breath that much more information is needed any more than the fact that the human factors engineer has many fine tools available to him does not mean that tool development can or should cease.

Some of the factors that are of extreme importance in systems operation have yet to be quantified, much less related to systems development. Morale and motivation might do for openers.

In addition, much relevant information is *not* presented so the engineer can use it. It was Fitts who first reminded us that unless the independent variable is one that the engineer can manipulate, an investigator really has nothing that can be of much assistance to design/ systems engineer. But that alone is not sufficient. We must never forget that the system that eventually emerges from the development process is, more than anything else, an integrated (hopefully!) assembly of *compromises*: compromises made chiefly by design and system engineers. This requires an appositive consideration by them of multiple variables and unless these independent variables are expressed in terms of a meaningful reference point or absolute zero (ratio scales), it is difficult to see how the engineer can use them to the full extent of their potential in trade-off situations.

Further, there is a basic incompatibility between the requirements and standards of good laboratory procedure and generation of data most useful to the design engineer. In the laboratory one generally carefully isolates variables so as to be able to define them more exactly and to control them more precisely in his manipulation of them. In addition, one often introduces highly artificial conditions (e.g. exposure times of one millisecond) apparently in order to magnify differences and to help establish significant differences. Unfortunately, very little is known about how all this relates to the real world of systems operation. It would be surprising, however, if the optimal selection of values and combinations of variables for good laboratory work were identical to those needed by the engineer who is about to make a design decision.

There is hope, however. People and machines are available to handle more complex (thus, hopefully, more realistic) experimental designs, models have been developed to handle original data for those cases where ratio scales are not available, etc.

Human factors personnel occasionally have criticized systems engineers for spending insufficient time following their designs into the field to learn firsthand of the problems encountered there. But human factors personnel are equally guilty. They have spent far too little time determining what operators and maintenance men actually do in the field (as contrasted to what a paper and pencil or computer analysis may have predicted they would do) and finding out how well laboratory findings generalize to field situations. These problems of currency and generality still are not being contended with to any significant extent.

This paper has been identified by Aerospace Medical Research Laboratory as AMRL-TR-69-82. Further reproduction is authorised to satisfy needs of the U.S. Government.

References

CHAPANIS A., 1959, *Research Techniques in Human Engineering* (Baltimore: The Johns Hopkins Press).

CHAPANIS A., GARNER W.R., and MORGAN C.T., 1949, *Applied Experimental Psychology: Human Factors in Engineering Design* (New York: Wiley).

CHRISTENSEN J.M., 1948a, Aerial analysis of navigator duties with special reference to equipment design and workplace layout. II. Navigator and radar operator duties on three arctic flights. *Memorandum Report No. MCREXD-694-15A, Air Materiel Command, Dayton, Ohio.*

CHRISTENSEN J.M., 1948b, Scientific methods for use in the investigation of flight crew requirements. *Flight Safety Foundation (Woods Hole, Massachusetts), Project* RP-1-F.

CHRISTENSEN J.M., 1949, A method for the analysis of complex activities and its application to the job of the arctic aerial navigator. *Mechanical Engineering,* **71,** January.

CHRISTENSEN J.M., 1959, The sampling method of activity analysis. *Personnel Psychology,* **3,** Autumn.

DEPARTMENT OF DEFENSE, 1968a, Human engineering design criteria for military systems, equipment and facilities. *MIL-STD-1472.*

DEPARTMENT OF DEFENSE, 1968b, Human engineering requirements for military systems, equipment and facilities. *MIL-H-46855.*

FITTS P.M. (ed.), 1951, Human engineering for an effective air navigation and traffic control system. *National Research Council, Washington, D.C.*

FITTS P.M., and JONES R.E., 1947a, Analysis of errors contributing to 460 'pilot-error' experiences in operating aircraft controls. *Memorandum Report No.* TSEAA-694-12, *Air Materiel Command, Wright-Patterson Air Force Base, Ohio.*

FITTS P.M., and JONES R.E., 1947b, Psychological aspects of instrument display I. Analysis of 270 'pilot-error' experiences in reading and interpreting aircraft instruments. *Memorandum Report No.* TSEAA-694-12A, *Air Materiel Command, Wright-Patterson Air Force Base, Ohio.*

FLANAGAN J.C., 1954, The critical incident technique. *Psychological Bulletin,* **51,** 327–358.

FUCIGNA J.T., 1965, The role of human factors in system development. *Unnumbered Report, Dunlap and Associates, Darien, Connecticut.*

HEILAND R.E., and RICHARDSON W.J., 1957, *Work Sampling* (New York: McGraw-Hill).

McFARLAND R.A., and MOSELEY A.L., 1954, Human factors in highway transport safety. *Harvard School of Public Health, Boston, Massachusetts.*

MITCHELL M.D., and KOPPEL C.J., 1965, Air Force Systems Command: Systems management. *San Diego, California: Paragon Design Company.*

MUNDEL M.E., 1947, *Systematic Motion and Time Study* (New York: Prentice-Hall).

NIEBEL B.W., 1955, *Motion and Time Study* (Homewood, Illinois: Richard D. Irwin, Inc.).

UNITED STATES AIR FORCE, 1963, Systems program documentation: Systems management. AFR 375–4, *Department of the Air Force.*

UNITED STATES AIR FORCE SYSTEMS COMMAND, 1964a, Systems Program Management Manual. *HQ United States Air Force Systems Command, Andrews Air Force Base, Washington, D.C.*

UNITED STATES AIR FORCE SYSTEMS COMMAND, 1964b, Systems Engineering Management Procedures Manual. *AFSCM-375-5, HQ United States Air Force Systems Command, Andrews Air Force Base, Washington, D.C.*

VASILAS J.N., FITZPATRICK R., DUBOIS P.H., and YOUTZ R.P., 1953, Human factors in near accidents. *Air University, United States Air Force School of Aviation Medicine, Report No. 1, Project No. 21-1207-0001.*

WOODSON W.E., and CONOVER D.W., 1964, *Human Engineering Guide for Equipment Designers* (Los Angeles: University of California Press).

Dr. Burrows holds psychology degrees. After working
with the Royal Air Force Institute of Aviation Medicine,
he joined Douglas Aircraft Company, Long Beach,
California, in 1959. He has extensive experience with
military and commercial aviation problems, and is now
Director of Research (Sciences).

The structure of an effective human factors effort

A. A. Burrows

1.   The problem

The question of criteria in the business of human factors (or ergonomics) cannot be explored
with any realism unless it is clear what functional structure a characteristic project has. This
does not change the fundamental criteria for the *quality* of experimental studies within any
discipline, physiological, psychological, engineering or others, which provide base data. These
must always be tied to good experimental design, correct statistical application and valid
inference.

The confusion as to the correct criteria in this field is, in reality, what valid data to apply
where and what new data to search for, having accurately diagnosed a lack of it.

We may find an interesting relationship between eminence in our field, the field's size and
stature, and the fact that the discipline is now formally *required* as a design function rather than
a consulting service. *Eminence* is almost purely confined to those individuals and small groups
who have maintained academic connections and connotations and have experimentally solved
problems in a generalized manner. Some eminence, but not very much, is present in individuals
and groups who have solved critical problems in the same way within large, evolving systems
and have either got those systems out of trouble or have improved their operating effectiveness.

A much lower status level is attached to the practitioner who applies or supplies data to
evolving systems. Being good at doing this either *avoids* problems (by correctly steering the
system at an early point away from a gap in relevant knowledge) or identifies the critical problem for some other person to solve. Our effective human factors practitioner, who is politically
astute enough to arrange for himself to be early enough in the programme and influential
enough to steer it well, tends to disappear (except in the eyes of perceptive management) the
better he is at the job.

Most of these chapters talk about the best and most current methods and criteria for *solving*
problems and not about diagnosing or avoiding them. Most of the work discussed is judged
through perpetuation in learned journals and not in its effectiveness within the project it
supports. This is not to criticize such work, but an attempt must certainly be made to put it
in perspective. We talk mainly about research and problem solving. In the development of
many systems in the last ten or more years, it has become clear that such activity is but one
necessary effort in the success or failure of a project.

One description (Burrows *et al.* 1967) shows a breakdown of a project into three sequential
phases and sixteen steps (only one of which is 'research' to provide fresh data) (Table 1). In
fact, the manpower aspects of such a realistic procedure show that the ratio of the research
function to 'applied practice' is about one to ten or greater. The fifteen other steps shown have
evolved or have been designed to provide the least problems for solution and to identify those
which are critical and unavoidable at the earliest time. This is to some extent because research
and problem solving of the sort we are concerned with have been accepted as so indefinite in
result and longevity that the complex project should be least contingent upon such activity.

Why then is status and eminence so tied to the experimental, problem-solving activity, and
should it legitimately be so? There is perhaps a persistent feeling among the professional, and
not necessarily eminent, problem solvers, that questions about the place of human performance

in a tightly temporal and cost-controlled project must always be tainted by compromise or by pragmatic specificity. Sufficient ties between problem solvers and academia maintain the trend toward immortal, elegant, generalizable and publishable studies. Furthermore, there are very few academic institutions which teach the related disciplines found in projects, that there should *be* relationships and what the structure should be; furthermore, there is a lack of emphasis on problem identification in comparison with problem solving.

Table 1. Human factors engineering functions in system development

Phase I: System analysis
1. Preliminary system planning
2. Flow charting
3. Estimates of potential commander/operator/ maintainer processing capabilities
4. Allocation of functions
5. Operational sequence diagramming
6. Definition of commander/operator/maintainer information requirements
7. Definition of control, display, and communication requirements
8. Human factors research
9. Commander/operator/maintainer task description
10. Personnel planning information

Phase II: System design and equipment development
11. Mockups
12. Dynamic simulation
13. Work station layout
14. Human factors engineering aspects of detailed design
15. Design follow-through

Phase III: System test and evaluation
16. Human factors engineering design verification

This is a very difficult state of affairs, because more and more projects are becoming formalized in the way described, less and less problems identified and solved, and the gulf of status and communication, of application and generalization, grows larger.

What does the non-experimentally oriented practitioner do, in fact? Very simply, he considers the proposition that in any developing system, he must confine himself to the extent to which the human operator is extending himself. The operator is with very few exceptions commanding, operating or maintaining the system, that is, telling it what to do, helping it to do it, ensuring or repairing its performance. The system is meant to extend the human capability in its purpose (move faster, go further, lift more) and man's functions within it may be extended by aids within each of the three functions. Our practitioner basically must decide the *extent* of the extension. But he lacks many tools. He has not yet a firm taxonomy of human capability with which he can even identify clearly the processes, behaviours and activities. Some reasonable attempts have been made to assist him (Christensen and Mills 1967, Berliner *et al.* 1964), as we see in Tables 2 and 3.

Even with a taxonomy available to him, the distributions, limits and, most of all, the interactions of human ability elements are not usefully designated. Bishop and Guinness (1966), after a long scrutiny of what happens, summarize, '. . . The final design was evolved through a series of compromises between the human factors requirements and the designer's functional concept. . . .' It is difficult to illustrate this process, except by example. As Meister *et al.* (1966) consistently point out, '. . . designers trained in the engineering disciplines tend to work from laid down rules and specifications, from tabulations and simple graphical descriptions about how components function. The process of design in mixed disciplines is always a matter of

Table 2. Classification of behaviours. From Christensen and Mills (1967) and Berliner *et al.* (1964)

Processes	Activities	Specific behaviours
Perceptual processes	1. Searching for and receiving information	Detects Inspects Observes Reads Receives Scans Surveys
	2. Identifying objects, actions, events	Discriminates Identifies Locates
Mediational processes	1. Information processing	Categorizes Calculates Codes Computes Interpolates Itemizes Tabulates Translates
	2. Problem solving and decision making	Analyzes Calculates Chooses Compares Computes Estimates Plans
Communication processes		Advises Answers Communicates Directs Indicates Informs Instructs Requests Transmits
Motor processes	1. Simple/discrete	Activates Closes Connects Disconnects Joins Moves Presses Sets
	2. Complex/continuous	Adjusts Aligns Regulates Synchronizes Tracks

matching different sets of descriptors with the engineer tending to replace the human element with something which is as (apparently) predictable, reliable and less variable and . . . with which he is familiar'.

Our experimenter, then, is losing a battle to have problems both identified and presented to him for solution, the less he cooperates with the non-experimental practitioner. Compromises are made, familiar components are substituted where human extension may be possible. Furthermore, we find new sorts of experimental studies beginning to appear; the human factors man, with little lead time and few guidelines except limited data, often will say 'let's run it' and carry out a very pragmatic study indeed. Most often, these studies are designed only to leave the door open to human element function in opposition to total automation.

Table 3. Percentage of time allocated to activities by three typical personnel. From Christensen and Mills' (1967) data

	Command (Approach control)	Operate (Pilot)	Maintain (Checkout)
Perceptual			
1	11%	42%	43%
2	9%	4%	24%
Mediational			
1	9%	0	0
2	9%	0	0
Communication	29%	7%	0
Motor			
1	0	9%	32%
2	33%	40%	0

Percentages to nearest per cent.

An early study in the lunar landing problem (Abbott 1962) was critical to design, where it was necessary to ask if an operator (when all else failed far from earth) could move several hundred pounds of mass at about one cycle a second for a minute. Tracking literature is replete with any but the solution to this problem. An answer to the question was found by a very limited but very good experiment, precisely to the point. With public and government interest high in the nature of passenger evacuation in the case of crash in commercial aircraft, some very good experimental studies were carried out in an extremely short time using hundreds of typical passengers in specific aircraft configurations, which not only led to rule-making and greater safety, but point to some previously unnoticed characteristics of a very basic nature in crowd behaviour while in hazard (Johnson 1969, Altman 1969).

These are but two of many examples which show a drift away from the *status quo* of our discipline and one which, it is certain, is obvious to many of us. Problem solvers seem clearly to be moving into more and more slender areas, aligning with academic institutions and becoming less rather than more interdisciplinary in their outlook and approach.

2.   Solutions

There is no reason to suppose that the large, complex projects or their cost or technical management systems are going to go away; if anything, they are going to get bigger and more complex. The growing pains of management of these programmes have to some extent passed and many lessons have been learned. One of these lessons is the correct diagnosis of necessary research areas and the avoidance of others by accepting state-of-the-art solutions.

Our place in this scene may be dictated by the trend. Not all of our work will be associated with these sort of projects, but nearly all of it will be affected by their process. Our growth and health as a discipline will depend upon accepting that support of the functions which identify and delegate problems will strengthen the discipline as a whole.

The basic controversy between physiological and psychological interests at a technical level within the ergonomic community, seems not to be a national one. Some surface indications are often misleading. The author's contention is that *psychological* interests internationally (with a few exceptions) by virtue of their evolution and formal training, lean heavily upon quite complex methods of experimental design and statistical inference. It has not been traditional to teach these methods within the medical or physiological communities, or, at least they are not given such priority. In consequence, there is a gap in communication and credibility between the two groups. Most noticeable is the trend within the physiological community not to approach research areas with multivariate techniques or to explore interaction effects of a high order, as do experimental psychologists. That this alleged 'gap' appears wider in the European–U.S.A. dichotomy is perhaps an accident in *who* in the two disciplines is dealing more with ergonomics issues. A second point is that very apparently, in the United States, there is a trend towards multi-disciplinary problem solving (necessary because of the larger size and complexity of projects) in which methodology 'rubs off' on the uninitiated. European workers often operate alone or in very small groups.

A trend away from this, however, is apparent. In combined behavioural, physiological and biochemical approaches (Carpenter *et al.* 1969, Fraser *et al.* 1967), it is becoming apparent that where little or no consistency in results is to be found (in the study of fatigue) the application of quite advanced design and inferential techniques results in solid patterns across the measures, linking them. It is interesting to note that in some of these studies, different orders of inference were experienced. In examining the physiological changes intimating activation in subjects against increasingly demanding tasks, individual pattern differences required inference at the ordinal level where behavioural shifts could be dealt with at a higher order. Non-parametric techniques appeared more appropriate when attempting general conclusions rather than describing individual subject trends. This is perhaps an emergent situation where the gaps of approach, methodology and background will narrow as the gains become apparent. Another good example of this shift to combined measures with increasingly sophisticated analysis, can be found in Kalsbeek (1968) where the two disciplines are found to mesh well without the stimulus of a large, complex programme or many workers providing cross fertilization.

We must ask, then, in conclusion, if we have a pseudo problem in physiological/psychological conflict in ergonomics, and whether we have or not, if it is a problem of the highest priority to the survival and development of our discipline. It is submitted here that there are more urgent resolutions to be made and more fundamental disputes to be settled.

Perhaps the question of criteria should first be addressed to defining the needs of the total effort, the criteria which are needed and which should be common to all steps within a project involving human extension. First of all, the old, but still pressing need for a realistic task taxonomy would have an extremely high priority. Second, a clear feeling for which parts of such a taxonomy can be given values which are unequivocal and can be safely applied. Doing this will define those areas which must be treated experimentally at an early point. Another need is a clear statement of those values attached to such a taxonomy which are very probably interactive and which will never or very seldom be quoted *without* experiment or at least expert study.

Another requirement which has been identified but not met is the agreement between investigators concerning the methodology and equipment necessary for evaluation of 'bread and butter' elements. Commonly accepted and repeatable determinants of error probability of displays commonly used, for example, would be immensely valuable.

Organizationally, the human factors of ergonomics operation must show these elements of application, experiment, standardization of commonly used procedures and equipment, basic data and data criteria for application; and it must show them and its personnel in close cohesion, without the stigma of disciplinary status differences. Without these elements present or in proper cohesion, we shall be showing a greater downhill trend than we do today.

References

ABBOTT P.E., 1962, An investigation of displacement error and control stick force as a function of mode, forcing, frequency, spring gradient and inertia. *DAC Report 31034c, Douglas Aircraft Company, Long Beach, California.*

ALTMAN H.B., 1969, Emergency evacuation aids in adverse visual environments. *Paper to Western Psychological Association Symposium in Aviation Safety, Vancouver.*

BISHOP E., and GUINNESS G.V., 1966, Human factors interaction with industrial design. *Human Factors,* **8,** 279–289.

BURROWS A., et al., 1967, The why and how of human factors. *U.S. Office of Naval Research Contract 0014-66-C0141, Douglas Aircraft Company, Long Beach, California.*

CARPENTER D., CREAMER L., GABRIEL R.F., and BURROWS A.A., 1969, Individual and crew fatigue in a simulated, complex airborne weapon system, human error research and analysis program. *DAC 67798, Contract No. 189-68-C-0565, Final Report—Part III, U.S. Naval Station, Norfolk, Virginia, March.*

FRASER H., ROLFE J.M., and SMITH E., 1967. Pilot response under ground simulation conditions and in flight. *R.A.F. Institute of Aviation Medicine Report, Farnborough. October.*

JOHNSON D.A., 1969, Passenger behavior in survivable aircraft accidents: Inaction under stress as a significant behavior. *Paper to Western Psychological Association Symposium in Aviation Safety, Vancouver.*

KALSBEEK J., 1968, Measurement of mental workload and of acceptable load: possible applications in industry. *International Journal of Production Research,* **7,** 1, 33–45.

MEISTER D., and FARR D., 1966, The utilization of human factors information by designers. *U.S. Navy, Nonr-4974-00.*

Dr. Hartman is a psychologist, has worked in the U.S. Army Medical Research Laboratory, and is now with the U.S. Air Force School of Aerospace Medicine, Brooks A.F.B., Texas. His work has been both clinical and experimental, with emphasis on environmental stress.

Some experiences in conducting interdisciplinary studies in a medical research institution

Bryce O. Hartman

1. Introduction

In early correspondence concerned with the focus of this symposium on physiological and psychological criteria for studies on man–machine systems, Professor Chapanis identified four problem elements:

- guidelines for choosing among the two measurement domains;
- comparability of results from the two approaches;
- utility of results for managers and designers;
- agreement between principles derived from the two approaches.

For the past 12 years, I have worked in a military medicine/physiology institution and have directed a small group of psychologists chartered to function as team members in interdisciplinary studies in support of operational aviation and space programmes. Major research areas have included the effects of drugs in common clinical use upon flying efficiency, physiologic and environmental stress in airborne and space systems, problems involved in work/rest and related time-anchored stressors, personality and behavioural factors in crew performance (including some elements of personnel management systems) and human engineering factors in aeromedical equipment. From our research, I have selected 15 papers which provide material for evaluating one of the problem elements listed above: comparability of results from the two approaches.

2. Review of the studies

As an indicator of the diverse nature of the studies, there are four involving simulated space flight (McKenzie *et al.* 1961, Morgan *et al.* 1963, Cutler *et al.* 1964, Rodgin *et al.* 1966), three directed at questions of space physiology (Graveline *et al.* 1961, Dunn 1962, Glatte *et al.* 1967), three in the drug area (McKenzie *et al.* 1965, Hartman *et al.* 1968, Cantrell *et al.* 1969), and five where psychologic variables were primary (Hartman *et al.* 1964, Hartman *et al.* 1967, Little *et al.* 1968, Buckley *et al.* 1969, McKenzie 1969). As an indicator of the interdisciplinary nature of the studies, only five (Dunn 1962, Hartman *et al.* 1964, McKenzie *et al.* 1965, Hartman *et al.* 1967, McKenzie 1969) did not involve a large multi-discipline research team. In all but two of the remaining studies (Little *et al.* 1968, Buckley *et al.* 1969), the psychologists on the team had essentially secondary functions, supporting the other disciplines with data which would make it easier to apply the medical and physiologic findings to specific operational systems. Overall, however, psychologic variables were primary in approximately half the studies.

Many of these studies have some unusual or historical feature which deserves to be mentioned. The first study in this series is, to my knowledge, the earliest experiment on the physiological effects of weightlessness, as well as the first to use the water-immersion technique to simulate weightlessness in the laboratory. Though there was only one subject (the principal investigator), the period of exposure was substantial (seven days), and the relatively benign effects of weightlessness during exposure and the cardiovascular problems upon return to 1g were confirmed in subsequent American and Russian space flights.

The second and fourth studies are among the earliest multi-man laboratory experiments involving simulated space flights. They were also the first to simulate space flights long enough to go to the moon and back (14, 17 or 30 days), at a time when American space shots were limited to one-man earth orbits, and were also the first to include a careful analysis of inter-personal effects. The fourth study also deserves special comment because it is the only simulated space flight experiment where several long-duration simulations accomplished over several months were organized into a formal experimental design permitting systematic manipulation of work/rest, day/night performance variables. Both the biomedical and psychologic changes were minimal, demonstrating the feasibility of long-duration space missions. A small decrement in performance was obtained for the combination of a low signal rate during the post-midnight work shift.

The third is one of the few papers reporting a statistically significant enhancement of performance when breathing oxygen-enriched air, though many athletes and aviators are convinced that such enhancement occurs.

The fifth study was conducted in a Launch Control Center at a Minuteman site, and appears to be the only study involving an extended period of confinement (30 days) and performance on operational tasks (simulated launches using operational equipment) in an operational missile installation, with the test focused exclusively on missile-crew problems. No psychologic or physiologic effects were seen. Many hardware-oriented problems in the crew-logistics and life-support areas were revealed.

The sixth and ninth studies had an engineering orientation. Designers of life-support systems for space vehicles were finding published tolerance limits on CO_2 concentrations too restrictive, and wanted less demanding tolerance limits in order to save weight and cost. The central question was the biomedical acceptability of less stringent standards. A team was assembled to evaluate the significant biomedical dimensions. Though physiologic changes were seen, they were in the category of 'adaptive' and posed no biomedical threat. Limits as high as three and four per cent (in contrast to one per cent in the published standard) were studied.

The seventh study involved an unusually close approximation in the laboratory of a three-day 'physiological conditioning' regime used to prepare fighter pilots for flight across the Atlantic, including giving a sedative to promote sleep the night before and a stimulant to sustain performance during a 12 hour simulated flight on the third day. A 'hang-over' effect from the sedative was found, and impaired performance long after the drug had been excreted. The facilitation of performance due to the stimulant was like that reported by others. We found this sort of 'faithful simulation' to be rather demanding on the staff.

The eighth study is the longest simulated space flight of which I am aware, and required assembling an exceptionally large interdisciplinary team numbering around 30 staff members representing over a dozen specialties. It also had a very practical orientation. Two-gas systems offer some advantage to life support and safety engineers of space vehicles, and helium as a dilutent permits some savings in vehicle weight. The study was directed at the biomedical effect of prolonged exposure to helium. Though several effects in the physiologic domain were seen, a helium-diluted two-gas system was found to be biomedically acceptable. It is worth noting that the 56-day duration puts this study in a time-frame which is now only beginning to be discussed in relation to orbiting space stations.

The tenth study was a replication of a work/rest study reported by another USAF laboratory. The other study used a 4/4 and 4/2 schedule. We added a 16/8 schedule. The study consisted of a series of 12-day test sessions, with three days of sleep deprivation midway. Performance changes were seen only during sleep deprivation and the first day of recovery, and were smallest for those men working on the 16/8 schedule. The use of sleep deprivation as a provocative procedure (an innovation for work/rest studies first applied by Alluisi and Chiles) is the unique feature of this study. Air Force researchers on work/rest problems have been rather routinely frustrated by their difficulty in attaching indices of merit to various work/rest schedules. Some major provocation, such as three days without sleep, facilitates this greatly.

The eleventh and fourteenth studies are neither unique nor of historical interest. They do have a practical orientation. They were done as part of a long-range programme to identify drugs which could be prescribed for aviators without grounding them. A malarial-prophylactic regime was studied by Hartman *et al.* (1968). An anti-hypertensive agent was studied by Cantrell *et al.* 1969. No psychomotor decrements were obtained.

The twelfth study was a parametrically-oriented study of acceleration stress with a very simple design, conducted by a master's degree-level graduate student doing a thesis problem, without any previous research experience. Nine subjects made repeated centrifuge runs at 5, 7 and $9G_x$ while performing a psychomotor task. Performance decrement increased as acceleration levels increased. The physiologic data were obtained from our routine medical monitoring procedures, which are always carried out for subject safety on the centrifuge. This approach is clearly not optimal for obtaining physiologic data.

The thirteenth study was unusual in several ways. Two USAF helicopters flew non-stop across the Atlantic, with nine in-flight refuellings. They set several world's records. The flight was also an operational test of a new operational capability for helicopters. The timing was such that they arrived in Europe during the Paris Air Show, thereby demonstrating a national technologic achievement. A flight surgeon (who was also the principal investigator) was a member of one crew. He collected urine samples, recorded ECGs, obtained fatigue and performance data. A linear increase in fatigue and endocrine signs of stress were obtained. Transient ECG changes were seen in response to the stress of refuelling.

In the fifteenth study, subjects were conditioned to one in a list of ten nonsense syllables and then went to bed on a normal sleep schedule. They were wired for the conventional psycho-physiologic measures. The nonsense lists were read at intervals, and the tracings were scored for signs of responses to the key syllable. We found that they responded in a non-differentiated way to any disturbing stimuli. As the reader can see, the experimental question was straightforward. Do people respond in a special way to special stimuli while sleeping?

These brief descriptions of each of the fifteen studies are intended to give the reader a better feel for the kind of material to be discussed in the remainder of this essay. I should state specifically that my group has functioned as part of the team on most of these studies, and that in most cases the unique characteristics of the research was due to the leadership of the principal investigator. The next and more important step is to summarize and integrate these 15 papers in a manner which addresses itself to the relationship between the physiologic and psychologic outcomes. The initial step will be taken in the first table. Table 1 summarizes each of the studies, focusing on procedures, outcomes and relationships. There is some detail in this table which augments the descriptive paragraphs above, and permits the reader to reconstruct the studies more fully.

Table 2 is a simplification of results from the 15 papers, categorizing the results in the clinical medical ('symptoms'), physiologic and psychologic domains. As Table 2 shows, procedures appropriate to exploring the relationships were not done in four of the studies (Dunn 1962, Hartman *et al.* 1964, Hartman *et al.* 1967, McKenzie 1969); they have been reported here because they reflect some element of the problem, either by orientation (Dunn 1963 and Hartman *et al.* 1964) or by procedural limitations (Hartman *et al.* 1967, McKenzie 1969). Three involved psychologic variables which were, at best, only indirectly oriented towards physiologic concerns (McKenzie *et al.* 1961, Hartman *et al.* 1964, Rodgin *et al.* 1966). All three involved prolonged confinement as a major aspect of the study. The decision to direct psychologic manipulations away from physiologic and towards psychologic questions was based in part on our earlier experiences with the essentially unstressful nature of confinement when the subjects had a meaningful job to perform. The remaining papers fit more comfortably into the focus of this review.

Table 3 is a further simplification of the 15 papers. (It becomes necessary to assume that this simplification is not a gross departure from the papers being reviewed.) Table 3 suggests that:
- more than half of the studies involved some meaningful or presumptive stress;
- several studies yielded a general kind of relationship between physiologic and psychologic outcomes;

Experiences in interdisciplinary studies

Table 1. Summary of papers related to physiologic/psychologic relationships

Abbreviated title	Physiol focus	Psychol focus	Psychol outcomes	Physiol outcomes	Physiologic/Psychologic relat.
1. Water-immersion (weightlessness) effects (*Aerospace Med.*, 1961).	Cardiovascular, neuromuscular, and work-capacity effects.	Simple psychomotor tests to measure changes across days of immersion. Sleep changes.	Small performance decrement across days. Gross performance loss post-immersion. Major changes in sleep quality (stage) and quantity.	Extreme physiologic changes approaching incapacitation, post-immersion.	Psychomotor and physiologic disruptions post-immersion agree well and could be assumed to be causally related.
2. Performance factors: simulated space flight (*Aerospace Med.*, 1961).	Wide range of classical medical and biomedical effects.	Psychomotor test battery: signal rate, work/rest schedule, circadian effects, effects across 17 days, altitude.	Scattered decrements, note related systematically to psychologic variables.	No systematic physiologic decrements of operational significance.	Physiol. effects which are essentially benign do not provide a setting for revealing relationships. Psychol. variables not chosen primarily to explore relationship.
3. Psychomotor functioning with O_2–N_2 (SAM-TR-62-82, 1962).	Psychomotor concomittants of known physiologic stress (nitrogen narcosis). No direct physiol. measures.	Performance changes on a multi-element tracking task.	No significant psychomotor decrement re N_2. Some O_2 improvements, unexpected.	Not evaluated except by observation.	Structure of study does not permit evaluating relationships.
4. Increased O_2 at altitude (*Aerospace Med.*, 1963).	Wide range of classical medical and biomedical effects.	Psychomotor test battery: general psychomotor effects across days, with secondary emphasis on circadian and work/rest effects.	No operationally important decrements.	No operationally or medically important changes.	Essentially benign physiologic effects do not provide a setting for evaluating relationships.
5. Extended confinement: Minuteman launch centre (SAM-TR-64-62, 1964).	Diaries reviewed for symptomatic complaints.	Simulated launch equipment, continuously manned, with occasional launch exercises.	No significant psychomotor changes; some sleep changes (stage and duration) of theoretical interest.	No major observations.	Structure of study does not permit evaluating relationships.
6. CO_2 at altitude (*Aerospace Med.*, 1964).	Wide range of classical medical and biomedical effects.	Psychomotor test battery: performance measures as an explicit element of physiologic status.	No significant psychomotor changes.	Physiologic changes primarily reflected appropriate compensation and acclimatization to environmental stress.	Physiologic changes essentially benign so far as functional status of subjects is concerned do not provide a setting for evaluating relationship.
7. Drug effects: tactical air mission (*Aerospace Med.*, 1965).	Psychophysiologic changes (EKG, respir.). Concurrent urine sampling, with separate report.	Simulated flying task: psychomotor changes re drugs and across time.	Significant drug 'hangover' and significant stimulant effect; typical fatigue effects.	No significant psychophysiologic changes; urinanalyses showed moderate to strong stress responses in general.	A time-locked relationship between (urine) stress measures and performance was not obtained but a general relationship was shown.

Study	Measures	Method/Task	Performance effects	Physiologic changes	Relationship conclusion
8. 56-day simulated space-flight (*Aerospace Med.*, 1966).	Wide range of medical and biomedical effects, classical and space-specific.	Psychomotor test battery: changes across days. Sleep and temperature effects across days.	No operationally significant effects.	Physiologic changes primarily reflected adaptation to the atypical environment.	Adaptive physiologic effects do not provide a setting for evaluating the relationship.
9. Hypercapnia (Lectures, 1967).	Medical variables specific to the physiologic mechanisms provoked by hypercapnia.	Psychomotor test battery: simple performance tasks reported to reveal hypercapnic stress (RT, vigilance, etc.).	No significant psychomotor effects.	Physiologic changes were typically adaptive except for small reduction in exercise tolerance.	Adaptive physiologic changes are basically functionally benign and do not provide a setting for studying the relationships.
10. Work/rest (SAM-TR-67-99, 1967).	None, though the study was directed at showing a disturbance in 'physical reserves'.	Many different performance tests.	Differential effects related to work/rest schedules were obtained during sleep-deprivation and recovery.	Not studied.	This study was a prime candidate for a physiologic/psychologic effort; no physiologic measures were obtained.
11. Acetazolamide (SAM-TR-68-65, 1968).	Blood studies, respiratory and exercise measures selected to reveal physiologic accommodation to altitude.	Simple psychomotor functions reported to reveal hypoxia (RT, hand-steadiness, etc.).	No systematic drug or altitude effects.	Minimal physiologic accommodation which was not seen on replication.	Small physiologic insult yields minimal physiologic changes without concommitant psychomotor changes. This was one of the better tests of the relationship.
12. Acceleration stress (*Aerospace Med.*, 1968).	EKG, blood pressure, and EEG; these measures served as both physiologic and safety procedures.	Discontinuous tracking task performed before, during, and after acceleration.	Significant psychomotor decrement, differentially related to acceleration level; mechanical load and distractions seemed to be primary cause.	Significant but not completely systematic physiologic effects; the psychological factor of anxiety seemed to play an important role.	The relationship was not time-locked, and physiology appeared to be driven by the psychologic mechanism of anxiety, appropriate to the stressor.
13. Helicopter flight (*Aerospace Med.*, 1969).	EKG and urine to assay operational stress.	Observations and rating scales to evaluate fatigue and performance.	Progressive increase in fatigue over time (32 hours), no major performance decrement observed.	EKG revealed 'acute' stress effects of refuelling. Urinary measures showed flight to be a 'non-specific' stressor.	A general but not time-locked agreement between subjective fatigue and urinary (endocrine-metabolic) measures. No agreement between EKG and the other measures.
14. Anti-malarial drug (SAM-TR-in press).	Wide range of clinical medical procedures to evaluate drug effects.	Simulated flying task supplemented by choice-reaction time task to evaluate drug effects.	No significant psychomotor effects relative to treatment. A learning curve was seen.	No significant clinical changes.	Drug treatments which are essentially benign do not provide a setting in which to evaluate the relationship.
15. Autonomic responses during sleep (SAM-TR-69-30).	Psychophysiologic measures were primary (heart rate, respiration, EEG, GSR/BSR).	Classical conditioning to shock to establish differential responses prior to experimental sessions.	Behaviourally, Ss responded in a non-specific way to any stimulation during sleep.	Psychophysiologically Ss showed altered respiration patterns and shifted to a lighter stage of sleep, non-specific.	Study not really structured to permit a test of the relationship.

■ half of the studies resulted in no significant differences in physiologic or psychologic data, analysed independently.

Concerning this latter point, one can speculate about the physiologic/psychologic relationship revealed in these studies only if one accepts the meaningfulness of 'no significant differences': e.g., accepts the null hypothesis.

Table 2. Categorization of results on physiologic/psychologic relationships

Reference	Symptoms	Decrements Physiol	Psychol	General relationship
1. Zero g	yes	yes	yes	Gross disruption appeared in all areas
2. Sim. space flt.	scattered*	no	scattered	Psychol. variables were not really oriented to concurrent physiol. concerns
3. H_2 effects	not done	not done	yes	No concurrent data
4. O_2 effects	no	no	no	Minimal stress levels
5. Confinement	from diary	not done	no	Minimal stress levels
6. CO_2 effects	scattered*	compensatory	no	Minimal stress levels†
7. Drugs/fatigue	not done	urine—yes psychophy—no	drugs—yes fatigue—yes	Urine—grossly related but not time-locked Psychophysiol.—not sensitive to other than 'acute stress'
8. Sim. space flt.	scattered*	no	no	Minimal stress levels
9. CO_2 effects	scattered*	compensatory	no	Minimal stress levels
10. Work/rest	not done	not done	yes	Important relationships are implied by the emphasis on depletion of physical reserves
11. Drugs/altitude	scattered*	compensatory for altitude	scattered	Minimal stress levels
12. Acceleration	no	scattered	—	The mechanical impediments to efficient performance had no systematic effect on physiol., which reflected only the anticipatory anxiety
13. In-flt. fatigue	EKG—yes?	urine—yes	subj. fat.—yes puf.—no	Urine—grossly related but not time-locked to subjective changes EKG—reflected a different (acute) aspect of stress
14. Anti-malaria drug	scattered*	no	no	Minimal stress levels
15. Autonomic responses	not done	psychophy—yes	not done	Psychophysiol. effects reflected in a gross sense the 'acute stress' of a disturbing stimulus during sleep

*Scattered clinical changes of medical interest could not usually be explained by the stressor or the related physiologic responses and/or mechanisms.

†Though these stress levels are described here as minimal, they exceeded CO_2 concentrations predicted from the literature to be definitely stressful.

Table 3. Further categorization and simplification of results on the physiologic/psychologic relationships

Reference no.	Meaningful stress Yes	No	Meaningful relationship Yes	No	Accept null hypoth.
1	×		×		
2		×			×
3	×		implied		
4		×			×
5		×			×
6	implied				×
7	×		grossly		
8		×			×
9	implied				×
10	×		implied		
11	×				×
12	×		grossly		
13	×		grossly		
14		×			×
15	grossly		grossly		

Some support for accepting a mixture of relationships between physiologic and psychologic outcomes in physiologic stress studies is shown in Figure 1. This figure proposes that the time-course of physiologic effects differs considerably from psychologic, that psychologic changes will lag behind physiologic or will require more insult before they are elicited, and that mixed relationships are probable when sets of physiologic/psychologic studies are examined. (However, an alternative and critical possibility is that there is indeed a stable relationship, which will be revealed whenever the correct procedures and instrumentations are used. I have some reservations of this sort about the studies being reviewed here, particularly with regard to 'multi-element, general purpose' test batteries. This problem, however, calls for a separate essay.)

Figure 1. Relationship between many physiologic and psychologic variables

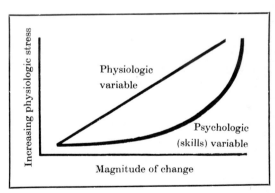

3. Final comment

This review has led in a step-wise fashion to the proposition that there is a mixture of relationships between outcomes from physiologic and psychologic data obtained in joint studies on the effects of physiologic stress. Fifteen studies done by my group have served as a vehicle for presenting this proposition. In addition to the reservation about measurement tools just expressed, I should add that perhaps the mission of this organization limits the scope of the problems studied (medically-oriented applied problems from the operational environment, frequently involving reasonably low intensities of the stressors being investigated). The time/intensity characteristics of many stressors places our studies at that point on the horizontal axis of Figure 1 where we would expect only small changes in the physiology and performance of man. This incidental limitation may be the major shaping function for the conclusions one can reach from these studies.

References

BUCKLEY C.J., and HARTMAN B.O., 1969, Aeromedical aspects of the first non-stop transatlantic helicopter flight: Part I. *Aerospace Medicine*, **40**, 710–713.

CANTRELL G.K., and HARTMAN B.O., 1970, *Psychomotor Effects of a Malarial Prophylaxis Regime*. SAM-TR-(in press), USAF School of Aerospace Medicine, Brooks AFB, Texas.

CUTLER R.G., ROBERTSON W.G., HERLOCHER J.E., McKENZIE R.E., ULVEDAL F., HARGREAVES J.J., and WELCH B.E., 1964, Human responses to carbon dioxide in the low-pressure, oxygen-rich atmosphere. *Aerospace Medicine*, **35**, 317–323.

DUNN J.M., 1962, *Psychomotor Functioning While Breathing Varying Partial Pressures of Oxygen-Nitrogen*. SAM-TR-62-82, USAF School of Aerospace Medicine, Brooks AFB, Texas, June.

GLATTE H., HARTMAN B.O., and WELCH R.E., 1967, *Nonpathologic Hypercapnia in Man*. Lectures in Aerospace Medicine, SAM Special Report, USAF School of Aerospace Medicine, Brooks AFB, Texas, February.

GRAVELINE D.E., BALKE B., McKENZIE R.E., and HARTMAN B.O., 1961, Psychobiologic effects of water-immersion-induced hypodynamics. *Aerospace Medicine*, **32**, 387–400.

HARTMAN B.O., and CANTRELL G.K., 1967, *MOL: Crew Performance on Demanding Work/Rest Schedules Compounded by Sleep Deprivation*. SAM-TR-67-99, USAF School of Aerospace Medicine, Brooks AFB, Texas, November.

HARTMAN B.O., and CRUMP P.P., 1968, *Psychomotor Effects of Low Doses of Acetazolamide to Aid Accommodation of Man to Altitude*. SAM-TR-68-65, USAF School of Aerospace Medicine, Brooks AFB, Texas, July.

HARTMAN B.O., FLINN D.E., EDMUNDS A.B., BROWN F.D., and SCHUBERT J.E., 1964, *Human Factors Aspects of a 30-day Extended Survivability Test of the Minuteman Missile*. SAM-TR-64-62, USAF School of Aerospace Medicine, Brooks AFB, Texas, April.

LITTLE V.Z., LEVERETT S.D., and HARTMAN B.O., 1968, Psychomotor and physiologic changes during accelerations of 5, 7 and 9 $+G_x$. *Aerospace Medicine*, **39**, 1190–1197.

McKENZIE R.E., 1969, *A Study of Autonomic Response to Auditory Stimulation During Sleep*. SAM-TR-69-30, USAF School of Aerospace Medicine, Brooks AFB, Texas.

McKENZIE R.E., and ELLIOTT L.L., 1965, *The Effects of Secobarbital and d-amphetamine on Performance During a Simulated Tactical Air Mission*. SAM-TR-64-79, and *Aerospace Medicine*, **36**, 774–779.

McKENZIE R.E., HARTMAN B.O., and WELCH B.E., 1961, Observations in the SAM two-man space cabin simulator. III. System operator performance factors. *Aerospace Medicine*, **32**, 603–609.

MORGAN T.E., Jr., CUTLER R.C., SHAW E.G., ULVEDAL F., HARGREAVES J.J., MOYER J.E. McKENZIE R.E., and WELCH B.E., 1963, Physiologic effects of exposure to increased oxygen tension at 5 p.s.i.a. *Aerospace Medicine*, **34**, 720–726.

RODGIN D.W., and HARTMAN B.O., 1966, Study of man during a 56-day exposure to an oxygen–helium atmosphere at 258 mm. Hg total pressure. XIII. Behaviour factors. *Aerospace Medicine*, **37**, 605–608.

Dr. Michon is a psychologist, and has pursued
fundamental and applied research in the Institute for
Perception RVO-TNO, Soesterberg, Netherlands, and
also in England and the U.S.A. He is now head of the
Department of Road User Behaviour at Soesterberg.
Dr. Fairbank was a visiting member of the Department
in 1969 and is now at New Mexico State University.

Measuring driving skill

J. A. Michon and B. A. Fairbank Jr.

1. Introduction

The experimental approach to human factors in the driving situation has been very unsystematic. As an area of investigation it has in the past attracted only the occasional attention of a
few outstanding research workers, and unlike some tasks that are predominantly military, and
where therefore research was stimulated by military resources, traffic research has mostly
remained at the implemental level of state and county traffic departments. Only now, as traffic
becomes, like other ecological pollutants, a major threat to the human species, can we see a
gradual increase in the amount of effort and funds being spent in traffic research.

Even though many countries have one or more special bodies to deal with the planning and
coordination of this research, few general principles emerge. We lack a systems analysis of the
traffic situation and of the factors that contribute in one way or another to the phenomenon. To
some extent this is understandable. The sudden threat of overcrowding and neglect of the
problem in the past have necessitated 'crash' programmes to give an immediate answer to
questions that are often scientifically trivial. Under this pressure too little time has been
devoted to fundamental questions about the conceptual frame of references in which research
findings ought to be placed. Only if the workers in the area are given ample funds and time to
develop and chart the area, may we expect future research projects to contribute to a firm body
of fundamental knowledge about traffic.

In a few areas within this conceptual chaos we find sporadic patches of knowledge that seem
to rest on a sound basis of prior knowledge. As a result an excess of research effort is spent in
these areas, which exceeds their relative importance in the whole. The attention given to
specification of street lighting, visual acuity and legibility are examples of such overvalued
areas. This is not to discredit the work that has been going on. On the contrary, we feel that
workers, who have been plodding around in a mass of phenomena that just will not fit nicely
together, may envy those colleagues who stand on such solid ground. In the future, however,
we ought to reassign weights to various factors to avoid a concept of the driving situation that
reduces practically everything to 'eyeball factors' (Figure 1).

Our task is to review some methodological points that are currently of importance in the
analysis of driving skill. While realizing the necessity of considering it in the light of traffic flow,
legislative, technological, social and other factors, we may nevertheless study driving behaviour in its own right. In doing so we are dealing with the psychology of a modern centaur:
man and car form a unit that cannot be separated functionally, because man's reactions are
largely determined by the possibilities of the car he is steering.

This sets the stage for a classification of studies in this area. While the perceptual apparatus
of the centaur is strictly human, the input presented to it is not: the scale of some of the input
dimensions like velocity, lateral displacement, or persistence of visual information are changed
in many respects. An analysis of these specific requirements has led to research in dynamic
visual acuity, reflectivity of road signs, symbolic indications, peripheral cues, and various
others. Many of these factors derived from an interest in flight performance.

The control operations which result from decisions about a probabilistic situation which has
many invariants, and which seems to surpass human capacity, may become problematic

Figure 1. The 'eye-ball' approach to driving skill, showing an excessive concern for the input aspect (From Allen 1967, p 118).

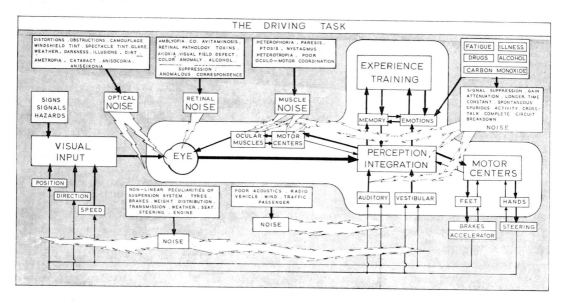

primarily through their speed of succession rather than their expectancy. In driving, these control operations form a well specified environment however : the number of response alternatives is vastly smaller than it is in flying, and it should therefore be fairly easy to define the range of possibilities that are in any given situation. The means of analyzing the immediate control manipulations have become available by extensive research in various control situations. Little however is known about the physiological functioning of the human part of the centaur during the driving task, or even about some of the more complex psychological functions such as attention, anxiety and arousal in general.

The constraints placed on the motor behaviour of the human agent are largely determined by the dynamic properties of the vehicle. The vehicle is the best understood part of the centaur. The ease with which the driver assimilates the dynamics of the vehicle, and his sensitivity to its parameters is another aspect of the driving skill. Included in the effect of handling the vehicle according to its dynamic properties is the feedback that derives directly from the vehicle's response. Velocity, for instance, is found to be directly related to the centrifugal force experienced in curves.

2.   Research methods

The methods used to investigate these various details have been derived from psychological and physiological practice. Generally speaking there are three main lines of empirical data collection : first, the laboratory experiment, which may range from a test battery to a sophisticated simulator study ; and second, the field study, ranging from a test track, rigid control experiment to full traffic measurements obtained in instrumented cars of greater or less sophistication. The third category is the unobtrusive traffic observation, usually carried out with the aid of slow movie cameras or electronic counting equipment. In the latter case we touch upon traffic flow studies, and interpretation of such data in terms of individual behaviour may be somewhat difficult, if it is attempted at all.

In reviewing the available literature the most sensible way of summarizing the data is to collect them in a table (see Appendix), in which the entries in one direction refer to the aim of the study, placed in the framework of the preceding exposition. The other, orthogonal, entries refer to the methods used. Thus in the appended table the rows represent the 'ends' and the columns represent the 'means' of the various studies. The numbers in the cells refer to the list of references. Of course stating means and ends, is only half the story. Where does a means-end analysis lead us?

We might try to specify some of the basic problems which arose in the studies referred to. Such specification is hardly feasible however, within the framework of the present paper. We shall try instead to sketch a research typology.

What becomes immediately clear is the overall lack of insight into what constitute reliable criteria of driving skill. Many of the variables that have been studied do not differentiate between conditions that, at least to the naive investigator, appear to be highly distinct. An example is the negligible effect of fatigue during prolonged driving (Brown 1967). Another point, essentially the inverse of the previous one, is that it is in no way clear what some of the significant differences that are observed here and there, mean in terms of driving. Is, for example, an increase in velocity when the driver is listening to Tijuana Brass music bad or good (Konz and McDougal 1968) and what about observed changes in heart rate or other psychological measures? Are they positive or negative signs?

We are led to the conclusion that the literature does not provide a generally accepted procedure for determining whether or not a driver is skillful in the sense of being a 'good' driver. In fact most of the existing studies of driver behaviour have not even been carried out with the object of measuring driver skill. Instead, many arbitrarily selected variables more or less related to driving behaviour have been measured, controlled, evaluated, or predicted. In sampling the literature related to driver behaviour, the task of one interested in driver evaluation must obviously be to determine how (or even if) each of the available studies of driving behaviour relates to driving skill. The question of whether or not a given driver is skilful is one of obvious practical importance.

Let us therefore consider driving skill from this point of view. How does one quantify the driving skill? To do this an *a priori* criterion for skilful driving is required, or at least an operational definition of 'skilful driving'. Unfortunately neither such a criterion nor such a definition exists. The difficulties in establishing an appropriate criterion are many, but the most obvious one is that the skill of a driver cannot be measured adequately unless one considers the interaction of the driver's car with the total traffic situation.

But here the problem amounts to the difficulty of specifying performance in real task conditions. We cannot expose our subjects deliberately to the hazards of the road; so we can never push the system really to its limits, and hence we have to satisfy ourselves with a somewhat less lethal approach.

Accident records suggest themselves; but as such are hardly adequate. A driver whose record indicates no accidents in the last ten years may nevertheless have caused accidents in which he was not involved. Furthermore, a driver may create a dangerous situation which does not actually cause an accident, yet which is detrimental to good traffic flow. As a further example, consider the typical New York City taxi driver: his accident record may be good, but few who have driven in New York would want every driver to behave the way some New York taxi drivers do.

Thus any measurement or combination of measurement which is confined to what is happening within the automobile will be insufficient to quantify driver skill fully, because we must also take into account what effect the driver's behaviour has on traffic in general. One consequence of this is that the development of an automatic or mechanical driving skill evaluator is not going to be a simple accomplishment.

The best way to find a criterion for 'good' driving would seem to be at first to establish several criteria with 'face validity' (i.e. criteria which seem to an experienced traffic worker to be valid), then select a number of drivers who meet these criteria, and finally measure as many aspects of the 'good drivers' behaviour as possible, and attempt to find differences between those measures and the same measures taken on a randomly selected group of drivers. As will be seen presently, a somewhat similar technique has recently been used to identify the characteristic behaviour patterns of 'bad' drivers (Quenault 1967). There is already a great deal known about the behaviour of randomly selected drivers; research into 'good' drivers could be compared against this mass of data.

The criteria for selecting the good drivers must be somewhat arbitrary: accident-free driving, no traffic citations, a reasonable number of kilometres driven per year, a mixture of city and highway driving, the agreement of, say, two referees that a subject's driving seemed 'good' might be a first approximation of such criteria. Given a subject population which meets these criteria, a series of measurements could be taken during driving to gather as much information as possible on the driving habits of the population. Until such a study is undertaken, however, research which claims to measure 'good driving', or which makes value judgments of driver behaviour must be seen as little more than educated guesswork.

Bearing these cautions in mind, it is nonetheless possible to comment on some of the relevant literature. The reports selected for comment here are only a few of the many which are available, but they do represent four fairly well-defined kinds of study. For a more inclusive list of papers see the Appendix.

Ritchie *et al.* (1968) examined the relationship between a vehicle's forward speed and lateral acceleration on curves. One of their hypotheses is that 'lateral force produced will be an inverse function of the speed'. This hypothesis is supported by the data. Their other hypothesis 'that when a driver chooses a speed he is predicting the lateral force which he will feel in the curve' is not tested in the examination of the data, nor are the data sufficient to permit such a test. This study is representative of many in that it selects a well-defined area of driving for examination, gathers data relevant to what drivers actually do in a driving situation, and reaches some conclusion which adds to our knowledge of driver behaviour. Unfortunately it is also typical in that it provides no clue as to how the behaviour described is related to the evaluation of driving skill. Furthermore, the authors conclude by suggesting further corollary hypotheses which are of greater relevance to driver evaluation than are the hypotheses tested. Unfortunately this too is a common feature of such studies.

In a quite different research tradition, attempts have been made to correlate personality variables or clinical psychological conditions with driving performance. Two such studies may be mentioned here, although both suffer from the weakness that lack of accidents is an implicitly accepted criterion of 'good driving'. In one, by Selzer *et al.* (1968), drivers who caused fatal accidents showed significantly more psycho-pathology and social stress than did a control group, a finding which clearly implies that it may eventually be possible to identify a 'high risk' category of drivers on the basis of personality variables. In a similar study, Spangenberg (1968) showed that existing standard psychological tests may be interpreted in ways which permit significant correlations between test scores and traffic accident likelihood. Again, the idea is supported that a safe (and thus presumably a 'good') driver may be detected by measuring behaviour at times other than when actually driving.

In a third representative type of driver behaviour investigation, a laboratory study is carried out, the findings are verified in a real traffic situation, and recommendations for drivers are made. Such studies may be somewhat more useful in driver evaluation and education than those just discussed, since it should be possible to determine whether or not drivers perform in accordance with the recommendations. A study of this type is reported by Crawford (1963). He investigated the behaviour of drivers overtaking and found that drivers should not attempt to overtake when they doubt the safety of the manoeuvre. In the cases where doubt exists, not only is the margin of the safety smaller, but the driver also takes longer to make his decision, thus further reducing his margin of safety. The advice not to overtake in a marginal situation is good advice, and might possibly be useful in the evaluation of drivers, but the large variability of situations judged 'safe' for overtaking means that a very large sample of a driver's judgments would be needed to evaluate this performance.

A fourth class of studies encompasses those which investigate psychological variables and relate them to driving performance. Occasionally these studies demonstrate propositions which are contrary to common expectations and thus are extremely valuable. Brown (1967) showed that driver fatigue is a relatively unimportant factor in traffic accidents, despite considerable speculation to the contrary. He did not show that fatigued drivers were no more likely to have accidents than unfatigued ones, but rather that in accident statistics fatigue is an unimportant causal factor compared with inexperience, excessive speed, and adverse road conditions.

These studies illustrate the approaches currently adopted in studies of driving behaviour; it is evident that the results, regardless of whether they are interesting and useful in their own right, are not easily applied to measurement of 'driver proficiency'. Two recent studies, however, illustrate the direction that research might take if experimental work is to contribute to what we might call the technology of driver evaluation. The first one (Herbert, 1963) submitted driving to factor analysis and identified five factors contributing to 'driving skill', while the second (Quenault 1968) compared ratings made of drivers who had and who had not been convicted for careless driving. In the Herbert experiment drivers were required to perform twelve complex driving tasks, none of which involved interaction with any other vehicle. The tests were actually of driving precision, rather than the more vaguely defined concept of driving skill. Herbert found that five factors contributed to driving precision: multilimb coordination, spatial orientation, proprioception, response orientation, and reaction time. No attempt was made in the report to define skillful driving operationally, rather it was assumed that those who scored high on the twelve tests are skillful, and the skill thus measured was analyzed. The ability to drive precisely may be a prerequisite to driving 'skilfully'; thus Herbert's report may be a good starting point in the establishment of a criterion for skilled driving.

Quenault examined driving on a level quite different from that of Herbert's analysis. The author selected two populations of drivers, the general population and a population of drivers who had been convicted of careless driving. He then took samples of fifty drivers from each population, and measured their driving behaviour. He found that those with convictions took significantly more risks, had a higher overtaking/being overtaken ratio, and made more unnecessary manoeuvres. Thus if we assume that those who have been convicted of careless driving are bad drivers, we now have a first indication of what may go into the making of bad drivers, or how bad driving differs from ordinary driving.

The task ahead is to determine, not how bad driving differs from the norm, but rather how good driving may be identified taught, and promoted. Studies similar to that of Quenault, that is, careful ethological observation of the centaur, offer the best chance of identifying good driving.

The driving situation offers a fascinating challenge. It is a man-machine symbiosis, in which man is fully and permanently engaged, and in which his life depends on his performance. The behaviour of the system is not yet well defined, and certainly the external influences are not.

Since however, a large proportion of the population is engaged in driving, the cost—benefit balance is very favourably biased toward the research side. This balance however, is constrained by some negative factors. More than most other areas of human factors work, traffic is a political affair: press, politicians, and quite powerful pressure groups all interfere with traffic research, and usually cause a long delay in the implementation of practical measures. Even more unfortunate is that their immediate needs also interfere with the basic build-up that is required in order to overcome the current highly piecemeal approach to traffic as a major ecological problem.

References

ALLEN M.J., 1967, The visual environment in the modern automobile. In *The Prevention of Highway Injury*. (Ann Arbor: Highway Safety Research Institute), pp. 118–121.

BROWN I.D., 1967, Car driving and fatigue. *Triangle*, **8**, 131–137.

CRAWFORD A., 1963, The overtaking driver. *Ergonomics*, **6**, 153–170.

HERBERT M.J., 1963, Analysis of a complex skill: vehicle driving. *Human Factors*, **5**, 363–372.

KONZ S. and McDOUGAL D., 1968, The effect of background music on the control activity of an automobile driver. *Human Factors*, **10**, 233–244.

QUENAULT S.W., 1963, Driver behaviour and unsafe drivers II. *Road Research Laboratory Report LR* 146.

RITCHIE M.L., McCOY W.K., and WELDE W.L., 1968, A study of the relation between forward velocity and lateral acceleration during normal driving. *Human Factors*, **10**, 255–258.

SELZER M.L., ROGERS J.E., and KERN S., 1968, Fatal accidents, the role of psychopathology, social stress, and acute disturbances. *American Journal of Psychiatry*, **124**, 1028–1036.

SPANGENBERG H.H., 1968, The use of projective tests in the selection of bus drivers. *Traffic Safety Research Review*, **125**, 118–121.

Appendix

The list of references in this appendix is not intended to be exhaustive, but rather representative of the kinds of work that have been done in the field of driving skill in the last few years. For easy reference the items of the list have been tabulated in a table in which the 'means' used as the principal method, and the 'ends' of the various studies appear as column and row entries respectively. The 'ends' of a study determine the 'means' to a certain extent; for example one would expect controlled studies to be popular in alcohol investigations, and it would be difficult to study visual factors by observing traffic patterns. Within these broad restrictions, however, the table shows a wide diversity of both ends and means of investigation. To some extent the classification of the various studies is arbitrary, and of course the choice of twelve ends classifications and eleven means is completely arbitrary. The studies which overlapped more than one area are listed in all applicable cells of the table. Thus although there are only 98 references, the table has 160 entries. Finally, it should be noted that a few papers listed in the references were unavailable to the authors in their complete form, and abstracts had to be utilized for classification purposes. In such cases, minor errors of classification may have occurred; the authors apologize to anyone thus misrepresented.

1. ALLEN M.J., 1964, Automobile dash panel visibility impairment due to wearing fit-over sun glasses. *American Journal of Optometry*, **41**, 592–598.
2. ALLEN M.J., and CARTER J.H., 1964, Visual problems associated with motor vehicle driving at dusk. *Journal of the American Optometric Association*, **35**, 25–30.
3. ARMED FORCES–NRC COMMITTEE ON VISION, 1958, *The Visual Factors in Automobile Driving*. Symposium Summary. NAS–NRC Publication no. 574, Washington.
4. ARTHUR D. LITTLE, INC., 1966, *The State of the Art of Traffic Safety* (Cambridge, Mass.: A. D. Little, Inc.).
5. BARRETT G.V., KOBAYASKI N., and FOX B.H., 1968, Feasibility of studying driver reaction to sudden pedestrian emergencies in an automobile simulator. *Human Factors*, **10**, 19–26.
6. BETZ M.J., and BAUMAN P.D., 1967, Driver characteristics at intersections. *Highway Research Record*, **195**, 34–51.
7. BJÖRKMAN M., 1963, An exploratory study of predictive judgments in a traffic situation. *Scandinavian Journal of Psychology*, **4**, 65–76.
8. BLANCHE E.E., 1965, The roadside distraction—how big a rôle does it play in accidents. *Traffic Safety*, **65**, 24–25 and 36–37.
9. BOWLES T.S., 1969, Motorway overtaking with four types of exterior rear view mirror. *Institute of Electrical and Electronics Engineers. Conference Record no. 69 C 58-NNS*. Papers from International Symposium on Man-Machine Systems, Cambridge 8–12 September.
10. BROWN I.D., 1967, Car driving and fatigue. *Triangle* (The Sandoz Journal of Medical Science), **8**, 131–137.
11. BROWN I.D., 1967, Decrement in skill observed after seven hours of car driving. *Psychonomic Science*, **7**, 131–2.
12. BROWN I.D., 1968, Human factors in the control of road vehicles. *Electronics and Power*, 275–279.
13. BROWN I.D., 1968, Some alternative methods of predicting performance among professional drivers in training. *Ergonomics*, **11**, 113–21.
14. BROWN I.D., SIMMONDS D.C.V., and TICKNER A.H., 1967, Measurement of control skills, vigilance, and performance on a subsidiary task during 12 hours of car driving. *Ergonomics*, **10**, 665–673.
15. BURNS N.M., BAKER C.A., SIMONSON E., and KEIPER C., 1966, Electrocardiogram changes in prolonged automobile driving. *Perceptual and Motor Skills*, **23**, 210.
16. BRYAN W. E., and HOFSTETTER H.W., 1958, A statistical summary and evaluation of the vision of drivers. *Journal of the American Optometric Association*, **29**, 112–117.
17. CALIFORNIA DEPARTMENT OF HIGHWAY PATROL, 1965, *The Roles of Carbon Monoxide, Alcohol and Drugs in Fatal Single Car Accidents*.
18. CRANCER A.jr., DILLE J.M., DELAY J.C., WALLACE J.E., and HAYKING M.D., 1969, Comparison of the effects of marihuana and alcohol on simulated driving performance. *Science*, **164**, 851–854.
19. CRANCER A.jr., and McMURRY L., 1968, Credit ratings as a predictor of driving behavior and improvement. *Washington (State) Department of Motor Vehicles Report no. 010.*
20. CRANCER A.jr., and QUIRING D.L., 1968, Driving records of persons arrested for illegal drug use. *Washington (State) Department of Motor Vehicles, Report no. 011.*
21. CRANCER A.jr., and QUIRING, D.L. 1968, The mentally ill as motor vehicle operators. *Washington (State) Department of Motor Vehicles, Report no. 013.*
22. CRANCER A.jr., and QUIRING D.L., 1968, Driving records of persons hospitalized for suicide gestures. *Washington (State) Department of Motor Vehicles, Report no. 014.*
23. CRANCER A.jr., and QUIRING D.L., 1968, Driving records of persons with selected chronic diseases. *Washington (State) Department of Motor Vehicles, Report no. 015.*
24. CRAWFORD A., 1963, The overtaking driver, *Ergonomics*, **6**, 153–170.
25. CRAWFORD A., and TAYLOR D.H., 1961, Driver behaviour at traffic lights. *Traffic Engineering and Control* **3**, 473–478, 482.
26. DARLINGTON J.O., 1967, Road accidents, human causes and general remedies. *Journal of the Institution of Highway Engineers*, **14**, 19–22.

Table 1.   References to investigations on driving skill

Ends \ Means	personality, social and legal variables	post-accident investigation; violation record	paper and pencil tests	observation of driver	medical and physiological techniques	lab and simulator studies	observation of traffic	controlled experiment	instrumented car and similar	mathematical models etc.	miscellaneous
general driver behaviour	41	8, 19	41, 94			41, 42, 57, 96	6, 48, 74	57	59, 83, 90	92	19
medical, including drugs	22	17, 22, 23, 31			17, 18, 67	18, 67		18, 67	81		
alcohol studies		17, 44, 95			17, 18, 95	18, 61, 70		18, 61, 70, 91	91		63
driver evaluation, safety criteria	20, 21, 42, 77	19, 20, 21, 31, 89, 97	13, 97	13, 78, 79, 80	21				39, 53, 75, 91		13, 19
effects of fatigue		10	43	11, 14, 43	15				14, 15, 86, 91		10
judgment, psycho-physics, visibility		97	72, 97			55		27, 30	2, 7, 27, 45, 53, 84, 85, 87		
visual factors					16, 19	51, 52, 70		70	2, 52, 81, 87, 98		3, 35
overtaking decision studies						55	24	24, 29	37, 93		
simulation and model building	89			28		5, 28, 92	6, 33, 34, 36, 40, 46, 60, 68, 88	60	85	54, 58, 60	
equipment and methodological			38	76		5, 69	73	1, 25, 73	9, 45, 49, 50	73	32, 73, 82
bibliographies, summaries, reviews		26									4, 12, 63, 64, 66
miscellaneous	56, 62	56		41, 71							65

27. DENTON G.G., 1966, A subjective scale of speed when driving a motor vehicle. *Ergonomics*, **9**, 203–210.

28. EDWARDS D.S., HAHN C.P., FLEISHMAN E.A., 1969, Evaluation of laboratory methods for the study of driver behavior: the relation between simulator and street performance. *American Institute Research Report*, R 69–7, *Washington D.C.*

29. FARBER E., and SILVER C.A., 1968, Behavior of drivers performing a flying pass. *Highway Research Record*, **247**, 51–56.

30. FARBER E., SILVER C.A., and LANDIS D., 1968, Knowledge of closing rate versus knowledge of oncoming car speed as determiners of driver passing behavior. *Highway Research Record*, **247**, 1–6.

31. FINESILVER S.G., 1969, *The Older Driver: A Statistical Evaluation of Licensing and Accident Involvement in 30 States and the District of Columbia* (Denver University: COLORADO COLLEGE OF LAW).

32. FORBES T.W., 1969, Factors in visibility and legibility of highway signs. *Paper presented at Annual Meeting of the National Academy of Sciences-National Research Council Committee on Vision*. May, 1969.

33. FORBES T.W., and SIMPSON M.E., 1968, Driver-and-vehicle response in freeway deceleration waves. *Transportation Science*, **2**, 77–104.

34. FORBES T.W., ZAGORSKI H.J., HOLSHOUSER E.L., and DETERLINE W.A., 1958, Measurement of driver reactions to tunnel conditions. *Highway Research Board Proceedings*, **37**, 345–357.

35. FRY G. A., 1967, The use of the eyes in steering a car on straight and curved roads. *American Journal of Optometry and Archives of American Academy of Optometry*, **45**, 374–391.

36. GEORGE E.T.jr., and HEROY F.M.jr., 1966, Starting responses of traffic at signalized intersections. *Traffic Engineering*, **36**, pp. 39, 40 and 43.

37. GORDON D.A., and TRUMAN M.M., 1968, Driver's decisions in overtaking and passing. *Highway Research Record*, **247**, 42–50.

38. GRAY P.B., 1964, Drivers' understanding of road traffic signs. *Traffic Engineering and Control*, pp. 6, 49, 50, 51, 53 and 65.

39. GREENSHIELDS B.D., and PLATT F.N., 1968, Objective measurements of driver behavior. *Highway Vehicle Safety, Collected SAE papers*, 1961–1967, (New York: SOCIETY OF AUTOMOTIVE ENGINEERS).

40. HARMS P.L., 1968, Driver following studies on the M4 motorway during a holiday and a normal weekend in 1966. *Road Research Laboratory, RRL Rep. LR 136.*

41. HEIMSTRA N.W., ELLINGSTAD V.S., and DE KOCK A.R., 1967, Effects of operator mood on performance in a simulated driving task. *Perceptual and Motor Skills*, **25**, 729–735.

42. HERBERT M.J., 1963, Analysis of a complex skill: vehicle driving. *Human Factors*, **5**, 363–372.

43. HERBERT M.J., and JAYNES W.E., 1964, Performance decrement in vehicle driving. *Journal of Engineering Psychology*, **3**, 1–8.

44. HIGHWAY SAFETY RESEARCH INSTITUTE, 1969, Alcohol safety. *Five Page Executive Summary in Highway Safety Literature*, HSL no. 69–17.

45. HOFFMAN E.R., and JOUBERT P.N., 1968, Just noticeable differences in some vehicle handling characteristics. *Human Factors*, **10**, 263–272.

46. HUBER M.J., and TRACY J.L., 1968, Operating characteristics of freeways. *National Cooperative Highway Research Program Report* 60, Part I HRB.

47. HURST P.M., 1964, Errors in driver risk taking. *Division of Highway Studies, Institute for Research, State College, Pennsylvania. Report* no. 2.

48. HURST P.M., PERCHONOK K., and SEGUIN E.L., 1968, Vehicle kinematics and gap acceptance. *Journal of Applied Psychology*, **52**, 321–324.

49. JANI S.N., and MENZES D.F., 1962, A comparison of seeing times using plane and convex mirrors. *British Journal of Physiological Optics*, **19**, 103–109.

50. JOHANSSON G., BERGSTROM S., JANSSON G., OTTANDER C., RUMAR K., and ORNBERG G., 1963, Visible distances in simulated night driving conditions with full and dipped headlights, *Ergonomics*, **6**, 171–172.

51. JOHANSSON G., and JANSSON G., 1964, Smoking and night driving. *Scandinavian Journal of Psychology*, **5**, 124–128.

52. JOHANSSON G., and OTTANDER C., 1964, Recovery time after glare. *Scandinavian Journal of Psychology*, **5**, 17–25.

53. JOHANSSON G., and RUMAR K., 1966, *Visible Distances and Safe Speeds during Night Driving Car Meetings.* (Uppsala University Department of Psychology).

54. JONES C.J. 1962, A theory of traffic collisions. *Medicine, Science, and the Law*, **3**, 489–499.

55. JONES H.V., and HEIMSTRA N.W., 1966, Ability of drivers to make critical passing judgments. *Highway Research* no. 122, 89–92.

56. KAESTNER N., and SYRING E.M., 1968, Follow-up of brief driver improvement interviews in Oregon. *Traffic Safety Research Review*, **12**, 111–117.

57. KAO H.S.R., 1969, A feedback analysis of eye-head angular displacements in human vehicular guidance. *Institute of Electrical and Electronics Engineers Conference Record no. 69 C 58–MMS*. Papers from International Symposium on Man-Machine Systems, Cambridge, 8–12 September 1969.

58. KIDD E.A., and LAUGHERY K.R., 1964, *A Computer Model of Driving Behavior; the Highway Intersection Situation.* (Buffalo, N.Y.: CORNELL AERONAUTICAL LABORATORY, INC.), *Rep. no.* VJ-1943–V–1.

59. KONZ S., and McDOUGAL D., 1968, The effect of background music on the control activity of an automobile driver. *Human Factors*, **10**, 233–244.

60. LAUGHERY K.R., and KIDD E.A., 1968, *Urban Intersection Study*, Volume 1, (Buffalo, N.Y.: CORNELL AERONAUTICAL LABORATORIES, INC.), *Rep. no.* VJ-2120–V–1.

61. LEWIS E.M.,Jr. and SARLANIS K., 1969, The effects of alcohol on decision making with respect to traffic signals. *Injury Control Research Laboratory Research Report* ICRL–RR–68–4 *Providence, Rhode Island.*

62. LITMAN R.E., and TABACHNICK N., 1967, Fatal one-car accidents. *Psychoanalytic Quarterly*, **36**, 248–259.

63. McFARLAND R.A., 1964, Alcohol and highway accidents—a summary of present knowledge. *Traffic Digest and Review*, **12**, 30–32.
64. McFARLAND R.A., 1965, *Publications in the Field of Highway Safety*. (New York: HARVARD SCHOOL OF PUBLIC HEALTH)
65. McFARLAND R.A., 1968, Psychological and behavioral aspects of automobile accidents. *Traffic Safety Research Review*, **12**, 71–80.
66. McFARLAND R.A., and MOSELEY A.L., 1954, Harvard School of Public Health. *Human Factors in Highway Transport Safety*.
67. MILLER J.G., 1962, Objective measurements of the effect of drugs on driver behavior, *Journal of the American Medical Association*, **179**, 940–943.
68. MONSEUR M., 1969, Driver's decision making at road intersection. *Institute of Electrical and Electronics Engineers Conference Record no. 69 C 58 MMS*. Papers from International Symposium on Man-Machine Systems, Cambridge, 8–12 September 1969.
69. MORROW I.R.V., and SALIK G., 1962, Vision in rear view mirrors. *The Optician*, **144**, 314–318, and 340–344.
70. MORTIMER R.G., 1963, Effects of low blood-alcohol concentrations in simulated day and night driving. *Perceptual and Motor Skills*, **17**, 399–408.
71. PERCHONOK K., 1964, The measurement of driver errors. *Division of Highway Studies. Institute for Research, State College, Pennsylvania. Report no. 1*.
72. PERCHONOK K., and HURST P.M., 1968, Driver apprehension. *National Cooperative Highway Research Program*, Rep. 60, Part IV HRB.
73. PERCHONOK K., and HURST P.M., 1968, Effect of lane-closure signals upon driver decision making and traffic flow. *Journal of Applied Psychology*, **52**, 410–413.
74. PERCHONOK K., and SEGUIN E.L., 1964, Vehicle following behavior, a field study. *Division of Highway Studies, Institute for Research, State College, Pennsylvania. Report no. 5*.
75. PLATT F.N., 1968, Operations research on driver behavior. *Highway Vehicle Safety, Collected SAE Papers 1961–1967*. (New York: SOCIETY OF AUTOMOTIVE ENGINEERS).
76. QUENAULT S.W., 1966, Some methods of obtaining information on driver behaviour. *Road Research Laboratory, RRL Rep. no. 25*.
77. QUENAULT S.W., 1967, Driver behaviour—safe and unsafe drivers. *Road Research Laboratory, RRL Rep. LR 70*.
78. QUENAULT S.W., 1967, The driving behaviour of certain professional drivers. *Road Research Laboratory, RRL Rep. LR 93*.
79. QUENAULT S.W., 1968, Driver behaviour safe and unsafe drivers II. *Road Research Laboratory, RRL Rep. LR 146*.
80. QUENAULT S.W., 1968, Dissociation and driver behaviour. *Road Research Laboratory, RRL Rep. LR 212*.
81. RAY A.M., and ROCKWELL T.H., 1967, An exploratory study of automobile driving performance under the influence of low levels of carboxyhemoglobin. *Unpubl. Master's thesis, Ohio State University*.
82. REILLY R.E., and PARKER J.F., 1966, *Visual Information Related to Vehicle Control During Overtaking*, (Arlington Va.: BIOTECHNOLOGY, INC.).
83. RITCHIE M., McCOY W.K., and WELDE W., 1968, A study of the relation between forward velocity and lateral acceleration in curves during normal driving. *Human Factors*, **10**, 255–258.
84. ROCKWELL T.H., and LINDSAY G.F., 1968, Driving performance. *National Cooperative Highway Research Program*, Rep. 60, Part II HRB.
85. ROCKWELL T.H., and SNIDER J.N., 1965, An investigation of variability in driving performance on the highway. *Systems Research Group, Ohio State University, Project RF 1450, Final Rep.*
86. SAFFORD R.R., and ROCKWELL T.H., 1967, Performance decrement in twenty-four hour driving. *Highway Research Record*, **163**, 68–79.
87. SALVATORE S., 1968, The estimation of vehicular velocity as a function of visual stimulation. *Human Factors*, **10**, 27–32.
88. SEGUIN E.L., and PERCHONOK K., 1965, Vehicle interactions on an urban expressway. *Division of Highway Studies, Institute for Research, State College, Pennsylvania, Report no. 6*.
89. SELZER N.L., 1969, Alcoholism, mental illness, and stress in 96 drivers causing fatal accidents. *Behavioral Science*, **14**, 1–10.
90. SENDERS J.W., and KRISTOFFERSON A.B., 1967, The attentional demand of automobile driving. *Highway Research Record*, **195**, 15–34.
91. SHAW W.J., 1957, Objective measurement of driving skill. *International Road Safety and Traffic Review*, **5**, 21–28.
92. SHERIDAN T.B., Vehicle handling; mathematical characteristics of the driver. In: *Highway Vehicle Safety, Collected SAE Papers 1961–1967*. (New York: SOCIETY OF AUTOMOTIVE ENGINEERS).
93. SILVER C.A., and FARBER E., 1968, Driver judgments in overtaking situations. *Highway Research Record*, **247**, 57–62.
94. SKELLY G.B., 1968, Aspects of driving experience in the first year as a qualified driver. *Road Research Laboratory, RRL Report LR 149*.
95. SMART R.G., and SCHMIDT W., 1967, Responsibility, blood alcohol levels, and alcoholism. *Traffic Safety Research Review*, **11**, 112–116.
96. SPURR R.T., 1969, Subjective aspects of braking. *Automobile Engineer*, **59**, 58–61.
97. VERHAEGEN P., HUYSKENS S., and VAN DER ELST A., 1969, Habitual decision making and accidents in private car drivers. *Institute of Electrical and Electronics Engineers Conference Record no. 69 C 58–MMS*. Papers from International Symposium on Man-Machine Systems, Cambridge, 8–12 September 1969.
98. WHALEN J.T., ROCKWELL T.H., and MOURANT R.R., 1968, A pilot study of drivers' eye movements. *Systems Research Group, Ohio State University, Rep. no. EES 277–1*.

Mr. Fox has a psychology degree. He has worked on industrial ergonomics problems since 1960, and is now Lecturer in Ergonomics, Engineering Production Department, University of Birmingham. His current interests are in industrial inspection, and the provision of ergonomics information for industry.

Ergonomics in production engineering

J. G. Fox

1. Introduction

There can be little doubt that ergonomics should have a role in the production (or industrial) engineer's approach to the adequate use of human and material resources in production systems. However, as in other areas of ergonomics, the identity of its contribution has only recently begun to emerge.

Dudley (1968) gave some direction, focussing attention on the need to develop research and application in the areas measurement, design and organisation of work. The benefits of an ergonomics contribution in these traditional areas of work study are clearly seen by comparison of columns 1 and 2 of Table 1 (Singleton 1969). An example of such development was demonstrated by Morrison *et al.* (1965). They compared work study raters' assessments of physiological load with a direct physiological measure, O_2 consumption. The very high correlations found between the volume of O_2 consumed by each subject and the Basic Minute Value for the task rated by a highly experienced work measurement practitioner suggested better ways of training and calibrating work study raters. This would be a not unimportant step in work measurement practice : for raters will continue for some time to estimate load via subjective estimates of performance where tradition or inconvenience prevents the use of more objective measures.

Table 1. Interaction of ergonomics with other behavioural technologies (Singleton 1969. By permission of W.H.O.)

	Ergonomics	Work study	Operational research	Cybernetics
Origin	Experimental human sciences	Shop-floor production problems	Mathematics and economics	Control system theory
Favoured procedure	Inferences from theory and experimental evidence	Generalizations from experience and observation	Construction of models	Simulation
Validation	Statistically controlled experimentation	Before/after comparisons	Check of model predictions against real occurrences	Compare simulated behaviour with real behaviour
Special interests	Human operator based systems	Human activity	Total systems	Biological/engineering analogies
Special techniques	Acquisition of behavioural evidence	Estimation by human observers	Various standard mathematical models	Comparisons of control systems
Advantages and limitations	Systematic and slow	Fast and superficial	Elegant but often oversimplified	Theoretically interesting but difficult to apply

The field having been claimed it had to be recognised that the nature of its problems was different from, for example, those of design or consumer ergonomics. Closer inspection would of course reveal that the boundaries between these areas were indeed fluid. But if the production ergonomist was to come to terms with his task, he had to realise that the problems of his design or consumer colleague were not his ; and that the difference in approach was not trivial. In the present context, techniques and criteria suitable in design of consumer evaluations were often not appropriate to production systems. The central difference between ergonomics in design and consumer satisfaction on the one hand and ergonomics in production on the other was that the latter had to be an on-line activity. Thus, for example, a meaningful and worthwhile question for the design ergonomist is the allocation of functions between man and machine based on known human capacities : but for the production ergonomist the question may very often be meaningless. He deals on-line with a system where functions evolved in the distant past ; and if they have to be reallocated, the criteria will be based almost entirely on such factors as traditional practice, engineering convenience or social constraints.

The second fundamental point which ergonomics practitioners had to elucidate before they could make an unequivocal contribution to the production equation was to set in perspective the relationship between experimental and systems criteria. In the event the relationship seems to have been easily established. In contrast to other areas where the claims of ergonomics may have been overstated, the relationship has resolved itself without conflict in a hierarchical structure. Using techniques, and experimental criteria, from the basic sciences of ergonomics (psychology, physiology and biomechanics), general data on human capacities have emerged, either in the laboratory or on the shop floor : and these data in turn have provided, in part, the criteria for systems evaluation. They can of course provide only part of the list of criteria in the evaluation, for in dealing with the production system the contribution of ergonomics has to take its place with those from other relevant disciplines, including production technology, quality control, operations research and so on. Human capacities must therefore find their place in a list of criteria which may also include 'initial and operating cost', 'acceptable level of reliability or quality', 'available training capacity', 'maintenance arrangements' or 'shortest critical path'. The weight given to human factors data in this list seems to depend on a third order set of criteria, briefly described under two main headings : 'systems effectiveness' and 'operator well-being'. Evaluations in this context are succinctly dealt with in this book by de Jong.

The qualitative aspects of ergonomics in production being established in some measure, what of the quantitative?

There is unfortunately a dearth of research in production ergonomics. Murrell's paper elsewhere in this book and Morrison *et al.* (1965) are rare exceptions. This in turn has meant that no experimental techniques or criteria specific to the area have been developed : and in application there is no specific production ergonomics data available as systems criteria.

This omission perhaps reflects the neglect by ergonomics workers in general of their fundamental concern—'man at work' and 'man-machine systems'. The advance in philosophy from the early days of interface problems has been slow : often it seems that there has been a preoccupation with the basic disciplines or with making human factors, or ergonomics, simply the technology of consumer convenience or the like. Thus, there has been no real development in the 'science of man-machine systems'.

A not atypical result of not grasping the nettle arises in industrial quality control. Attempts at using human factors data to increase the low levels of efficiency (often 40–50%) recorded when industrial quality control is attempted by the use of human 'overlookers' are rarely noted for their success. This failure, despite the volume of experimental data apparently available, is in a large measure due to the fact that such data relate to vigilance phenomena, identifying them as the principal parameter in effective inspection performance. Such a view is questionable. Little or no evidence exists indicating that the oft-found decrement in the laboratory has a parallel effect in the industrial setting. Further, many characteristics of inspection tasks bear little resemblance to those of vigilance tasks. The general implication is that if the ergonomist is to understand industrial inspection and expect to generalise his results, then he must redirect his research to look at the influence of other parameters and their interaction with one another in this activity.

Taking the point more generally: if ergonomics applications are to be successful here and in other areas, then ergonomics research must really look at 'man-machine systems' and develop techniques and criteria appropriate to multifactor investigations.

With no specific experimental techniques or criteria of his own, the production ergonomist draws on the whole range currently used in ergonomics research : and in applications draws on the general pool of ergonomics data for his systems criteria. He is catholic in his choice showing no preference for psychological over physiological measures or vice versa ; or even matching psychological and physiological techniques to the solution of psychological and physiological problems respectively.

Thus, the excellent reviews in the earlier sections of this book satisfy the need for a discussion of techniques and criteria in production ergonomics. However, in the next section some studies are highlighted, with some risk of redundant discussion, since they review the use of general techniques in production situations.

2. Ergonomics in engineering production practice

2.1. *Validity, reliability and convenience*

Drawing on the general pool of data, it would seem that the production ergonomist can forcefully say something about the measurement and design of work, particularly in relation to physiological and biomechanical loads ; and something less surely about mental loads. It would seem somewhat premature to feed the current experimental data on work organisation into systems criteria.

With this, perhaps, eclectic view of the current role of ergonomics in production systems the question of criteria and techniques to evaluate load seems somewhat simplified. At first sight they can be related to measures of performance : in the straight transformation of energy and information inputs into desired outputs by the operator. It is tempting to say that we can base our human factors criteria on such relationships : to say if one set of inputs allows the operator to produce x pieces per minute and another $x + y$, then the latter is ergonomically better.

The question is of course a little more complex. Measures of direct performance are not valid or reliable indicators of load. Such measures are obviously important in a systems evaluation but in strictly human factors considerations they fail since they give no indication of the cost to the operator of maintaining the overt level of performance.

In dealing with the physiological load imposed intrinsically by the job, input or output descriptors or measures alone are clearly not acceptable. Responses of different human organisms to the same input vary widely. Output measures may be completely erroneous, being affected adversely by decrements in social satisfaction or being maintained at an acceptable level by virtue of the introduction of compensatory physiological mechanisms. Direct measures of the appropriate function of the organism are necessary.

In considering psychological loads it is equally unsatisfactory to use simple input/output relationships. Attempts to measure them in terms of inputs (Conrad 1954) have shown the problems. The number of different input conditions possible suggests an infinite number of combinations which have to be accounted for in any scale of mental load. Further, the mathematical relationships between the simplest combinations cannot be elucidated with our present state of knowledge. Looking at the output side offers no solution to our problem of measuring mental load. It is even more clear from everyday experience that outputs can be maintained at the required level even under severe informational loads : and that equally they are prone to fluctuation under the same load when, for example, motivational factors change. Again direct measures of the appropriate functions are necessary to assess mental load. Unfortunately, lacking the physical definition of their physiological counterparts, psychological functions currently only lend themselves to 'black box' descriptions which by their nature preclude direct measures.

With only a 'black box' we must look again to performance measurement: and if direct measures of performance are unsatisfactory then we must turn to indirect measures. This does leave us dependent on the accuracy of our theories and the descriptions of our hypothetical constructs and intervening variables within the 'box' for our success in measuring psychological load. The measures will only be as accurate as the theories and descriptions are proved to be valid and reliable in varying contexts. It is in this respect that we might appraise the utility of the work on measuring mental load by the secondary task method discussed elsewhere in this volume: and recognise the importance of the work of Brown (1967, 1968) and Michon (1966) in using the concept in extensions of normal load conditions, such as during training.

Another approach to the measurement of mental load may be to look at the accessible physiological functions as indicators. Their value can be illustrated by an experiment by Berman and Pettitt (1960). The study called for subjects to undergo high G loads while in a centrifuge. Two trials were carried out according to the expected experimental programme and each showed rises in the excretion of adrenaline and noradrenaline as reliable indicators of increased psychological load. On the third trial the centrifuge stayed at rest but nevertheless analysis showed that the catecholamine excretion rose as high on this mock trial as on the others, showing that anticipatory anxiety had been present. This could not have been defined by the energy input since none existed and it is unlikely to have been demonstrated in an analysis of performance output such as tracking since it had been previously shown that subjects maintained their performance on this task despite increases in their psychological load. It came to light only through the evidence of a physiological response to a psychological condition.

Perhaps the most recent evidence of the value of this approach comes from the Apollo 13 flight. In the grossest way, perhaps, we have one measure of the combined psychological and physiological load carried by its crew during their hazardous flight, in the 14lb body weight loss experienced by Lovell, the space-ship commander.

The obvious limitation is the lack of convenience of the technique. Engineering production is concerned with on-line evaluations and it is unlikely that criteria based on the above technique, or rectal temperature, for example, would find much favour with the ergonomist, or the operator, on the shop floor.

The available techniques for evaluating operator mental load are, indeed, few when they have to meet the requirements of validity, reliability and convenience.

2.2. *Physiological evaluations*

The large number of studies carried out in the area of physiological load has given rise to practical knowledge which can readily be applied by the ergonomist to the evaluation of an industrial situation. Measures of heart rate, O_2 consumption, etc. are easily recorded, and translated into direct energy costs for the job to be evaluated against criteria such as those shown in Figure 1 (Müller 1962).

Of the techniques available, heart rate must be considered the foremost against the demands of convenience, reliability and validity.

A review of the literature shows that it has been used as a measure for at least the following four purposes.

Work load comparisons, where a previously 'calibrated' man has worked at two different jobs and the cardiac cost of the two jobs compared.

Rest requirements, where the condition has been specified that heart rate should not rise throughout the working day. More directly, this type of study has given a measure of the physiological strain so that precise relaxation allowances can be estimated to ensure adequate physiological recuperation.

The effects of thermal environment, both as a means of estimating the work decrement due to heat and as a control over men at work in very hot surroundings.

Matching men to jobs, where jobs have been calibrated, simple step tests have been used to estimate an individual's work capacity for any given job.

Typical of these types of study is one by Pratt (1969). Full day studies were undertaken of operators using vertical boring machines during which their heart rate was continuously monitored. The work loads were evaluated against known human performance limits and desirable changes in work design and organisation for matching the machine to the man were able to be recommended.

Figure 1. Normal values of maximal and occupational work capacity of men ($\male$) and women ($\female$) at different ages. (Müller 1962)

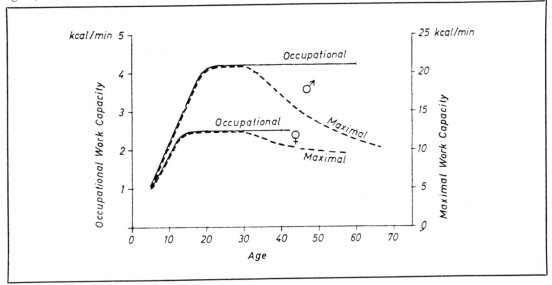

The other principal technique available is, of course, O_2 consumption. It does not stand up well against the demand for a convenient measure : the ancillary equipment which must be attached to the operator for the collection of samples makes it less than satisfactory for use on the shop floor. Grieco *et al.* (1968) successfully used this method in an ergonomics study in an iron foundry to evaluate the working conditions of a group of men assigned to moulding opera-tions using an old manual method and a new method with a 'speed sandslinger'. Energy cost as measured by the partial O_2 pressure in expired air showed the use of the 'sandslinger' to be an ergonomic improvement on the hand moulding operation.

Bonjer and others have given many more examples in this volume. Despite possible prejudice against this technique by virtue of its inconvenience, it must be noted that it has high claims to reliability and validity for energy cost measures in working conditions : see for example Wyndham *et al.* (1966).

2.3. *Psychological evaluations*

With as yet no well validated, reliable or convenient techniques for direct measurement of psychological load there are few reports in the literature which can be discussed in the context of production ; though as has been reported by Wisner earlier in this book there have been many industrial evaluations of their potential.

Critical flicker fusion measurement held the field for some years as a possible technique : but doubts on the nature of the basic phenomenon seem to have undermined its position. However, it is worth noting that Grandjean (1958) achieved satisfactory results, whatever the underly-ing rationale, both in the theoretical and production sense using flicker fusion levels as criteria. He investigated a process in which needles which had to be straightened were projected on to a screen and were manipulated by the worker until they corresponded to a standard. The contrasts in this job were high and it was found that the flicker fusion rate of women on this work was lowered at the end of the working day ; a change not shown by other women in the same shop on different work. Changes were made in the visual environment to reduce contrast and as a result there was a smaller drop in flicker fusion rates and at the same time the output increased. The implication was that changing the environmental conditions reduced fatigue. This must surely be a classic example of performance and load being used in conjunction to assess the ergonomics of a situation.

Evoked potentials have in recent years been suggested as a more meaningful measure of mental load but clearly they have yet to undergo rigorous appraisal : and as yet there would appear to be no references to their use to make definitive statements on the demands made on an operator or to justify changes in work design or organisation to increase production.

Similarly with the use of related physiological measures of mental load. The recent literature has been full of hopeful signs that levels of sinus arrhythmia are valid and reliable criteria. Kalsbeek has already reported that its use is feasible with workers on assembly lines. However, again there would appear to be no report of its use in a strictly production evaluation.

Thus measures through performance scores either directly or indirectly seem to be the continuing tool of the production ergonomist. Techniques using indirect measures, e.g. the secondary task method, are clearly on a more satisfactory and rigorous theoretical footing than are direct performance measures with regard to performance changes; being consistent measures of fluctuations in load, uninfluenced by other factors. However, the scientific and pseudo scientific literature seems to show little use of the former and much use of the latter in industry. This is only to be expected since they are the most meaningful to short term production goals. Often the use of the direct performance measure can be defended in this context by the suggestion that after even a little experience on a task the worker will settle his performance at an acceptable constant load level which he maintains regardless of external conditions. Thus significant differences in conditions are reflected by load shedding via reduction in speed and increase in errors. See for example Whitfield (1964) ; whose evaluation of two computer interfaces used speed of operation as a criterion and showed meaningful and significant differences between the two interfaces. Streimer has earlier shown that the corner stone of this argument— constancy of acceptable operator load—is true for physiological load; but it would seem only intuitively supported for psychological load.

In summary, it appears that on the question of psychological load we have only gone forward a short way since Murrell (1965) said, in the broader context of the ergonomic contribution to all work measurement and methods : 'techniques have been almost entirely used by academic research workers on research projects ; what is now needed is further development work on the shop floor in order to perfect these techniques for industrial use and to define areas in which they can most usefully be applied'.

2.4. *Biomechanical evaluations*

Thus far the assessment of biomechanical load has been ignored. This is because it does not pose the same questions. Criteria based on inputs to and outputs from the system are possible. Statements can be made about the maximum and optimal pushing and pulling forces which the operator should be asked to develop ; or the posture he should be asked to adopt.

Less certain seems to be how the criterion values change over time : what is the cumulative load and the eventual result of maintaining, even intermittently, over a prolonged period, a specified biomechanical output or an unenviable posture? Dentists constrained by the design of their equipment and their task are forced to adopt a posture which apparently does not degrade their immediate effectiveness as measured against any current ergonomics criteria. However, the statistics for low back morbidity in the U.K. after 15–20 years of practice in this profession (B.D.A. 1963) clearly suggest that the dentist at work has an excessive biomechanical load when it is measured against criteria which include a time integration function.

Problems of posture and prolonged biomechanical load are deceptive. How far advanced we are in dealing with them is well summed up by Branton (1969) who when dealing with a special aspect—comfort—gives our problems here as twofold. 'We have yet to determine what is to be measured and how it is to be done'.

Yet in the layout of working areas, posture and biomechanical load are of primary considera-tion. The only reasonably established techniques for evaluating the biomechanical load induced by continuous effort or habitual postures, are, however, electromyography and subjective ratings. The former's use is discussed elsewhere in this volume but it is worth re-affirming that with modern techniques it provides a convenient method.

Fox and Jones (1967) report its use in the evaluation of dentists' postural load using R.A.F. dentists under operational conditions: neither the dentist or the patients appeared to suffer inconvenience.

Its validity and reliability in evaluating postural load is less sure. In the study by Fox and Jones there was certainly a high correlation (·72) between integrated E.M.G. values and sub-jective assessments of postural comfort by the dentists. But Floyd and Ward (1969) after extensive studies using E.M.G. are only willing to say 'our studies lead us to believe that further studies in myography may well be most valuable in determining the postural behaviour, in addition to the basic anthropometric requirements, that should be considered in . . . design'.

The technique most commonly recorded is subjective assessment using rating scales. Normally the rating is done by an unselected population. Shackel *et al.* (1969) tried a variation on this theme by having the ratings of seats done by a team of experts on the ergonomics aspects of seating.

The experts, it is reported, were not significantly better than the general population in giving accurate comfort assessments and these authors conclude that assessment by opinion cannot be considered a valid method for assessing sitting comfort.

In the study by Fox and Jones (1967) yet another variation on the theme was used. Postural load was evaluated subjectively: not by the dentists who were the subjects, but by a team of experts in postural disorders (two physical medicine specialists and five physiotherapists). Their assessments suggested a reliable technique was being used in that the correlation between judges was ·87. There were also strong hopes for its validity in that the subjective judgements had a correlation of ·64 with the E.M.G. evaluations of load. But since it has been noted that the validity of E.M.G. is far from established in this context, there is some measure of begging the question here.

3. The validity of laboratory criteria

Clearly the doubtful relevance of laboratory criteria to real life situations must be of major concern to the production ergonomist. Particularly when circumstances force him to transfer laboratory results to the shop floor.

Chapanis (1967) warned of the difficulties. Yet the outlook may be less pessimistic than he suggests.

The problem associated with the application of laboratory studies would seem to be one of three:

the Hawthorne Effect;

the phenomenon demonstrated in the laboratory is masked by the numerous other variables operating in the real world;

the parameters manipulated in the laboratory are not those relevant to real world criteria.

The Hawthorne Effect can appear in any experimental situation. The associated drop in performance is certainly prone to appear when we go from the laboratory to a prolonged industrial evaluation. This need not mean that the data has no real-world significance for productivity.

Fox (1963) reported the evaluation of performance of an industrial inspection system. In the laboratory evaluation the mean operators' performance was approximately 1300 articles/min inspected with an efficiency of 92%. In the evaluation of *exactly* the same system on the shop floor, the performance measured at the beginning of the study approximated to the laboratory figures. Measured again after a prolonged period of familiarization the inspectors' mean perfor-mance had dropped to 870 articles/min inspected with an efficiency of fault detection of 81%. Clearly the Hawthorne phenomenon had taken its toll of the very optimistic laboratory results.

However, it must surely be agreed that it had not made them valueless when it is realised that the system which was being evaluated was replacing one where the inspector's performance had been, for the previous 70 years, 700 articles/min inspected with a 51% fault detection efficiency.

Even more disheartening to the experimenter is the demonstration of some aspect of human behaviour in the laboratory which cannot be repeated in the real world by virtue of being masked by other variables, perhaps held constant, in the laboratory. However, there can be little argument that much of human behaviour is sufficiently well established to survive its transfer from the laboratory. The laws of visual perception hold equally well whether we are tracking in the laboratory or playing tennis. Thus, it would be foolish to under-rate the value of laboratory results for the real world. Fox and Haslegrave (1969) took Colquhoun's (1961) laboratory experiment on the relationship between the probability of a defect appearing and inspection efficiency and repeated it in an industrial situation. The efficiency of detection for the various conditions in the industrial task correlated highly ($\cdot$87) with the efficiency predicted from Colquhoun's laboratory experiment.

Finally, there is the question of the parameters manipulated in the laboratory not being those relevant to the real world relationship. This, of course, is a matter of the experimenter's rigour and creativity. This might not be the complete answer however. It might well be that the question is being viewed from the wrong end. Considering some unsuccessful attempts at applying laboratory derived ergonomics data it is not inconceivable that it is the practitioner rather than the original laboratory experimenter who must carry the burden of failure. All too often he does not analyse fully his real world situation to determine all the significant parameters included in it. His resultant lack of success in applying laboratory data to the problem is due not to the data's irrelevance but to insufficient data being applied.

The point has already been taken in another context in the introduction to this paper, but is worth repeating in the present argument specifically. Wilkinson's (1961) experimental results on the lack of effect of pacing on inspection are often quoted in ergonomics practice but rarely found in *ad hoc* industrial studies of inspection stations. This does not imply a criticism of Wilkinson's laboratory study which was essentially the investigation of a vigilance effect. In industrial inspection this is but one of the parameters influencing total performance and, while paced and unpaced conditions may not have significantly different effects on vigilance, the possibility of pacing affecting some other parameter, for example pattern recognition, is not precluded.

4. Human factors in production design

In his primary role the outlook for the ergonomist in production engineering is reasonably bright. Validated against the criteria of the basic disciplines of ergonomics his experimental data allow him to make tactical advances in his problem solving activity. However, such data must be incorporated in the overall strategy which is the production equation. Here his horizon is rather darker. Currently he has no strategy for making his contribution amidst the other parameters of this equation.

To be most effective he must have a strategy : and in the same way as the impetus for a system for design ergonomics (Singleton 1967) came from the ergonomist rather than the designer it will surely be necessary for the production ergonomist to take the initiative rather than his production colleagues.

Fox (1967) evaluated the systems design procedure with respect to its utility for production systems. While clearly the framework is of immense value to the design ergonomist it has only limited utility for his production counterpart. The limitation essentially is that it demands that the ergonomist is involved at the earliest stage in a system as yet uncreated. The production engineer is concerned with ongoing systems and is engaged in what can be conveniently described as regenerative ergonomics. Thus the systems design procedure poses inappropriate questions for the production engineer and does little to assist him in generating the correct solutions.

It is the lack of this procedural framework rather than of criteria which currently provides ergonomics in production engineering with its greatest problem.

References

BERMAN M.L., and PETTITT J.A., 1960, The effect of stress and anticipation to stress of urinary levels of a catecholamine catabolite. *Aerospace Medicine*, **31**, 297.

BRANTON P., 1969, Behaviour, body mechanics and discomfort. *Ergonomics* **12**, 316–327.

BRITISH DENTAL ASSOCIATION, 1963, *Memorandum on Fatigue in Dentistry* (British Dental Association: London).

BROWN I.D., 1967, Measurement of control skills, vigilance and performance on a subsidiary task during twelve hours of car driving. *Ergonomics*, **10**, 665–673.

BROWN I.D., 1968, Some alternative methods of predicting performance among professional drivers in training. *Ergonomics*, **11**, 13–21.

CHAPANIS A., 1967, The relevance of laboratory studies to practical situations. *Ergonomics*, **10**, 557–577.

COLQUHOUN W.P., 1961, The effect of 'unwanted signals' on performance in a vigilance task. *Ergonomics*, **4**, 41–51.

CONRAD R., 1954, In *Human Factors and Equipment Design* (Edited by W. F. FLOYD and A. T. WELFORD) (Lewis: London).

DUDLEY N.A., 1968, *Work Measurement: Some Research Studies* (Macmillan: London).

FLOYD W.F., and WARD J.S., 1969, Anthropometric and physiological considerations in school, office and factory seating. *Ergonomics* **12**, 132–139.

FOX J.G., 1963, The ergonomics of coin inspection. *The Quality Engineer*, **28**, 165–169.

FOX J.G., and JONES J.M., 1967, Occupational stress in dental practice. *British Dental Journal*, **123**, 465–473.

FOX J.G., 1967, The use of checklists in regenerative ergonomics. *Ergonomics*, **10**, 718.

FOX J.G., and HASLEGRAVE CHRISTINE M., 1969, Industrial inspection efficiency and the probability of a defect occurring. *Ergonomics*, **12**, 713–721.

GRANDJEAN E.A., 1958, In *Fitting the Job to the Worker: a Survey of American and European Research into Working Conditions in Industry.* (Paris: OEEC), p. 125.

GRIECO A., SARTORELLI E., and TALAMA L., 1968, The ergonomic evaluation of two jobs in an iron foundry. *Ergonomics*, **11**, 467–472.

MICHON J.A., 1966, Tapping regularity as a measure of perceptual motor load. *Ergonomics*, **9**, 401–412.

MORRISON J.F., BROWN A., and WYNDHAM C.H., 1965, A comparison of work study assessments and physiological measurements of men at work. *The South African Mechanical Engineer*, **14**, 234–238.

MULLER E.A., 1962, Occupational work capacity. *Ergonomics*, **5**, 445–452.

MURRELL K.F.H., 1965, *Ergonomics* (Chapman Hall: London).

PRATT FIONA M., 1969, An investigation of the layout of machine controls on vertical boring machines. *Report No. B/SR/3412. Department of Engineering Production, University of Birmingham.*

SHACKEL B., CHIDSEY K.D., and SHIPLEY PAT, 1969, The assessment of chair comfort. *Ergonomics*, **12**, 269–306.

SINGLETON W.T., 1967, The systems prototype and his design problems. *Ergonomics*, **10**, 120–124.

SINGLETON W.T., 1969, *Introduction to Ergonomics* (World Health Organization: Geneva).

WHITFIELD D.J.C., 1964, Validating the application of ergonomics to equipment design: a case study. *Ergonomics*, **7**, 165–174.

WILKINSON R.T., 1961, Comparison of paced, unpaced, irregular and continuous display in watchkeeping. *Ergonomics*, **4**, 259–267.

WYNDHAM C.H., MORRISON J.F., WILLIAMS C.G., HEYNS A., MARGO E., BROWN A.N., and ASTRUP J., 1966, The relationship between energy expenditure and performance index in the task of shovelling sand. *Ergonomics*, **9**, 371–378.

Professor Shackel has a psychology degree. After skills
research at A.P.R.U. Cambridge, he became head of the
Ergonomics Laboratory at E.M.I. Electronics Limited,
working on several advanced man–machine systems. He
is now Professor of Ergonomics at the University of
Technology, Loughborough, England.

Criteria in relation to large-scale systems and design

B. Shackel

1. Ergonomics in relation to systems—three general problems

On this subject of ergonomics criteria and the study, design and validation of systems, I would like first to suggest three of the general problems which seem rather important, and then to offer some comments and deductions from experience.

1.1. *Timescale*

The first problem is the timescale involved. Table 1 shows the typical sequence of activities. There is always considerable overlap of successive phases, with many iterations within and sometimes between each, so that it is not of course a simple linear process; and some phases may sometimes be omitted, particularly the production prototype. But this is the general form, for ordinary equipment design as well as systems. The distinguishing characteristic in system design is the long duration of each phase, because of the scope and complexity of the system to be created.

Table 1. The life cycle of system design

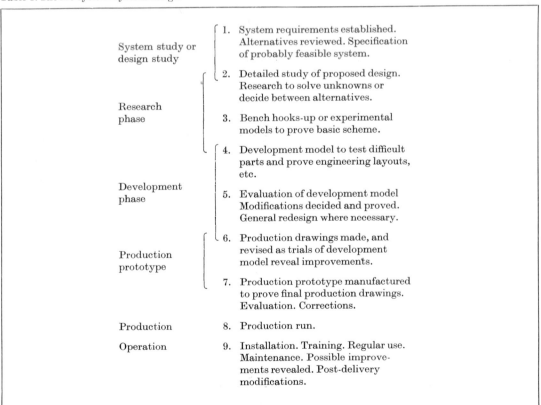

System study or design study	1. System requirements established. Alternatives reviewed. Specification of probably feasible system.
Research phase	2. Detailed study of proposed design. Research to solve unknowns or decide between alternatives.
	3. Bench hooks-up or experimental models to prove basic scheme.
Development phase	4. Development model to test difficult parts and prove engineering layouts, etc.
	5. Evaluation of development model Modifications decided and proved. General redesign where necessary.
Production prototype	6. Production drawings made, and revised as trials of development model reveal improvements.
	7. Production prototype manufactured to prove final production drawings. Evaluation. Corrections.
Production	8. Production run.
Operation	9. Installation. Training. Regular use. Maintenance. Possible improvements revealed. Post-delivery modifications.

The study phase may take typically between about three months and two years, the research and development phases about fifteen months to six years or more, and the production proto-type, or trials stage before acceptance or before production starts, about six months to two years. Thus, the total design timescale, before acceptance or production, can seldom be less than two years and may often extend up to ten years or more. Production and regular use may vary even more widely in duration, from zero, if the system is not accepted or put into production (as happens not infrequently with military systems and has been known to occur even with costly non-military ones) up to more than thirty years for basic structures such as railway and telephone networks.

This long timescale of the system design process seems to require attention, on the human factors side as much or even more than on the engineering side, to such aspects as continuity, documentation and consistent methodology.

Continuity is needed in the ergonomics staff who join the system design team. Thus they learn by full involvement over a long time the many ancillary facts and factors which are relevant for critical design decisions. Moreover, there are stages when the ergonomics work is only intermittent; it is important to bridge the possible gaps which might grow, leading to ergonomics re-entry too late, by occasional visits from the same ergonomics staff on an informal basis.

Apart from any formal requirements, *documentation* is needed to overcome memory difficulties and ensure successive development based upon precisely agreed foundations at each stage. Meticulous recording of both decisions and reasons is needed to avoid re-reasoning and re-making decisions, in the short term, and to enable valid questioning and re-thinking of decisions in the long term, e.g. reappraisal in the development phase of aspects provisionally decided in the study phase.

By *consistent methodology* I mean the adoption of an agreed framework, general pattern of approach and detailed plan for tackling the human factors aspects. When stated, this seems obvious; but the complexity of system design situations makes it less easy in practice, and deliberate attention is needed to establish and maintain this approach. There are two main reasons for this need, apart from the obvious value of a planned programme for ensuring that ergonomics work matches the system timescale and provides results and advice when needed: a most important factor.

The first reason is that it helps to ensure a comprehensive view of the system, so that all significant aspects are considered thoroughly instead of only the obvious ones. This enables the relevant knowledge from all the human sciences to be brought to bear as appropriate, and also gives the best insurance against the risk of not finding an important detail until too late to avoid costly redesign, or worse, rebuilding. The second reason is that it considerably assists the joint work with the design team to be able to show the general pattern and detailed plans, for the ergonomics aspects, and how they relate to the whole system design process.

The type of framework and pattern of approach I have in mind are shown in Figure 1 and Tables 2 and 3. Guidelines such as these are no doubt familiar to many and have been proposed in similar form by several authors. However, we should remind ourselves of two things. First, these outlines are as yet only general guides, from which a detailed plan may be developed for the specific system study concerned. Second, even as general guides they have received little study or validation. Surely, we should decide that much further study is needed on this aspect of system design methodology, to see how far it is possible to develop a more detailed plan for the ergonomics work, and to validate that and these conceptual frameworks which we have already.

1.2. *Criteria at different stages*

The second problem is that there seem to be at least three general levels, or stages, in the human factors work, and that the main criteria for each are to some extent different and conflicting.

Figure 1. Human factors related to system design (after Singleton 1967)

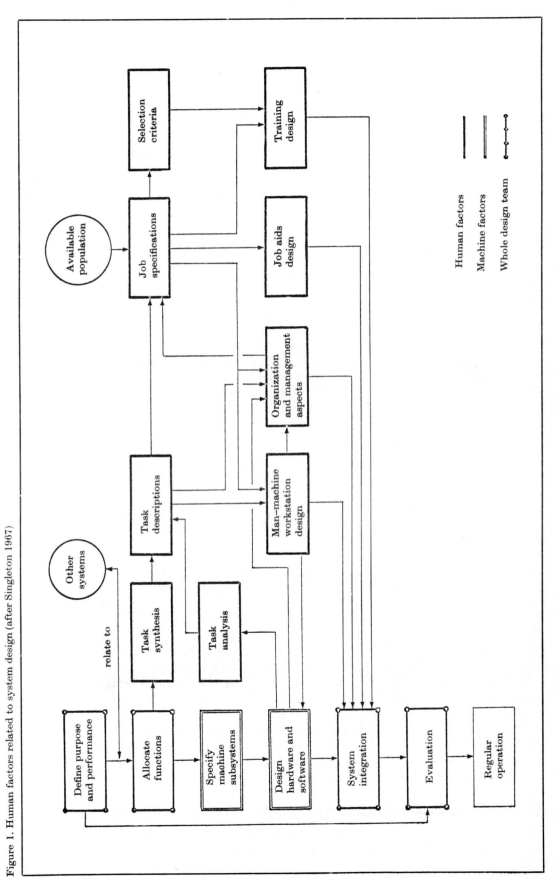

Table 2. Systems design analysis outline concentrating on ergonomics aspects

---

System level

1. Define system purpose and performance requirements. (General indoctrination and first data gathering—involves going through the main procedures up to No. 7 below, but broadly and briefly to get over-view; later reiterate in detail. Develop system pictogram.) Thus provide criteria for evaluation (at 13 and 17).
2. Study possible solutions to achieve required purpose and develop concept of a feasible system design.
3. Identify next larger and smaller systems and analyse interactions.
4. Define possible sub-systems within the design envisaged; function analysis and first allocations of functions. Establish research and trials as necessary to develop possible new solutions and to evaluate alternatives. Specify broadly the input–output requirements and components for each sub-system.

Sub-system level—reiterate the following for each sub-system

5. Flow or sequence analysis (sequence of events, with or without time-line).
6. Activity, decision and task/job analyses; develop manning policy.
7. Analysis of critical design requirements.
8. Revision of function allocations as necessary.
9. Revision of flow, activity, decision and task analyses as necessary. Assist with preparation of draft operating and maintenance procedures.
10. Design workstations, analysing each in all sectors (man, machine, workspace, environment) using workstation analysis.
11. Analyse all man–man and man–machine communication links.
12. Evaluate each workstation with mock-ups and on prototype when available.

System level

13. Evaluate whole system as far as possible on prototype when available.
14. Assist with completion of operating and maintenance manuals.
15. Using prototype as tool if possible, develop selection procedures and training schedules.
16. Do or assist with selection and training as necessary.
17. Liaise during installation, commissioning, final training and early regular operation, to evaluate finally and learn lessons from faulty design decisions.
18. Evaluate post-delivery modifications proposed to ensure compatibility with functions of and loads on system operators.

First of all, at the system level and early design stages, ergonomics can assist with system analysis, function allocation and setting limits within which humans can be used. The main criterion here would seem to be *system efficiency*.

Secondly, at the workstation level in the development stage, ergonomics can turn the analysis inside out, as it were, and concentrate on the design of each separate workstation around the man so as to fit the jobs to him. In this case, the main criterion would seem to be *individual well-being*, with a secondary criterion of individual efficiency.

In the third stage, installation and regular operation, the aspects of selection and training are studied to achieve an optimum fit of the men to the jobs. However, it is an observable fact that fitting the men to the jobs is not the only important factor in this stage ; as men become more concerned with job satisfaction, in the widest sense, so *operator acceptance* becomes a vital criterion. Certainly system efficiency is still an important criterion, but if various conditions are not acceptable then the operators may refuse to work and efficiency is zero.

Now, of course, this does not mean that these three stages are separate or dealt with in isolation ; the last two are considered during the first (e.g. see Table 2), and the design process is essentially iterative and interacting. But there do tend to be natural breaks between these stages : cost reappraisal and R & D contract procedures between the first two, delivery onto site before the third. Therefore, conflict could arise, and the criteria are certainly different. There is no obvious way of resolving this problem. It is an important aspect to remember for specific attention within each separate system design.

1.3. *Criteria for different groups involved*

The third problem to be borne in mind is the importance of the interactions with and criteria of the other main groups concerned. Four main groups are involved with the Human Scientist in the process of system design and operation : Manufacturer, Designer, User and

Table 3. Workstation analysis outline

---

1. **Man**
   Consideration of   sex   physique   training
                            age   intelligence   motivation
                            size   experience
   Definition of operational modes, e.g.   searching   monitoring
                                                tracking   decision-taking
   required in final situation and thus consideration of abilities and limitations of human operator for all aspects of the task.

2. **Man-Machine Interaction**
   Influence, on operator and his decisions, of
                        displays—sensory input to operator
                        controls—motor output from operator
                        panel layouts—display-control compatibility
   based upon study of human information—decision—action patterns of human, equipment and task operational sequences.

3. **Man-Workspace Interaction**
   Influence, on operator's position, posture and reach, of
                        machine size
                        chairs, desks, etc.
                        adjacent machines, structures and material, etc.

4. **Man-Environment Interaction**
   Influence, upon behaviour and performance, of
                        physical aspects
                        chemical aspects
                        biological aspects
                        psychological aspects
   Physical: light and colour, noise, heat, ventilation, gravity, movement, electromagnetic and nuclear radiation.
   Chemical: gas or liquid, composition, pressure, smell.
   Biological: microbes, insects, animals.
   Psychological: workteam, command structure, pay and welfare, shift conditions, discomfort or risk, socio-psychological aspects of the particular factory, neighbourhood, town and type of industry concerned.

5. **Special Questions**
   Consideration of non-standard conditions, such as errors, exceptional circumstances, or similar factors not included in the previous analysis of normal operation. Consideration of problems peculiar to the specific case under investigation.

Buyer. Each will be involved to some extent with the aims of the system to be designed, each will also tend to view the system in the light of his own scale of values and criteria. We assume, of course, that contracts have been placed and the system design has begun.

So the manufacturer will have as his prime criterion simplicity in manufacture. The designer, in the engineering sense, will be concerned primarily with engineering performance, the buyer with price (the risk of increases), and the user with acceptability, in the sense of a combination between ease of operation, convenience, comfort, appearance, and familiarity. The ergonomist, also, will have his own aims and criteria as a professional worker and consultant.

Having to interact extensively, as he must, with each of these groups, and assist in resolving the joint decisions which must be achieved, certainly adds to the complexity of the ergonomist's task. The less homogeneous and the more diverse these groups, and the more they tend to be separate organizations (e.g. different industrial companies), the more likely it is that the resolving of joint decisions and the whole design process will take substantially more time.

2.   Some case study examples

With these three problems and the related suggestions in mind, let us now turn to two case studies as pointers to the deductions I propose to offer (these are not described in detail because reports are available elsewhere).

2.1. *Port of London Authority—meat automation*

This project was concerned with all human factors aspects of the automation plant commissioned by the Port of London Authority (PLA) for unloading meat from refrigerated ships (Shackel *et al.* 1967, Shackel 1969). We provided much design work (but not the basic system configuration), a rapid applied research project, and a comprehensive training programme.

Particular points to note, with regard to this discussion, are as follows.

- From the comprehensive planned approach to the system analysis adopted it was possible to identify an unresolved detail which proved vitally important and would have cost considerably more to deal with later.
- To study the keyboard operating task as a function of different types of input, and to test dockers' performance handling cartons in a particular way, laboratory simulations were used. Scope and timescale were limited by cost constraints.
- Simulation in the classroom was also used successfully in the training programme ; both for the keyboard task and for the clerks and supervisors.
- Selection for keyboard operators, except for colour vision, was unacceptable to usual Docks practice.
- Experiments proposed to establish dockers' performance on the new task in the hold, as a basis for pay negotiations, were not acceptable to the employers and, in some aspects, to PLA (partly because the Devlin Report had not yet been implemented).
- At times the ergonomics team acted as a clearing-house for information between the many different groups involved, who did not have to see each other as often as the ergonomists saw each of them.

2.2. *Esso—London Airport refuelling*

This project involved the analysis of the whole operation, and the redesign and validation of the control centre, for aircraft refuelling at Heathrow (Hatfield *et al.* 1969, Pritchard and Shackel 1970). The analysis, aided by interviews with and suggestions from the Esso staff, led to redesign proposals which were validated by a detailed laboratory simulation before implementation. Subsequent attitudes by Esso staff and management, and validation results, suggest a satisfactory conclusion.

Some of the task problems for the Esso controller are :

- many information sources,
- complex decisions in scheduling men and trucks,
- time pressure adding to load and stress,
- complex interactions with airlines, airlines' staff and Esso operators.

Some of the items built into the redesign are :

- state-board showing essential details about resources, including past allocations, to assist scheduling,
- logical layout of information sources, radio and telephone channels, etc.,
- sit or stand desk arrangement with swivel chairs on castors,
- space layout for reach based on anthropometric and mock-up studies.

The aims of the validation experiment were :

- to study controller performance (i.e. decision speed and quality not to be worse, and better if possible),
- to test it as a function of loads expected over the next two to four years,
- to test it as a function of one or two men in the control room,
- and to test it with and without the new state-board method.

Particular points to note, with regard to this discussion, are as follows :

- Only the ergonomists and Esso personnel in various departments were involved in this project, until it came to manufacture and installation of the new system.
- The Esso staff made many suggestions of value, and also took major roles in the laboratory simulation.
- To test the proposed new job method, a very full simulation was made in the laboratory of the whole operation, as it affected the Esso controller, involving the setting up of the full task in real time for three hour spells.
- When aided by the new method, the controller was found to change his strategy in line with the real world situation as he saw it, that is, he did not service aircraft any sooner (for the most part this was not needed) but he gave his men more five-minute rest pauses.
- In the simulation we could and did add stress, without an 'artificial' secondary task, by increasing the extra telephone calls to the controller (which were typical of his normal work).
- The controller could be overloaded at times, leading to late decisions and aircraft delay, which amount to 'failure', so we were able in effect to 'test to destruction', as far as our system was concerned ; it is often said that one cannot do this with the human component in a system.

3. Some deductions from these examples

Some deductions can be made from these two case-studies and other systems work.

- The question has been posed whether field problems and experiments can be brought into the laboratory with any measure of success. With caution I would answer that it is possible, by using simulation. But the choice of the critical aspects to be simulated is still an art rather than a science. One particular point we found with the Esso project was that having involved the Esso supervisors and fuelling chiefs, they contributed several practical ideas and methods for the simulation.
- Usually, and especially with non-military systems, we do not design in a complete vacuum. There is always some system or operation, somewhat similar, which exists already, and improvement by comparison with that previous system is often a major aim and performance criterion.
- The choice of criteria and of methods depends ultimately upon the system design process itself and those involved in it. For instance, some of our measures in the Esso simulation were worked out in detail along with the senior Esso shift supervisor. While we can propose a general list of all the likely criteria and methods, we cannot specify exactly without close involvement in the actual system being considered.
- Criteria and methods must also depend on the implicit or explicit goals and criteria of the groups with whom we are working. Thus, we cannot specify exactly without also a close involvement with the actual manufacturers, designers, buyers and users concerned.
- The system process, choice of criteria and methods, etc, is likely to be easier, I suggest, if these groups are less diverse. For instance, in the PLA project all four groups involved different organizations, in some cases more than one: whereas in the Esso project separate organizations outside Esso were only involved in the manufacturing area (in both cases not counting the human factors group).

4. Some general conclusions

In turn some more general conclusions can be drawn with regard to criteria in large-scale systems and design.

- The limits of time and designing costs, in most system designs, constrain the range of criteria and of measures that can be used.
- At the least, ergonomics criteria can set design limits within which final equipment solutions can be selected using other criteria.

- Neither psychological nor physiological criteria alone are sufficient in systems design; the essential criteria of practical importance are those of the system itself, such as performance, cost, safety, etc, and those of the real world and its value systems.

- The criterion problem can perhaps best be considered in terms of a hierarchy of criteria, which may vary with different systems, and for some systems the human factors criteria are, or should be, the most important. What these criteria should be, and in what ranked or weighted order, for any given system is not so easy to decide.

- At present, of course, there is always a question of trade-off against system cost, and this must be accepted as a basic factor when considering all criteria.

- In system design work at present there is a continual risk of insufficient ergonomics involvement, at global system level and early enough in the design process, resulting in inappropriate function allocation. For example, there are certain trends in the engineering design process which, without ergonomics counteraction, may at times almost inevitably cause the use of humans in effect as machines (Thorne and Shackel 1969).

- There is also a continual risk of the ergonomics contribution, even at system level, being insufficiently wide and leading to sub-optimum use of humans (cf. Jordan 1962). This is particularly so with regard to the various social science aspects.

- Again, there is a continual risk of the ergonomics contribution being insufficient or inappropriate because our methodology is inadequate as well as our knowledge. There is no doubt that much research is needed to increase the knowledge available from the human sciences for system design work; but much research is also needed to develop and prove better techniques for making our contribution to this work.

- We must be especially aware of the norms and value systems of the real world. For example, it could be said that one of the important tasks for ergonomics is to set limits within which the bargaining of pay, working conditions, etc, can safely be done, i.e. to set limits within which no harm should arise for the operator. For instance, it used to be the situation in the Port of London that, for 'big' money, some men would literally work themselves to death with heavy loads by 40 years of age; we would agree, surely, that we should help to prevent anything like this but how far may we say 'thou shalt not harm thyself for high wages'? Take another, and more debatable example, the question of noise damage. Deafness used to be a 'badge' of status in the Royal Artillery and among boilermakers, and I suspect there are still many situations where operators would accept similar risks for higher pay. In pursuing our ergonomics philosophy, we must also be realistic about some of the very different attitudes and norms in the real world. In some cases we may be able to adopt criteria and set limits for safety, but in other cases we may have to accept that there is still much education to be done.

- Finally, and following from the last item, there is a whole gamut of criteria which we have not so far considered, but to which we shall need to give much attention in the future. It is easy, and only natural, for those in any system design to concentrate almost exclusively on the criteria within the system. But as systems grow in size, complexity, and influence upon the world at large, so of increasing importance are the criteria by which those responsible for the environment outside the system should decide whether a given system is acceptable or not (cf. the present problem of the location of the third London Airport). There is no doubt about ergonomics aspects forming an important part of the hierarchy of criteria relevant to these situations.

In conclusion, I should emphasize that the above opinions are very tentative, based upon detailed experience with only about ten major system designs. We need more gatherings of collective opinions and we need more research studies of this problem area. But at least we have begun.

References

HATFIELD M., HARRISON R.G., and PRITCHARD WENDY, 1969, Some control room methods evaluated by simulation. *Ergonomics*, **12**, 768.

JORDAN N., 1962, Motivational problems in human computer operations. *Human Factors*, **4**, 171–175.

PRITCHARD WENDY, and SHACKEL B., 1970, Esso refuelling control centre at London Airport—a case study. *To be published*.

SINGLETON W. T.,1967, The systems prototype and his design problems. *Ergonomics*, **10**, 120–124.

SHACKEL B., 1969, Human factors in the PLA meat handling automation scheme—a case study and some conclusions. Paper to NATO Advanced Study Institute on Human Factors. *To be published*.

SHACKEL B., BEEVIS D., and ANDERSON D.M., 1967, Ergonomics in the automation of meat handling in the London Docks. *Ergonomics*, **10**, 251–265.

THORNE P., and SHACKEL B., 1969, Man as an adaptive recognition component in an automation plant. *Proceedings of ERS & IEEE International Symposium on Man–Machine Systems*, Cambridge.

Section 3 -Applications

Epilogue

In practice there would appear to be no conflict either within the use of the traditional experimental criteria of the behavioural sciences, which are the grass roots of ergonomics, or between them and real world systems criteria.

Much more concern was expressed about the utility for ergonomics practice of the data generated by ergonomics off-line research. The general consensus would seem to be that the ergonomics practitioner in the future can only make a credible contribution to operational problems if ergonomics research provides data derived from experiments which reflect in some measure the multiple variable interaction which belongs to real life systems.

Data from classical ergonomics studies of the man-machine interface can give the human factors specialist only a modicum of credibility as a serious contributor to modern technology. Such data drawn from simple, formalized, and even artificial, situations in the laboratory or in the field are often too trivial to have economics significance : or, ignoring the complexity of man's performance in real life, often have little predictive value in setting up human factors criteria for a system.

Thus ergonomics must face a fundamental problem in its research effort and seek to make available information which is pertinent to the complex and multifactor questions on operator efficiency and well-being generated by modern complex systems. In the nature of systems development, the practitioner will always have a need for Christensen's 'quick and dirty studies' or Burrows' 'very pragmatic study'. But there can be little room in the future for the unsystematic basic research of the past two decades : little room for, to quote Michon and Fairbank, '*programmes* to give an immediate answer to questions that are scientifically trivial' ; 'sporadic patches of knowledge' ; or 'an excess of research effort in areas which exceeds the relative importance of these areas on the whole'.

Sophisticated tools will be required to support these more ambitious research aims. But already techniques and hardware are available which allow more sophisticated experimental designs.

A more significant need will be for an integrated framework to describe and predict complex human skilled performance : which will allow principles to be generalized from laboratory to operational task and from one task to another. The demand is, of course, for a task taxonomy and for the development of techniques in skills analyses and task synthesis and analysis.

Here, the need is, as yet, unsatisfied. Techniques originally conceived to provide human performance data for these purposes seem to have fallen short in their development.

It will come as no surprise to those actively engaged in ergonomics practice to find that in defining more worthwhile research aims the 'applied ergonomists' have also suggested that ergonomics findings should be set in the social context of man-at-work : and should be qualified with data on individual differences. To this might be added the need to set them in their cultural context. The man-machine system is embedded in a social setting with which it interacts continually. Often it is the social or cultural influence which gives rise to the ergonomics problem. A simple example is population stereotypes : but Sinaiko mentioned informally during the Symposium a more costly one in the transfer of the concept of maintainability from a

233

western to an eastern culture where its novelty precluded even a single appropriate word in the national vocabulary.

Whether this leads to a case for extending the scope of ergonomics is debatable. In terms of its current development ergonomics has still to formulate its requirements in terms which are meaningful to engineering. To overclaim the field has its attendant dangers for professional identification and subsequent credibility. The need to incorporate data from social or occupational psychology in systems design thinking is possibly best met by including these separately identifiable disciplines in the design group. Rather than incorporate them, ergonomics might seek to stimulate new approaches in its sister disciplines.

There is probably less emphasis in this Section, than might have been expected, on the ultimate criterion the ergonomics practitioner must face in promoting his viewpoint among his design or production colleagues: the cost-benefits of implementing ergonomics recommendations. The heuristic use of data in systems design decisions involves of necessity cost-benefit analyses: alternative designs are available each with their ergonomics advantages and disadvantages and differing predicted performance capabilities: in the decision as to which is developed, cost and permitted tolerances in the operators' performance and well-being have their place.

There are, however, remarkably few results of ergonomics intervention which have been reported in measures which allow conversion into tangible financial savings. This is perhaps because such measures do not easily sit in the scientific literature.

It could be argued as Beevis and Slade (1970) have recently suggested that 'It seems unnecessary that the value to a design team of a specialist in human performance should be evaluated on a financial basis when the advice of specialists in electrical, mechanical, metallurgical or dynamic performance is accepted as natural in order to achieve a given specification.' This is perhaps a naïve viewpoint. In the scheme of things technological specialist will have a 'reserved place' in a design team as their skills are overtly required, e.g. in the design of an electronics system, the place of the electronics specialist cannot be denied. The ergonomics specialist has no such 'divine right' and unless he can show his contribution to be obviously worthwhile his inclusion will be in doubt. Similarly management will only seek an ergonomics reappraisal of its plant and production systems if it can be shown to be clearly advantageous to the company. In the rude world of modern industry the contribution will only be seen as worthwhile if it can be interpreted in financial terms via the systems effectiveness.

Reference

BEEVIS D. and SLADE I.M., 1970, Ergonomics—costs and benefits. *Applied Ergonomics*, **1**, 79–84.

Dr. Sinaiko holds psychology degrees. He has pursued personnel research in industry, and engineering psychology research in university departments and with the U.S. Naval Research Laboratory. He is now engaged in analytic and experimental research with the Institute for Defense Analyses, Arlington, Virginia.

Some concluding remarks

H. Wallace Sinaiko

This paper summarizes the main problems and issues that arose during the meeting. These comments have the advantage of three months' hindsight and, at the same time, they suffer some omissions because of the time lapse.

It becomes less and less interesting to continue to describe differences in the approaches to ergonomics problems by Europeans and Americans: 'they' tend to be physiologically oriented and 'we' are behavioural. But all of us know about this difference and most of us have speculated about how things got to be that way. What is much more important is that there appears to be less parochialism among ergonomics practitioners and, one can predict, this trend will continue. Wisner indicated, among others, that his French ergonomics colleagues were becoming much more eclectic.

Much wider emphasis is needed on the dissemination of negative results of ergonomics studies. Too often we fail to write up for publication this type of information. One reason for this is that the traditional academic disciplines, e.g., psychology or physiology, and their associated journal editors accept only positive findings. Often such findings are published without regard to their practical consequences; if a difference between two conditions is shown to be statistically significant, it is accorded an importance that may not be relevant to the real problem at hand. Although neither idea is new, the two issues of negative results and the difference between practical and statistical significance were given appropriate emphasis by Chapanis in his opening paper.

Another important methodological point, addressed by Singleton among others, had to do with the obtrusiveness of our measurement techniques, particularly in recording certain physiological responses, such as O_2 expenditure. One wonders about the extent to which these techniques interfere with subjects' normal performance. (Physiologists with whom I raised this point in Amsterdam denied that there was a reason for concern. But I am less sanguine than they are about non-interference.)

Several participants echoed the need for ergonomics research to get closer to the real problems confronting men-at-work. I strongly endorse this notion. Singleton pleaded for less dependence on traditional laboratory methods, which he said brought about simple-minded and unrealistic experimental situations. (Is the over-worked bicycle ergometer a classic example?) Edwards proposed that ergonomics practitioners need to become much more involved in the problems they are investigating. Shackel's report on his studies of airport fueling systems was one reality experience; he correctly emphasized the importance of incorporating value systems and norms of the work situation into ergonomics research. Chiles touched on another 'reality' problem: since many actual tasks involve time-sharing, or divided attention, such details should also be included in ergonomics studies.

Throughout the conference there was a confirmation of a long-felt need: we have very little base-line data on human performance under a variety of normal conditions. Thus, ergonomics studies deal with man-in-stress of one sort or another but we have no 'normal' data with which to assess our observations. Rolfe, among others, commented on the need for much more physiological data on individual differences.

Related to this is another obvious need for data, although it was not mentioned directly during the Conference. I refer to a need for much more investigation of the ergonomics factors in critical but low frequency tasks. For example, millions of people use cameras but very few photographers are constantly involved with their equipment. Most of us are infrequent picture-takers, i.e., during holidays or at other special times. We expect our photographic results to be at least adequate but we often become over-committed to highly sophisticated equipment. Practically nothing is known about the ergonomics aspects of learning to use a camera or the retention of such skills as are infrequently practised. Similarly, there are other complex tasks that demand skilled performance, sometimes under stress, from non-professionals, consider the private and occasional airplane pilot. Ergonomics research should consider addressing this class of problems.

One issue arose during the meeting that was both new and, to me, an important recognition of a neglected problem: social and motivational aspects of human tasks. Although in complete agreement, I wondered about the backgrounds of most ergonomics specialists that would enable them to cope effectively. Kalsbeek suggested in passing that information about sinus arrhythmia could be used to improve social climates of work situations. Christensen, among many other important points, said we should be concerned with enriching the lives of systems users. I concur in this.

Several times during the conference, both during formal presentations and in comments by listeners, it became apparent that we are still too dependent upon subjects' verbal self-reports as a main data source. Psychologists are probably most prone to rely on interviews and questionnaires as their observational method. While operator attitudes and opinions are impor-tant, they are too subject to the usual distortions (not always deliberate) of human reporters. We need to place much more emphasis on the development of behavioural or performance indications of what people do rather than what people say they do. (Webb *et al.* 1966 provide an excellent source of ideas although they do not deal with ergonomic applications.)

Two papers (those of Rabideau and Burrows) illustrated the value of a technique my colleagues and I have called 'the indelicate experiment' (Sinaiko and Belden 1965). Very briefly, it has been shown that useful data can be obtained from trials involving typical operators doing representative tasks under realistic conditions. Usually these experiments depart from the more sanitary world of the laboratory and its tight controls; also, such studies do not produce results that are acceptable to editors of scholarly journals. But, to the ergonomics practitioner who is usually interested in solving practical problems, the technique has been found to be most valuable.

One wonders why so much ergonomics work, at least among Americans, has been limited to the more glamorous tasks such as that of the aircraft pilot or men-in-space. Michon, in reporting his very elegant studies of driver behaviour, suggested that there has been a lack of interest by government sponsors in the more mundane, but much more frequently occurring, human occupations. Michon also called for more orderliness in planning ergonomics research; we should eliminate, for example, the imbalances between the phenomena human factors special-ists study and their actual importance to society. Too often the laboratory worker is guided by inertia or the simple ease of continuing to do what he has been doing. (Are 25 years of human tracking studies an example of this?)

I would like to reiterate my own final remarks at the Symposium which cited two papers as most interesting and deserving of special recognition: Borg's correlational studies of perceived stress (e.g., ratings of exertion) and actual physiological response (e.g., heart rate): and Edwards' definitions of levels of ergonomics criteria and their relationship to the researcher's objectives.

Finally, here are some proposals for agenda items to be considered if a subsequent meeting is held. Much more needs to be said about ergonomics criteria, particularly in the realm of cost-effectiveness. Also, the users of ergonomics data should be heard; for example, Wisner men-tioned the French trade unions as among the field's best supporters. Equally important, there should be increased emphasis on the presentation of detailed case studies.

References

SINAIKO H.W., and BELDEN T.G., 1965, The indelicate experiment. In *Information System Sciences* (London: Macmillan and Co., Ltd.).

WEBB E.J., CAMPBELL D.T., SCHWARTZ R.D., and SECHREST L., 1966, *Unobtrusive Measures* (New York: Rand McNally).

Dr. Wilkins is a psychologist, and after war service he was
Professor of Psychology at the Universities of Notre
Dame and of St. Louis. He is now Scientific Director,
U.S. Navy Medical Neuropsychiatric Research Unit,
San Diego.

Some comments on the direction of the symposium

Walter L. Wilkins

This comprehensive symposium on the study, design, and validation of man–machine systems focused its attention on criteria, both physiological and psychological, and so we had new data, reviews, state-of-the-art and position papers from the relative orientations of psychology, of physiology, and of industrial engineering, as each is partially defined in ergonomics. The organizing committee's choice of areas for emphasis and of scientists to review the current situation is, in itself, a sort of definition of the present state of our knowledge in this challenging field.

In presenting some reflections on the directions these discussions have taken, I should like to attend mostly to those papers which explored the possibilities of new techniques, or new applications of well-known techniques, from the physiological side. Our own laboratory, as Singleton kindly mentioned, uses multiple approaches to the assessment of human performance which include biochemical, psychophysiological and behavioural measures while stressing on-the-job criteria and rejecting training criteria, when appropriate to do so. The physiological and the psychological must both be used in any valid approach to the kinds of problems which ergonomics concerns itself with.

De Jong summed up the arguments for an integrated approach by saying that such an approach was 'a condition for the correct interpretation of data collected by means of ergonomics techniques. This is true for the outcome of physiological measurements (e.g., of the sinus arrhythmia) as well as for opinions (gathered through personal interviews and the like) on the strenuousness of the work, the disturbing effect of the environment, etc.' Sound evaluation of any man–machine system, in de Jong's view, must include, in addition, information on physical conditions of work, the work load, the satisfaction in the job, the organization, and technical and economic analyses.

I should like to make somewhat more explicit a factor which I feel was implicit in de Jong's presentation. I refer to the usefulness of an epidemiological approach to the problems of ergonomics. The easy example, of course, is Michon's topic, driving safety, with all of its implications about the automobile and life and health. The epidemiology of automobile accidents is still in its beginnings, but has centred our attention on the areas where the most serious problems lie, and may re-direct our efforts to areas where some worthwhile change may be possible. Epidemiology solves few problems, but often high-lights those features of a problem which may be amenable to experimental study, or at least to a more appropriate analysis than had typified the situation. And it helps to define whether you do in fact have a problem which is researchable. Many things in life need only administrative decision—well-informed and wise decision, hopefully—but not decision which needs or must await definitive research.

The topic of Michon's paper also illustrates the difficulty of translating research results into practice and of modifying a complex situation. Michon points out that the automobile, with the traffic problems and health hazards it creates, involves largely matters of skill and safety, and indeed much of the variance is accountable by factors of skill. Yet, as Christensen illustrated in his discussion, there are often hidden costs, which are not readily identified in routine analyses. These costs in the case of the automobile may not seem very hidden, for we learn from the newspapers how many people are killed and wounded daily: many more than in that other public health scourge, war. The automobile, however, has additional costs which while not hidden are at least not attended to by policy makers who make decisions which affect health. It may be that the automobile's contribution to poor health and lessened efficiency through production of smog could be greater than its cost from accidents. The accidents we keep count of for purposes of insurance. The lessened efficiency and reduced health we have no adequate measures of. Epidemiology reminds us of these.

How do the papers in the present symposium address themselves to the problems presented? The evoked response, a very fashionable area in psychophysiology these days, was reviewed by Elisabeth Groll-Knapp and by Defayolle, and intriguing possibilities envisaged, especially for the eventual monitoring of readiness to react. We were unable, however, to translate these possibilities into any specific man–machine systems for ready application.

Borg's clever analyses of some of his data from studies of fatigue, and his formulae for allowing individual comparisons and for relating the subject's perception of his strain to actual measurements, suggest some new possibilities in other areas. His report of a correlation between heart rate and perceived strain of 0·8 and his experimental modification of the work/rest cycle to permit a more valid measure of effort are impressive.

Fox reviewed cogently the current practices in the use of heart-rate measures—the cardiac costs involved in work-load, the determination of rest requirements, the thermal effects, and the use in job selection—matching men to jobs. Immediate illustrations of Fox's points can be found in a wide variety of man–machine systems. I thought of the work of submariners and divers and of Philip Rasch's studies on marines doing prescribed exercises below decks in a ship crossing the Equator and the subsequent assessment of their fitness immediately upon landing in South-east Asia. Those who did daily exercises under hot, steaming decks proved to be much fitter and able to perform satisfactorily.

Fox called our attention to a very real and practical disadvantage of most of the physiological and psychophysiological measures—inconvenience. They tend to be burdensome, nettlesome, often a nuisance, and sometimes almost painful. Inconvenience for the man who has to be wired up for telemetering his responses or for recording leads to reluctance on the part of the subject, and frequently to refusal. Some refusal to cooperate is the result of more than inconvenience. It may be close to fear, as the discussion following Fox's paper illustrates. We heard Sinaiko's advocacy of unobtrusive measures in psychological assessment. I'm sure that unobtrusive measures in physiological assessment are equally desirable. Perhaps with recent advances in micro-miniaturization we may be close to unobtrusive physiologic measures.

More than one discussion raised the question of the fallibility of psychophysiological data. The EEG was cited as an example of a device which generates mountains of data, some part of which is amenable to computer reduction and analysis, but much of it has questionable validity. So much pessimism was expressed about prediction from physiological information, however, that I should like to make the point that there is a sizeable body of data with good predictions.

For example, in the training of underwater swimmers, data from our laboratory suggest that serum urate and cholesterol measures, as well as some measures of personality factors, can predict with fair accuracy which men are likely to drop out or to be dropped from the arduous training and even at what week of training such failure is likely to occur. The general area of sleep loss and performance is another example. Both the behavioural measures used by Wilkinson at the Medical Research Council's Applied Psychology Unit at Cambridge and the psychophysiological measures used at San Diego have shown that decrements in performance following sleep loss are predictable with some accuracy.

The position papers by Bonjer and by Wisner were both comprehensive and perceptive. Bonjer's emphasis on environmental physiology is fundamental to any study of the validation of man–machine systems and his concept of 'aerobic power' most useful. I was edified, too, by his analysis of man as an energy converter and the implications this concept has for allocating tasks between man and machine. The additional criteria for such decisions, as I infer them from this meeting, are simplicity, parsimony, and cost. Wisner's emphasis on the usefulness of physiological variables in low physical loaded work is an excellent state-of-the-art paper. I was impressed that he called our attention away from the exotic and glamorous jobs of astronaut, aquanaut, or explorer (even though many of the studies of our own laboratory, in the Antarctic, undersea, and in the hospital, have focused on unusual populations) and to the ordinary jobs in the world of work, where the overall usefulness of ergonomics must be manifested.

Singleton's paper too, in its very first sentence, called attention to both aspects of ergonomics: to improve efficiency of the system in which man works *and* to ensure that his health is not in jeopardy. His stress on the potential utility of physiological measures of arousal in assessing work efficiency and upon multiple approaches and multiple criteria for judging the worth of ergonomics studies is well taken.

Defayolle's use of the evoked response as a measure of a main task's performance allows some assessment of the effects when a second task or a distraction is added to the work load. It is of course possible, when a worker is overloaded, to obtain a behavioural measure of decrement in performance, but the monitoring through the evoked potential allows one to get at the effects of attitudes or expectations as these may affect the response, and thus adds a further dimension to our measures.

Streimer's report on oxygen needs, with its implicit assumption that a good many workers may be over-working, is timely. We have heard a good deal lately about unrealistic work loads for university students, as in medicine and engineering, and for intellectuals of many varieties, including university people, scientists, and managers, but the illustration of the costs of work and of over-work is a healthy reminder that persons in a variety of life situations may be facing daily loads which are physiologically and psychologically unhealthy.

Consideration of the costs of work leads to Kalsbeek's experiments utilizing the sinus arrhythmia. Here the behavioural effects of introducing a second demanding task when an initial one is already on-going is shown. Further, it is demonstrated that the degradation in performance may result not from fatigue or panic but from breakdown in thinking and reasoning when the rate of making binary choices is increased in the distraction task. It should be possible, therefore, to determine what sorts of error are likely in distraction situations and to build in arrangements to prevent errors.

It is clear from our discussions that the exact place of the EEG, electromyography, sinus arrhythmia, evoked potentials, galvanic skin response, steroids, catecholamines, etc., is far from settled. Some of these physiological or psychophysiological measures have already an accepted place in ergonomics research and a few in ergonomics practice. Some have what appears to be a limited usefulness.

Now there are some jobs in the world of work so critical that psychophysiological monitoring of on-going performance is mandatory. At present, the astronauts are so monitored, I presume, when they are out in space. A few other jobs may deserve such intensive monitoring. Some tasks would profit from periodic or aperiodic sampling of physiological costs, to assure safety and health of workers or to assure vigilance in tasks where alertness is critical to performance. Most of the jobs in the world, however, do not merit such psychophysiological monitoring.

There can be little doubt of the usefulness of the physiological approaches to the study and design of man–machine systems. The physiological studies should, just as the observational studies, lead to clues for further study and hopefully to possible means of increasing human effectiveness together with human well-being.

Ergonomics constantly stresses the well-being of the worker, but it cannot relax in its emphasis. The intellectual climate in the universities the last few years has been such that students have begun to feel that the world of work is hostile and unfeeling and unrelated to human concerns. Misperception and misunderstanding of the real thrust of ergonomics is quite possible in a university climate of opinion which is so near to being anti-intellectual. Constant emphasis on health and on well-being as well as on efficiency, on productivity, and on performance is well advised.

Final discussion

The frequently occurring problems and common themes across the papers in this book have been given sufficient emphasis by the various authors and in the epilogues. There remain for discussion the questions posed in the preface and a few general issues in relation to ergonomics.

1. Ergonomics techniques

Before attempting to answer these questions it is important to ensure that the terminology is as unambiguous as possible. The distinction between physiological and psychological techniques has already been made, admittedly somewhat arbitrarily, in the Preface. There is some additional potential confusion about the term 'criterion'. In the dictionary sense a criterion is a standard or a measure by which something is judged. For the human scientist it often means no more than the dependent variable in an experiment and thus has no necessary relevance to real systems. In ergonomics the problem is: what is being judged? The state of the man or the effectiveness of the system? The questions must be considered in relation to each of these purposes.

■ For what kinds of ergonomics problems is it more appropriate to use physiological criteria, psychological criteria, a combination of the two?

Physiological criteria indicate that something is happening to the state of the man. The decision as to whether or not this matters, either to the man or to the system, depends on the availability of adequate norms. These are available only for climatic stress, for heavy work and for the combination of the two. Thus in these situations physiological criteria are clearly the ones to use.

For the ordinary man–machine situations, where neither the environmental stress nor the physical workload is high, various measures can be taken. For example, cortical potentials, muscle potentials, biochemical changes, heart rate changes, speed of performance, errors, attitudes, opinions, and so on. For all of these measures, the few norms which are available still require interpretation by a specialist in the particular technique. It could be argued that this does not matter if the problem is merely to compare the two tasks, for any one of the above criteria will show a difference, and the preferred direction of difference is usually fairly obvious. This argument ignores the practical decision, which will depend not on whether there is a difference between the tasks but on how important the difference is in terms of health, system performance, cost etc. Thus we have to admit that in the present state of the art, the skills of interpreting criteria are much more relevant to the ergonomics problem than the criteria themselves. To put it another way, with the exceptions noted above, the skills of the investigators determine the utility of the techniques. It can be added that currently most skilled investigators are using batteries of criteria which transcend the physiology/psychology distinction. Furthermore, skills of interpretation are not static, but are constantly influenced by new research and new theoretical advance; thus, the criteria used presently by an investigator may well be abandoned by him within a year as his understanding and conceptualization develop.

■ When both physiological and psychological criteria are used in an experiment, how well do they agree?

Sometimes very well, sometimes not at all, depending on what is being measured at what time, and the number of intervening and uncontrolled variables. For example, physiological and psychological criteria of arousal usually correlate, but this correlation can be eliminated and even reversed by certain combinations of drugs. Since the two classes of criteria are sensitive to different and often complementary bodily systems, it is not surprising that results are not consistent. To take one general example, physiological criteria often reflect changes of effort and psychological criteria reflect changes of performance. Now other things being equal one would expect effort and performance to correlate and thus the physiological and psychological criteria would agree. But, for some extraneous reason performance might fall and the operator might increase his effort to restore the performance. In this situation the physiological and psychological criteria would not agree. In comparing two operators, one might be highly skilled and achieve a high performance with a low effort, while another operator of lesser skill might only achieve a poor performance with a high effort: again the physiological and psychological criteria would not agree. Often the physiological criteria relate to the man and the psychological criteria relate to the situation. This difference also can lead to superficial disagreement between the criteria.

In general whether the two classes of criteria 'agree' depends on how well the total situation and all the relevant parameters are understood. It follows that since the investigator's understanding of the situation is unlikely to be comprehensive he is often well advised to use both physiological and psychological criteria and to utilize the information he gains from their agreement or disagreement.

■ When principles of equipment design are based on findings achieved in experiments using physiological criteria, do they agree with principles arrived at using psychological criteria?

There is little direct evidence which would make it possible to answer this question unambiguously. It often happens that different principles rely on different kinds of evidence. For example, principles of population stereotypes rely entirely on speed/error criteria and one would not expect physiological criteria to be relevant. Psychological criteria are often used to validate physiological criteria and thus the two agree by definition, e.g. thermal comfort measured by skin temperature distributions, and ultimately validated by opinions. Where the two provide relatively independent assessment the agreement is often good, e.g. principles of chair design arrived at by opinions and electro-myography, and principles of bench space design arrived at by performance and by anthropometry.

It is sometimes argued that the physiologist concentrates on well-being and the psychologist concentrates on system efficiency and thus one would expect them to arrive at different principles. Both the premise and the deduction are probably unjustified. Although individual well-being is based on absence of bodily damage and disorder, which are mainly detected by physiological criteria, the concept extends to parameters such as job satisfaction to which psychological criteria are probably the most sensitive. If, in a given task, there appears to be a conflict between the demands of system efficiency and individual well-being, this is often simply because the criteria for system efficiency have been arrived at on a too narrow range or a too limited time scale.

2. Ergonomics as a science

The papers and discussions at the Amsterdam meeting reveal few traces of some of the traditional antagonisms and narrow attitudes within ergonomics. In particular there was no apparent suspicion between research workers and practitioners; it was accepted that both are essential to the furtherance of the subject. This may have been because few of the participants had spent their working lives at either extreme. Most, in fact, had done some original research and had also been involved in a variety of more or less direct applications. There was general feeling that these activities are symbiotic; application cannot progress without supporting research and research without the discipline of applications can easily become sterile. Similarly there was an attitude of mutual respect between those originally trained in psychology and physiology respectively. Neither side claimed to be central or to have the essential key to the general problems facing the ergonomist. As one expects in a rational rather than an emotional atmosphere it was the specialists in particular fields who were most vocal and perceptive about the limitations of their own approach. This modesty based on the security of real achievements is perhaps one of the healthiest signs that ergonomics is coming into maturity.

Beyond the physiology/psychology continuum there was an awareness that anatomical problems were not receiving the attention that their importance warrants. In particular the problems of dynamic anthropometry and the study of posture were regarded as worthy of further study. At the other extreme social psychological and sociological factors emerged at intervals. There was some unease about the way in which factors such as motivation and morale are left out of consideration both in the laboratory and in making predictions about performance in real systems.

On the other hand, it did emerge that ergonomics is still highly dependent on concepts which originate in the separate life sciences. There are few ideas which are clearly ergonomics and yet neither physiology nor psychology. To put it another way, ergonomics is still limited by the lack of general theories of man/machine systems.

The so-called 'systems approach' is essentially a philosophy and perhaps a methodology. The necessary supporting techniques for acquiring, manipulating and evaluating evidence come from the traditional disciplines. It can be argued that this position is at least as defensible as that of the engineer. His concepts and methods are no more valid in relation to the core topic of 'systems relevant criteria' than are those from ergonomics. The systems approach is valuable in relation to completely new systems. However, these are rare in industry where most system designs are dominated by the necessity to develop just one step further from previous practice. The systems approach lacks the historical dimension.

3. Ergonomics as a technology

Accepting the definition of a technology as the useful application of science, it is still not entirely clear how ergonomics is intended to be useful. To whom is the ergonomist responsible for what? The simple answer is that it depends on whom he is representing at the time; the designer, the manufacturer, the operator or the user. This is, of course, an oversimplification since there are bound to be overlaps and in any case the ergonomist has a general social responsibility as well as a specific technical one.

It was emphasized during the conference that the study of man at work does not make sense without the concept of purpose. There is always some reason or set of reasons why the man is doing what he is doing and the activity must be studied in this context. In addition, the criteria of success are inseparable from the definition of purpose. Looking at the problem as one of cost effectiveness, it is not clear what we mean either by cost or by effectiveness. Effectiveness is related to purpose which is not only multidimensional within the individual but also is different and sometimes conflicting between different individuals. For example, designers versus manufacturers, management versus workers and makers versus users. Cost has a biological as well as a financial interpretation but this is very difficult to define. Is the ergonomist always trying to reduce stress? It may be that some stress is not merely difficult to eliminate but that it is undesirable that it should be eliminated. High intensities of body functions do not necessarily indicate poor work design.

To each of these problems there is no simple answer and there may not be an unambiguous general answer. It depends on the circumstances. For example, acceptable stresses in combat conditions are obviously not the same as those for industrial conditions for a whole range of reasons: the kinds of personnel involved, the durations of exposure, the urgency and importance of the objectives. The ergonomist, in common with every other technologist, must deal with every situation on its merits. His expertise is not merely scientific; he must exercise that more elusive quality usually called judgment. This depends on long experience and precedent as well as on formal training. Uncertainty about validity and sometimes reliability also leads to the viewpoint that, within certain constraints, measurement during work is preferred to measurement before and after work and that measurement in the field is preferable to measurement in the laboratory. The constraints are that usually measurement implies some interference and this cannot always be accepted either for the system on the grounds of maintenance of efficiency or for the individual on the grounds of personal privacy. The realism of the field does not always compensate for the uncontrolled nature of relevant variables. On the other hand, the control possible in the laboratory has too often in the past been achieved at the expense of relevance and applicability to real problems.

Acceptance of the limitations and inherent probabilistic nature of even the best ergonomics advice and prediction leads to the need for flexibility in systems design. Flexibility which can be used to reallocate system functions or redesign human tasks as experience of the system 'on-line' increases for the designer and skills of operation increase for the operator.

4.   Ergonomics as a profession

An appraisal of the current status of ergonomics inevitably raises the issue of whether or not it should be regarded as a profession. Presumably the criteria are that there is a unique body of expertise which can be expected from any member of that profession and there is a set of immutable ethics which every member of the profession must accept and uphold.

On both counts further progress is needed. The body of knowledge expected of the trained practitioner is not yet clear to the point where all those concerned could agree. It can be argued that ergonomics is already too extensive for one person to encompass and yet the hierarchy of knowledge is not such that there is a basic core subject on which specialists in particular aspects could build. It does look as though this core will develop from the kind of material which is in this book. That is, the basic skills of the ergonomist rest in this expertise in the acquisition of evidence and the drawing of conclusions in relation to man in man/machine systems.

The ethical responsibilities of the ergonomist also are in need of further clarification. It is accepted that the twin objectives of individual well-being and system efficiency are always present but their relative weighting is not agreed. The preservation of individual health in the sense of absence of disease and damage requires highest priority, but when health is extended into the more general context of total well-being it is not obvious that this should always have priority over system efficiency. This latter ultimately reflects community needs as distinct from individual needs.

The avoidance of operator trauma extends beyond the problems of particular system design. The ergonomist has a responsibility to educate the designer towards this objective and also to the worker who needs increased awareness to decrease the likelihood of damage to himself or to others by inappropriate activities or strategies. This last point is becoming more important as the use of complex or potentially dangerous systems by untrained users increases. The most obvious case of this, at present, is the road vehicle. As hardware technology changes so also must ergonomics change in techniques and in objectives.

The search for a professional identity on the part of the ergonomist should not be misinterpreted as the activity of another specialist taking himself and his role too seriously. The measurement of man is much too difficult a task to ever be capable of neat completion and packaging. The ergonomist merely tries to help by contributing some systematically acquired data to the making of decisions about people and systems which in principle are impossible but in practice have to be made.

Author index

Subject index